Physlet® Physics

Interactive Illustrations, Explorations, and Problems for Introductory Physics

Wolfgang Christian and Mario Belloni

Davidson College

Contributing Authors

Anne J. Cox Eckerd College
Melissa H. Dancy Western Carolina University
Aaron Titus High Point University

Exploration Worksheets Author
Thomas M. Colbert Augusta State University

PEARSON EDUCATION, Inc.
Upper Saddle River, New Jersey 07458

Library of Congress Cataloging-in-Publication Data

Christian, Wolfgang.
 Physlet physics: interactive illustrations, explorations, and problems for introductory
 physics / Wolfgang Christian and Mario Belloni; with contributing authors, Thomas M.
 Colbert... [et al.].
    p. cm.
   ISBN 0-13-101969-4
   1. Physics—Computer-assisted instruction 2. Physics—Problems. Exercises, etc. 3.
 Internet in education. I.Belloni, Mario. II.Title

QC30.C477 2004
530'.078'5—dc21

                                        2003051722

Associate Editor: Christian Botting
Senior Editor: Erik Fahlgren
Editor-in-Chief, Science: John Challice
Vice President of Production and Manufacturing: David W. Riccardi
Executive Managing Editor: Kathleen Schiaparelli
Assistant Managing Editor: Beth Sweeten
Production Editor: Michael Drew
Assistant Managing Editor, Science Media: Nicole Bush
Media Production Editor: Debra Wechsler
Manufacturing Buyer: Alan Fischer
Manufacturing Manager: Trudy Pisciotti
Executive Marketing Manager: Mark Pfaltzgraff
Managing Editor, Audio Visual Assets and Production: Patricia Burns
AV Editor: Jessica Einsig
Art Director: Jayne Conte
Cover Designer: Bruce Kenselaar
Cover Photo: Getty Images, Inc.
Editorial Assistant: Andrew Sobel
Marketing Assistant: Melissa Berringer
Production Assistant: Nancy Bauer

© 2004 Pearson Education, Inc.
Pearson Education, Inc.
Upper Saddle River, NJ 07458

Printed in the United States of America
10 9 8 7 6 5 4 3 2 1

ISBN 0-13-101969-4

Pearson Education Ltd., *London*
Pearson Education Australia Pty., Limited, *Sydney*
Pearson Education Singapore, Pte.Ltd
Pearson Education North Asia Ltd., *Hong Kong*
Pearson Education Canada, Ltd., *Toronto*
Pearson Educación de Mexico, S.A.de C.V.
Pearson Education—Japan, *Tokyo*
Pearson Education Malaysia, Pte.Ltd.

About the Authors

Wolfgang Christian is the Herman Brown professor of physics at Davidson College where he has taught since 1983. He received his B.S. and Ph.D. in physics from North Carolina State University at Raleigh. He is the co-author of *Physlets: Teaching Physics with Interactive Curricular Material, Just-in-Time Teaching*, and *Waves and Optics Simulations*, Volume 9 of the Computational Physics Upper Level Software, CUPS, series. He has been books editor of the APS journal Computers in Physics. He is currently Chair of the American Physical Society Forum on Education and is a member of the Committee on Educational Technologies of AAPT. He has received a Distinguished Service Citation from the American Association of Physics Teachers and received a 2002 Multimedia Educational Resource for Learning and Online Teaching (MERLOT) Award for Exemplary Online Learning Resources for his work on Physlets. His research interests are in the areas of computational physics and instructional software design.

wochristian@davidson.edu

Mario Belloni is currently an assistant professor of physics at Davidson College. He received his B.A. in physics and economics from the University of California, Berkeley and his Ph.D. in physics from the University of Connecticut at Storrs. His research interests are in the areas of theoretical physics and interactive curricular material development. He is the co-author of *Physlets: Teaching Physics with Interactive Curricular Material* and the forthcoming *Physlet Quantum Mechanics*, to be published by Prentice Hall in 2004. He has received a 2002 MERLOT Award for Exemplary Online Learning Resources for his work on Physlets. He is the current Chair of the Committee on Educational Technologies of American Association of Physics Teachers, is the Section Representative of the North Carolina Section of the AAPT, and is a member of the ComPADRE Quantum Physics Editorial Board.

mabelloni@davidson.edu

Contents

CHAPTER 19   Heat and Temperature   163

CHAPTER 20   Kinetic Theory and Ideal Gas Law   169

CHAPTER 21   Engines and Entropy   181

CHAPTER 22   Electrostatics   189

CHAPTER 23   **Electric Fields   197**

CHAPTER 24   **Gauss's Law   205**

CHAPTER 25   **Electric Potential   213**

CHAPTER 26   **Capacitance and Dielectrics   221**

Preface

By now it is hard to imagine an instructor who has not heard the call to "teach with technology," as it has resounded through educational institutions and government agencies alike over the past several years. However, teaching with technology has often resulted in the use of technology for technology's sake and also often resulted in the development of tools that are not pedagogically sound. For example, consider PowerPoint lectures, which are a popular response to the "teach with technology" push. While PowerPoint lectures are more colorful, they are generally no more interactive than chalkboard lectures. The physics community has, to its credit, worked to use technology in a variety of highly interactive and effective ways including wireless classroom response systems that allow for in-class quizzing of students and MBLs (microcomputer-based laboratories) that free students from the drudgery of data collection so that they can spend more time understanding the underlying physical concepts. Into this we offer *Physlet Physics*, a collection of ready-to-run interactive computer simulations designed with a sound use of pedagogy in mind. The aim of *Physlet Physics* is to provide a resource for teaching that enhances student learning and interactive engagement. At the same time, *Physlet Physics* is a resource flexible enough to be adapted to a variety of pedagogical strategies and local environments.

Content

Physlet Physics contains a collection of exercises spanning the introductory physics sequence. These exercises use computer animations generated in Java applets to show physics content. We call these Java applets Physlets (**Phy**sics content simulated with Java app**lets**). Every chapter of *Physlet Physics* contains three quite different Physlet-based exercises: **Illustrations**, **Explorations**, and **Problems**.

Illustrations are designed to demonstrate physical concepts. Students need to interact with the Physlet, but the answers to the questions posed in the Illustration are given or are easily determined from interacting with it. Many Illustrations provide examples of physics applications. Other Illustrations are designed to introduce a particular concept or analytical tool. Typical uses of Illustrations would include "reading" assignments prior to class and classroom demonstrations.

Explorations are tutorial in nature. They provide some hints or suggest problem-solving strategies to students in working problems or understanding concepts. Some Explorations ask students to make a prediction and then check their predictions, explaining any differences between predictions and observations. Other Explorations require students to change parameters and observe the effect, asking students to develop, for themselves, certain physics relationships (equations). Typical uses of Explorations would be in group problem-solving and homework or pre-laboratory assignments. Explorations are also often useful as *Just-In-Time Teaching* exercises. To aid in the assignment of the Explorations, Exploration Worksheets are included on the *Physlet Physics* CD. The Worksheets provide students with extra structure to aid in the completion of the Exploration and provide instructors with an easy way to assign Explorations.

Problems are interactive versions of the kind of exercises typically assigned for homework. They require the students to demonstrate their understanding without as much guidance as is given in the Explorations. They vary widely in difficulty, from exercises appropriate for high school physics students to exercises appropriate for calculus-based university physics students. Some Problems ask conceptual questions, while others require detailed calculations. Typical uses for the Problems would be for homework assignments, in-class concept questions, and group problem-solving sessions.

Conditions of Use

Instructors may not post the exercises from *Physlet Physics* on the Web without express written permission from the Publisher for the English language and from Wolfgang Christian and Mario Belloni for all other languages. As stated on the Physlets website, Physlets (that is, the applets themselves) are free for noncommercial use. Instructors are encouraged to author and post their own Physlet-based exercises. In doing so, the text and script of Physlets-based exercises must be placed in the public domain for noncommercial use. Please share your work!

Authors who have written Physlet exercises and posted them on the Internet are encouraged to send us a short e-mail with a link to their exercises. Links will be posted on the Physlets page: *http://webphysics.davidson.edu/applets/Applets.html*.

More details can be found on the Conditions of Use page on the CD.

Web Resources

In addition to the interactive curricular material in this book and CD, instructors may also wish to view the *Physlet Physics Instructor's Guide* by Anne J. Cox and Melissa H. Dancy. The *Physlet Physics Instructor's Guide* as well as the *Exploration Worksheets* by Thomas M. Colbert are available for download from Prentice Hall's Teaching Innovations in Physics, *TiP*, website.

Instructors can access the official Prentice Hall Web page for *Physlet Physics* by visiting the *TiP* website at *http://www.prenhall.com/tiponline*, then click on the *Physlet Physics* link.

Before You Start

Assigning *Physlet Physics* material without properly preparing the class can lead to frustration. Although Physlet problems often appear to be simple, they are usually more challenging than traditional problems because novice solution strategies are often ineffective. In addition, small technical problems are bound to occur without testing. We use Physlets extensively in our introductory courses at Davidson College, but we always start the semester with a short laboratory whose sole purpose is to solve a Physlet problem in the way a physicist solves a problem; that is, to consider the problem conceptually, to decide what method is required and what data to collect, and finally to solve the problem. As a follow-up, we then assign a simple Physlet-based exercise that must be completed in one of the College's public computer clusters. This minimal preparation allows us to identify potential problems before Physlet-based material is assigned on a regular basis.

In response to these possible difficulties, we have written Chapter 1: Introduction to Physlets. This chapter provides students and instructors with a guided tutorial through the basic functionality of Physlets. After completing the exercises in Chapter 1, students and instructors alike should be in a position to complete the exercises in the rest of the book.

Before you begin, or assign material to students, you should also read the section on Browser Tests and System Requirements.

Acknowledgements

There are a great many people and institutions that have contributed to our efforts, and we take great pleasure in acknowledging their support and their interest.

We thank our colleague Larry Cain for the many hours he spent reading the manuscript and for providing many insightful comments and suggestions. We also thank our colleagues and our students at Davidson College for testing of Physlet-based material in the classroom and the laboratory. Mur Muchane and the Davidson ITS staff have provided excellent technical support. We would also like to thank the Davidson College

Faculty Study and Research Committee and Dean Clark Ross for providing seed grants for the development of Physlet-based curricular material. We also thank Nancy Maydole and Beverly Winecoff for guiding us through the grant application process.

The Physlets project has benefited tremendously from collaborations with non-U.S. universities. In particular, special thanks and recognition go to Francisco Esqucmbre and Ernesto Martin at the University of Murcia (Spain), to Sasa Divjak at the Universtiy of Ljubljana (Slovenia), and to Frank Schweickert at the University of Kaiserslautern (Germany) for translating Physlet-based material into their respective languages and for maintaining non-English-language Physlets websites.

W.C. would like to thank the numerous students who have worked with him over the years developing programs for use in undergraduate physics education. Some of our best Physlets are the result of collaborative efforts with student coworkers. In particular, we would like to single out Mike Lee, Cabel Fisher, and Jim Nolen.

M.B. would like to thank Mario Capitolo, Anne J. Cox, Edward Deveney, Harry Ellis, Kurt Haller, Bill Junkin, Ken Krane, Ken Krebs, and Steve Weppner for many useful and stimulating discussions regarding teaching and the incorporation of Physlets with existing curricular material.

Some people have been such frequent contributors of time and ideas that we have brought them in as contributing authors of this book. We would like to thank Anne J. Cox, Melissa H. Dancy, and Aaron Titus (whose work was supported in part by NSF DUE-9952323), both for their writing and for the many valuable ideas we have gained during our associations with each of them. In addition we would like to thank Thomas M. Colbert for his work creating Worksheets for the Explorations in this book.

Special thanks to Chuck Bennett, Scott Bonham, Morten Brydensholt, Anne J. Cox, Melissa H. Dancy, Dwain Damian, Andrew Duffy, Fu-Kwun Hwang, William Junkin, Steve Mellema, Chuck Niederriter, Evelyn Patterson, Peter Sheldon, Aaron Titus, and Toon Van Hoecke for their contributions of curricular material to this book. In addition we thank Harry Broeders, the CoLoS consortium, Fu-Kwun Hwang, Ernesto Martin, Toon Van Hoecke, and Vojko Valencic for the use of their applets in this book.

We would like to thank all those who reviewed material for this book. During the initial writing we received feedback from Rhett Allain (Southeastern Louisiana University), Cornelius Bennhold (George Washington University), Thomas. M. Colbert (Augusta State University), Edward F. Deveney (Bridgewater State College), Kevin M. Lee (University of Nebraska), Chuck Niederriter (Gustavus Adolphus College), and Steve Mellema (Gustavus Adolphus College). We also would like to thank Harry Ellis, Eduardo Fernandez, and Steve Weppner of Eckerd College for the feedback we received from their class testing of the exercises in this book.

Ranking tasks in this book are inspired by the ranking tasks in *Ranking Task Exercises in Physics*, T. O' Kuma, D. Maloney, and C. Hieggelke. Their Two-Year College (TYC) Workshops have been an especially fruitful arena for the give-and-take of ideas with fellow faculty. The Physlet strategy could not have grown and matured without these opportunities and the exchange of ideas that they afforded.

Both of us express our thanks to Erik Fahlgren, Christian Botting, Mark Pfaltzgraff, and their coworkers at Prentice Hall for supporting the development of *Physlets Physics* and for all of their hard work getting this book to press on an accelerated schedule. In addition, we thank Ruth Saavedra for her copyediting of the manuscript and Michael Drew and his coworkers at nSight for their work formatting and typesetting this book.

We also wish to express our sincerest thanks to those who have encouraged us the most: our spouses, Barbara and Nancy, and our children, Beth, Charlie, and Rudy and Emmy.

This work was partially supported by the National Science Foundation under contracts DUE-9752365 and DUE-0126439.

To the Student

We have found that using computers with our introductory physics students is neither trivial nor obvious and have developed *Physlet Physics* in response to our students' own educational needs. Our goal has been to develop computer-based material that is adaptable to a wide variety of student learning modes, uses a standard easy-to-understand interface, and covers a wide variety of topics. We hope to engage students in ways that are fundamentally different from traditional physics problems by requiring the use of multimedia elements in the course of solving the problem. When we ask students what they think of Physlets, or review end-of-semester evaluations, what is striking is how often they mention that Physlet-based exercises help them to visualize situations.

> *"Physlets do help me visualize problems at hand and are quite a bit more interesting than the text."*
>
> *"The [Physlet-based] problems show one what actually occurs and just doesn't force one to guess what may be happening."*
>
> *"[Physlets] help to see a practical use of physics and they do emulate real life situations."*
>
> *"The material in interactive problems tends to be more challenging, however because they can be seen it is slightly easier to do. Being able to actually see the problem is beneficial."*
>
> *"Use of Physlets on the computer is good for experimenting and observing."*
>
> *"In addition to class demonstrations, the Physlets help bring to life sterile mathematical equations."*
>
> *"As individuals learn in different ways, little harm is done by presenting a different perspective on a given problem."*

This appears to be the most obvious benefit to students.

Although Physlet problems often appear to be simple, they are usually more challenging than traditional problems because you must first figure out a problem-solving strategy before determining what to measure in a Physlet-based exercise. In addition, small technical problems are bound to occur if you are not familiar with the ways you can interact with the Physlet-based materials.

In response to these possible difficulties, we have written Chapter 1: Introduction to Physlets. This chapter provides students with a guided tutorial through the basic functionality of Physlets. After completing the exercises in Chapter 1, you should be in a position to complete the exercises in the rest of the book without technical difficulty.

Before you begin, you should read the following section on Browser Tests and System Requirements.

Browser Tests & System Requirements

Physlet Physics provides physics teachers with a collection of ready-to-run, interactive, computer-based curricular material spanning the entire introductory physics curriculum. All that is required is the *Physlet Physics* CD and a browser that supports Java applets and JavaScript to Java communication. This combination is available for recent versions of Microsoft Windows and most versions of Unix. Although we occasionally check Physlets using other combinations, Microsoft Windows 2000 and Windows XP with both Internet Explorer and the new Open Source Mozilla browser are our reference platforms.[1]

To check whether your computer already has Java installed, go to the Preface Chapter on the CD and navigate to the Browser Tests and System Requirements page. There you will find two buttons.

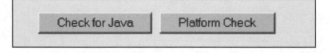

Figure 1. The "Check for Java" and "Platform Check" buttons.

Click the buttons to check for Java and your computer platform. If your browser fails the Java test, please continue reading for information on getting and installing Java.

Microsoft Java Most, but not all, versions of Windows include the Microsoft Java Virtual Machine (JVM). To test if any version of their JVM is installed on your computer, type the jview command at a DOS prompt. If a program runs, you have a Microsoft JVM. If you receive an error that no program by that name exists, you don't.

The Microsoft JVM is installed and updated on your computer with Windows Update. Previously it could be downloaded separately, but now Microsoft only uses Windows Update. The main Web page from Microsoft about Java is *http://www.microsoft.com/java*.

Sun Microsystems Java The Sun JVM is downloadable from the Java website: *http://java.sun.com*. After downloading the file to your hard drive, double-click on its icon to run the installer. Follow the instructions the installer provides.

You can check your computer for a properly installed Sun JVM and change its properties by clicking on the Java plug-in icon in the Windows Control Panel.

Figure 2. The Control Panel folder showing the Java plug-in icon.

---

[1]Physlets have been tested on Linux and various versions of Unix. The only major operating system vendor that does not support Physlets is Apple Computer since the standard Macintosh and Power PC browsers do not support JavaScript to Java communication.

The following dialog box will appear:

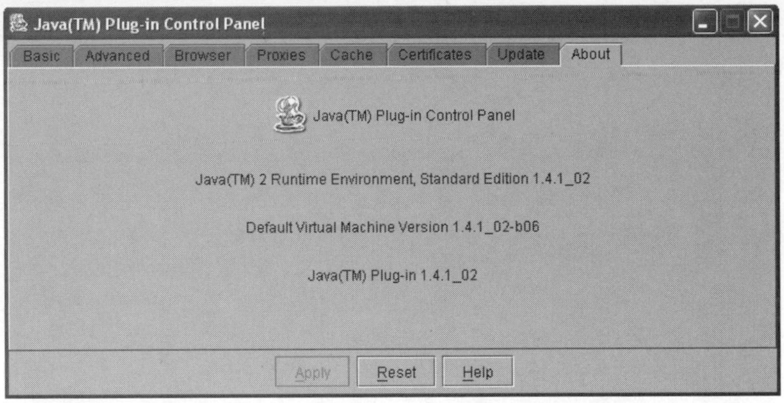

Figure 3.   The Java plug-in dialog box is accessed from the Windows Control Panel.

Although it is possible to simultaneously install Java VMs from Microsoft and Sun Microsystems on Windows computers, a browser can only run one VM at a time. You can switch between these two JVMs in Internet Explorer. Start Internet Explorer and click the *Advanced* tab under *Tool | Internet Options* from the menu bar. The following dialog box shown in Figure 4 will appear.

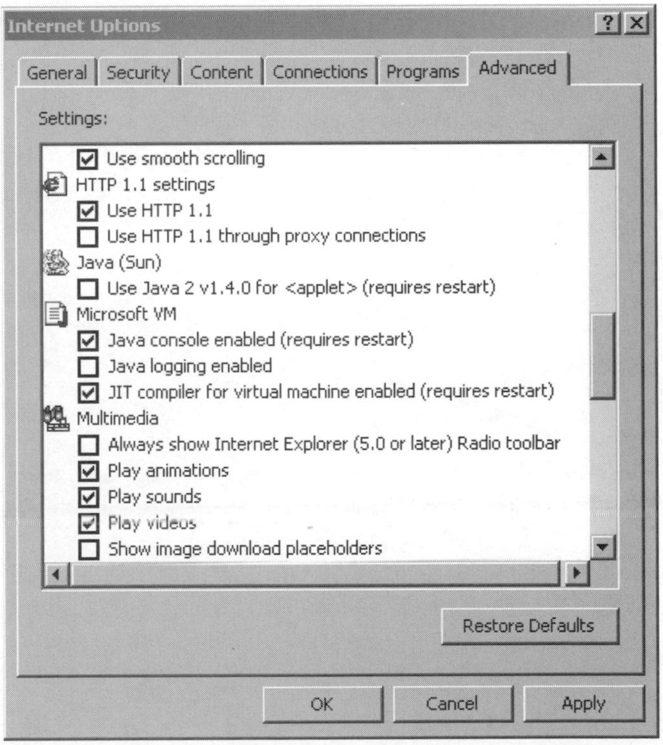

Figure 4.   The advanced Internet Options dialog box is accessed from within Internet Explorer.

Figure 4 shows that this computer has two Java VMs and that it is currently configured to run the Microsoft VM. The option for the Java (Sun) VM will not appear unless the Sun Java Run-Time Environment has been installed. You will need to close all browser windows if you decide to switch VMs. You do not, however, need to restart the computer.

Non-Microsoft Browsers   Netscape, Opera, and Mozilla offer alternatives to the Microsoft Internet Explorer on Windows operating systems. You can download this browser from the Mozilla website: *http://www.mozilla.org*.

After downloading the file to your hard drive, double-click on its icon to run the installer. Follow the instructions the installer provides. The Mozilla browser requires that the Sun JVM be installed on your computer.

Introduction to Physlets

Topics include animations, units, measurement, dragging, and getting data in and out.

Illustration 1.1: Static Text Images Versus Physlet Animations

Illustration 1.1 describes how to use Physlets. First of all, a Physlet is a **Phys**ics (Java) App**let**. We use Physlets to animate physical phenomena and to ask questions regarding the phenomena. Sometimes you will need to collect data from the Physlet animation and perform calculations in order to answer the questions. Sometimes simply viewing the animation will be enough.

The Physlet animations presented in *Physlet® Physics* will be similar to the static images in your textbook. There are differences that need to be examined, however, because we will be making extensive use of these types of animations throughout *Physlet® Physics*. First consider the image taken from Sir Isaac Newton's *Principia*. It is a static image depicting possible orbits of an object around Earth. We are supposed to imagine objects thrown from the mountain top with different initial velocities and imagine where they would land. (We are also supposed to imagine that for just the right conditions objects could orbit in the circles farther out from the center of Earth.)

Now consider the Physlet animation of a similar situation. Illustration 1.1 shows 10 identical balls about to be thrown off a mountain top. The initial positions of the balls are identical, but they have different initial velocities. **Restart**.

Press "play" to begin the animation. Note that the VCR-type buttons beneath the animation control the animation much like buttons on a VCR, CD, or DVD player. Specifically,

- **play** starts the animation and continues it until either the animation is over or is stopped.
- **pause** pauses the animation. Press "play" to resume the animation.
- **<<step** steps the animation backward in time by one time step (the size of the time step varies with the animation). In this animation, there is no "<<step" button.
- **step>>** steps the animation forward in time by one time step.
- **reset** resets the animation time to the initial time. Press "play" to start the animation from the beginning.

Make sure you understand what these buttons do, since you will need to use them throughout the rest of the book when you interact with the Physlets on the CD.

In addition to these buttons, there are hyperlinks on the page that control which animation is played. For example, on this page, **Restart** reinitializes the applet to the way it was when the page was loaded. On other pages there will often be a choice of which animation to play, but **Restart** always gets you back to the initial condition the animation was in when the page loaded.

So what is so neat about this animation compared to the static image? Plenty. Most of what you will study in physics is related to objects in motion. It is difficult to understand the details of the motion of an object if you are trying to describe it with a static picture. Since the examples in this book are interactive animations, you can actually see the details of the motion as the objects undergo their motions.

Restart (or reset) the animation and play it again. What do you notice about the motion of the balls? Specifically, what can you say about the motion of the balls that have orbits inside of the red ball? What can you say about the motion of the balls that have orbits outside of the red ball? First, all of the orbits are squashed circles (called ellipses) except for the red ball that moves in a circle. Second, all of the balls—except the red one—change speed throughout their orbits. The inner balls travel faster near the bottom of the screen as opposed to the top, while the outer balls travel slower at the bottom of the screen as opposed to the top. (Note that we are choosing to show the complete orbits of the balls, even the ones that would have hit Earth. We do this to compare all of the orbits.)

This is not something that is obvious from Newton's drawing from the *Principia*, but it is made clear by the animation. This effect is even easier to see with **only three balls**. Ghost images of the balls are placed at equal time intervals to further display this effect when you click **only three balls**. Don't forget to press "play" after selecting the hyperlink!

In the natural sciences, simulations are almost always deterministic. By deterministic, we mean that the simulation evolves in time according to a predefined mathematical model. The models we have built for this text may or may not represent physical reality. In fact, we will often present multiple models and ask you to determine which model is in agreement with experiment. Do not assume that every simulation obeys the laws of physics.

It is important not to confuse deterministic with predictable. Simulations that depend on random numbers, contain large numbers of parameters, or exhibit chaos are often not predictable in the sense that the exact behavior may depend on infinitesimal changes of initial conditions. However, even if the details of the dynamics cannot be determined, the model may still give useful information about the types of behavior that can occur.

Illustration 1.2: Animations, Units, and Measurement

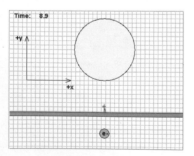

Most physics problems are idealizations of actual physical situations. In many physics problems, moving objects are set into motion before the problem even starts. This allows the problem to focus on one particular concept. Physlet animations are no different. The animations depict only a short period of time in the "life" of an object. Sometimes the objects start at rest, and when you press "play" they begin their movement. Other times, the object is already moving, even before you press "play," and pressing "play" just starts the animation. Look for visual clues as to whether the object starts at rest or is already moving when the animation begins. Both animations on this page (Animation 1 and Animation 2; do not forget to press "play") depict objects that are already moving when the animation starts (at $t = 0$). (Likewise, when the animation is over and you see the "End of Animation" message, the animation is over, but the motion of the object could continue. This continuing motion is just not depicted.)

Units are important to physicists. However, computer simulations store numbers and these numbers do not have units. Calculations are performed just as they are on a pocket calculator. **Restart**. This can cause confusion since the time and distance units shown on the computer display do not have an *a priori* relationship to the real world. In other words, we can assign the relationship to be anything we want it to be. In **Animation 1**, for example, the part of the animation that models the motion of an electron might define the distance unit to be 10^{-9} m, i.e., a nanometer, and the time unit to be 10^{-6} s, i.e., a μs. Another part of the animation, depicting a person walking, might define position given in meters and time given in seconds (the MKS system of units). Still another part of the animation depicts a star and might define position units to be 10^8 meters and time to be Earth years. In general, you should look for the units specified in the problem (whether from your text or from *Physlet® Physics*): **On the *Physlet Physics* CD all units are given in boldface in the statement of the problem. The units for Animation 2 are given in the following paragraph.**

Although computer simulations allow precise control of parameters, their resolution is not infinite. The numbers used to calculate the position and velocity of a particle have finite precision, and the algorithm updates these values at discrete times. Consequently, data is only available on the screen at certain predefined intervals. Whenever this data is presented on screen as a numeric value, it is correct to within the last digit shown. Start **Animation 2** and follow the procedures below to make position measurements **(position is given in meters and time is given in seconds)**.

Some problems require that you **click-drag** the mouse inside the animation to make measurements. Try it. Place the cursor in the animation and left-click and hold down the mouse button. Now drag the mouse around to see the x and y coordinates of the mouse change in the lower left-hand corner of the animation. Notice the way the coordinates change. As you drag the mouse around can you find the origin of coordinates? This is an easy question in these animations because the coordinate axes are shown. However, the coordinate axes will not always be shown. You can always find the origin of coordinates by click-dragging the mouse.

In addition, **these measurements cannot be more accurate than one screen pixel.** This means that depending on how you measure the position of an object you may get a slightly different answer than another student in your class.

For example, where is the man in **Animation 2** at $t = 10$ s? You could get anywhere between 19.4 m and 20.3 m depending on whether you are measuring the position of the man from his front, back, or center. In order to make good measurements you must be consistent.

You also must be careful to choose a problem-solving approach that does not depend critically on the difference between two numbers that are almost equal. It may not be possible to extract certain types of information from an animation if the changes in the relevant parameters are too small.

Illustration 1.3: Getting Data Out

In Illustration 1.2 you learned about units and how to click-drag in an animation to get the position of objects. In this Illustration we will discuss several other ways in which data is depicted in animations. **Restart**.

Select **Animation 1** to begin (position is given in meters and time is given in seconds). Shown in the animation is a red ball that, when you press the "play" button, will move across the screen in a predefined way. Along with the red ball are depictions of the object's position: an on-screen numerical statement of position, a data table, a graph, an arrow, and ghost images. You may of course click-drag in the animation to measure position as well.

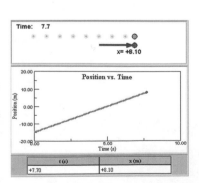

Why all of these different representations? Select **Animation 1** or reset the animation and play it again. Notice how the different representations of the motion change with the motion of the ball. With a lot of practice physicists are able to look at the motion of objects and tell you the various properties of the motion. How do we do that? Mentally we have different pictures in our heads. Specifically,

- **on-screen numerical statements of position** facilitate the measurement process, as the value is always given. These statements can be for any variable, not just position.

- **data tables** are used to compare two or more values that are changing as in the above animation where time and position are changing.
- **graphs** are used to summarize all of the data corresponding to the motion of an object that occurs over a time interval. The graph summarizes all of the data shown in the on-screen calculation and the data table. When you get a good-looking graph, you can always right-click on it to clone the graph and resize it for a better view. Try it!
- **arrows** show a vector quantity. In the above animation the arrow shows the position vector. Notice that unlike the on-screen calculation and the data table, the arrow depicts both a number and a direction (we call such things vectors).
- **ghost images** are used to represent what is called a motion diagram. We use motion diagrams to help facilitate a mental picture of the motion. In the above animations, the ghost images are placed at equal time intervals so you do not have to do this in your head.

We never set up an animation to give you all of these depictions of motion simultaneously as we have done above. We usually pick one or two representations that best depict the phenomena. Select **Animation 2** (position is given in meters and time is given in seconds) to see the depictions of velocity.

Note that this, and most, animations depict motion that started before the animation begins and continues beyond the time that the animation ends. In the animations on this page, the ball always has, and always will, move to the right at 3 m/s.

When you get a good-looking graph, right-click on it to clone the graph and resize it for a better view.

Exploration 1.1: Click–Drag to Get Position

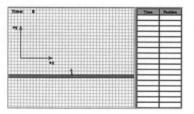

Some problems require that you **click-drag** the mouse inside the animation to make measurements. **These measurements cannot be more accurate than one screen pixel**. This means that depending on how you measure the position of an object you may get a slightly different answer than another student in your class. **Restart**.

Use the following techniques (position is given in meters and time is given in seconds) to measure the position of the man in the x direction as a function of time:

a. Pause the animation at $t = 0$ s (you may have to step back or reinitialize or reset the animation).

b. With the cursor in the animation, hold down the left mouse button and drag the cursor to the center of the man to measure his position in the x direction.

c. Step forward by 2 s and record the time and the man's new position in the x direction.

d. Repeat these measurements for $t = 4, 6, 8, 10,$ and 12 s.

Look at **show data in table** after you have finished part (d). Be sure to take a close look at the data table.

e. Do your answers agree with the table? Why or why not?

Exploration 1.2: Input Data, Numbers

Exploration 1.2 shows 10 identical balls about to be thrown off a mountaintop (position is given in arbitrary units and time is given in arbitrary units). The initial positions of the balls are identical, but they have different initial velocities. The difference in orbital trajectory, therefore, is due to the balls' initial velocities. **Restart**. We will explore how we can get numerical values into an animation and therefore change the animation depicted on the screen.

Click the "set value and play" button. Now change the value of the initial position, y_0, by typing in the text box and then click the "set value and play" button again.

a. Find the limits of the values you can type in the text box.

b. Why do you think these values have been chosen?

c. Now try typing in "abcd." What happens?

Exploration 1.3: Input Data, Formulas

In many animations you will be expected to enter a formula to control the animation (position is given in centimeters and time is given in seconds). **Restart**. In Exploration 1.3, you are to enter in a function x(t) to control the position of the toy yellow Lamborghini. There are a few important rules for entering functions. Notice that the default value in the text box is $3 * t$ and **NOT** $3t$. This is the way the computer understands multiplication. You must enter in the multiplication sign " $*$ " every time you mean to multiply two things together. Remove the " $*$ " and see what happens. You get an error and you can see what you entered. Division is represented as $t/2$ and **NOT** $t\backslash 2$. In addition, the Physlet understands the following functions:

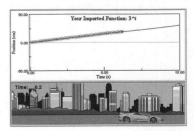

sin(a)	cos(a)	tan(a)	sinh(a)	cosh(a)	tanh(a)	
asin(a)	acos(a)	atan(a)	asinh (a)	acosh(a)	atanh(a)	
step(a)	sqrt(a)	sqr(a)	exp(a)	ln(a)	log(a)	
abs(a)	ceil(a)	floor(a)	round(a)	sign(a)	int(a)	frac(a)

where "a" represents the variable expected in the function (here it is t).

Try the following functions to control the Lamborghini (note that you are controlling x(t) of the red ball attached to the Lamborghini):

a. $0.3 * t * t$

b. $-20 * t + 3 * t^{\wedge}2$ (note that $t^{\wedge}2$ is equivalent to $t * t$)

c. int(t)

d. $10 * \sin(\text{pi} * t/2)$

e. $\text{step}(t - 2) * 3 * (t - 2)$

Try some others for the practice. Try to keep the Lamborghini on the screen!
When you get a good-looking graph, right-click on it to clone the graph and resize it for a better view.

Problems

Problem 1.1

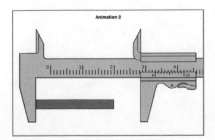

During the laboratory portion of your physics class, you will most likely have to measure objects rather precisely. A Vernier

caliper is a device that can precisely measure the length of small objects less than about 20 cm long **(position units on the caliper are given in centimeters)**. You read the caliper by placing an object in the caliper, closing the arm of the caliper, and reading off of the scale. **Restart**.

In the animations you are to drag the red circle on the caliper arm to move it. Your cursor will change into a little hand when it is over the circle. Once the cursor changes, left-click to drag the object around.

The 0 line on the movable arm tells you centimeters and tenths of centimeters. The zero line is almost always between two tenths on the scale. To figure out the hundredths place, you look to find which line on the movable arm matches up with a line on the fixed scale. The number on the movable scale tells you the hundredths place. The default position of the caliper after clicking **Restart** is 1.64 cm.

Now try it yourself for Animations 1 through 4. What are the lengths of these objects? Note: In Animation 4 you must drag the object in position before you can measure its length.

Problem 1.2

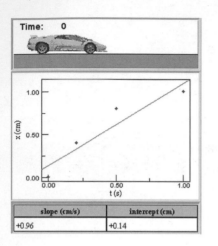

slope (cm/s)	intercept (cm)
+0.96	+0.14

A toy yellow Lamborghini moves across the screen as shown in the animation (position is given in centimeters and time is given in seconds).

Take data from the animation and create a position vs. time graph. To add data to the graph, type your (t, x) data into the text boxes and then click the "add datum" button. Use the "clear graph" button to start another graph.

Once you have taken enough data (you need more than five data points), use the "linear regression" button to calculate and plot the linear regression. If you later add data, the regression line and values from the table are removed and you must redo the regression. **Restart**.

a. Report the slope and intercept of the graph.

b. What do these two values represent regarding the motion of the toy Lamborghini?

When you get a good-looking graph, right-click on it to clone the graph and resize it for a better view.

Problem 1.3

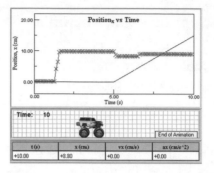

t (s)	x (cm)	vx (cm/s)	ax (cm/s^2)
+10.00	+8.80	+0.00	+0.00

Use the mouse to drag the rear bumper of the toy monster truck (the red ball). The goal of this exercise is to match the position/velocity/acceleration vs. time graphs as shown in the animation (position is given in centimeters and time is given in seconds). There is some smoothing for the velocity and acceleration matching animations. **Restart**.

Answer the following questions after trying to match the motion:

a. Which of the graphs was the easiest to match and which one was the hardest to match?

b. Why? Base your answer on physics and mathematics.

When you get a good good-looking graph, right-click on it to clone the graph and resize it for a better view.

One-Dimensional Kinematics

Topics include position, displacement, average and instantaneous speed, average and instantaneous velocity, acceleration, constant acceleration, and acceleration due to gravity.

Illustration 2.1: Position and Displacement

In physics we often talk about distance traveled and displacement to describe how the position of an object changes. Sometimes we even (appear to) use these terms interchangeably. However, they are not necessarily the same. Distance traveled is just that: the distance an object travels. The displacement of an object is a comparison of the final position to the initial position: $\Delta x = x - x_0$, the distance displaced during the object's motion. Can you think of an example in which distance traveled and displacement are the same or different?

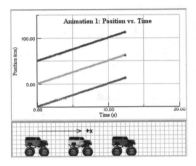

Shown are two animations, each with three toy monster trucks and their position vs. time graphs (position is given in centimeters and time is given in seconds). The arrow in each animation shows where the origin of coordinates is located. All measurements of position are with respect to this origin. **Animation 1** shows the three monster trucks starting at different positions at $t = 0$ s. During this animation, for each individual truck, the distance traveled is equal to the displacement.

Note that we are treating each monster truck as an idealized object, a point. We are always measuring the position of a convenient part of the monster truck and then describing the motion of this part of the monster truck. For this animation the part of each monster truck that we follow is its front bumper, but it could have been the middle or the rear bumper. What matters is that we are consistent in the measurement process. The position of the object will vary depending on where you take the measurement (the front bumper, the middle, the rear bumper of the truck), but the difference in position measurements, the displacement vector, always stays the same. Therefore, it is not position that is important in physics, but rather the change in position or displacement that is important.

When you get a good-looking graph, right-click on it to clone the graph and resize it for a better view.

What are the average velocities of the trucks in **Animation 1**? Even though the trucks start at different positions, all three trucks have the same average speed (distance traveled/time) and average velocity (displacement/time). This can be seen from the slope of the graph (note that the distance traveled and displacements for each individual truck are identical in this animation). In **Animation 2** the initial positions of the three monster trucks are the same, but each truck travels a different distance and also has a different displacement. The truck with the largest average speed and average velocity (for each individual truck, its average speed is still equal to its average velocity) is the one with the largest slope on the position vs. time graph.

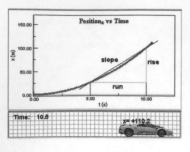

Illustration 2.2: Average Velocity

When the velocity of an object remains constant, its average velocity is equal to the instantaneous velocity, and both remain constant over time (position is given in centimeters and time is given in seconds). This is precisely why we can use the following definition of average velocity,

$$\mathbf{v}_{avg} = \Delta \mathbf{x}/\Delta t,$$

to describe the motion of an object moving at a constant velocity. We rewrite this equation as $x = x_0 + v(t - t_0)$. But what happens when an object is not moving at a constant velocity?

While we won't get into a full discussion of why things move until Chapters 4 and 5 (Newton's laws), we can still use the concept of average velocity to describe the motion of an object. The animation shows a toy Lamborghini traveling at a non-constant velocity.

What is the Lamborghini's average velocity, $\mathbf{v}_{avg}$, in the time interval between $t = 5$ s and $t = 10$ s? It is still the displacement divided by the time interval, but how can we see this graphically?

Click the "show rise and run" button and then the "show slope" button. During this time interval (between 5 s and 10 s) the rise is the displacement and the run is the time interval; therefore, the slope of the line segment joining the points [$x(5)$, 5] and [$x(10)$, 10] represents the average velocity in this interval. An object beginning at [$x(5)$, 5] would arrive at [$x(10)$, 10] if it moved with the constant velocity represented by the slope of the line connecting those two points. Note that the notation [$x(5)$, 5] describes the point on the graph at $t = 5$ s.

When you get a good-looking graph, right-click on it to clone the graph and resize it for a better view.

Illustration 2.3: Average and Instantaneous Velocity

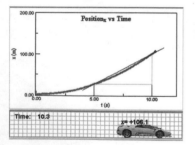

When an object's velocity is changing, it is said to be accelerating. In this case, the average velocity over a time interval is (in general) not equal to the instantaneous velocity at each instant in that time interval. So how do we determine the instantaneous velocity? Play the first animation where the toy Lamborghini's velocity is changing (increasing) with time (position is given in centimeters and time is given in seconds).

Click the "show rise, run, and slope" button. The slope of the blue-line segment represents the Lamborghini's average velocity, $\mathbf{v}_{avg}$, during the time interval (5 s, 10 s). What is the Lamborghini's average velocity during the time interval (6 s, 9 s)? It is the slope of the new line segment shown when you enter in 6 s for the start and 9 s for the end and click the "show rise, run, and slope" button.

When you get a good-looking graph, right-click on it to clone the graph and resize it for a better view.

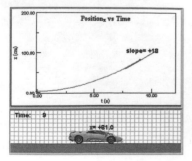

What is the Lamborghini's average velocity, $\mathbf{v}_{avg}$, during the time interval (7 s, 8 s)? How about the average velocity during the time interval (7.4 s, 7.6 s)? As the time interval gets smaller and smaller, the average velocity approaches the instantaneous velocity as shown by the following **Instantaneous Velocity Animation.**

The instantaneous velocity therefore is the slope of the position vs. time graph at any time. If you have taken calculus, you know that this slope is also the derivative of the function shown, here $x(t)$. The Lamborghini moves according to the function: $x(t) = 1.0t^2$, and therefore $v(t) = 2t$, which is the slope depicted in the **Instantaneous Velocity Animation.**

Illustration 2.4: Constant Acceleration and Measurement

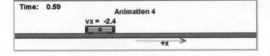

A 1.0-kg cart on a track experiences several different constant accelerations as shown in the animations (position is given in meters and time is given in seconds). The red dot shows you where position measurements are taken. But how can you tell what acceleration the carts have? There are several ways that we will discuss in this Illustration. Before we begin, play each animation *without* the velocity calculation (leave it unchecked, since that would be cheating). How would you characterize each cart's motion? How would you describe each cart's acceleration? How can you show that your assertions are correct?

Hopefully when you looked at Animations 1 and 2 you saw that the carts moved with a constant velocity (Animation 1 has a positive velocity and Animation 2 a negative velocity). The motion of the carts during each animation is uniform, and just by looking you can tell (with practice) that each cart undergoes the same displacement in the same time interval. This is easy enough to check. In Animation 1 the cart is at $x = 0$ m at $t = 0$ s, at $x = 0.5$ m at $t = 0.25$ s, at $x = 1.0$ m at $t = 0.5$ s, at $x = 1.5$ m at $t = 0.75$ s, and finally at $x = 2.0$ m at $t = 1.0$ s. The cart's motion is uniform at $v = 2$ m/s. The cart in Animation 2 has $v = -2$ m/s, which can be verified by taking data and showing the velocity calculation for the animation.

What about Animations 3, 4, and 5? Hopefully you saw that the motion was not uniform and that the carts were all accelerating. So how can we prove it and calculate the accelerations? It depends on the situation and the data given. Below are the three most used constant acceleration formulas:

$$v = v_0 + at,$$

$$x = x_0 + v_0 t + 0.5\, at^2,$$

and

$$v^2 = v_0^2 + 2a(x - x_0).$$

So which of these should we use for Animations 3, 4, and 5? We can rule out the first equation (unless we cheat and turn on the velocity calculation) because it requires instantaneous velocity, which we do not have. We can get position and time measurements from the animation, so that means we can use the second equation. In both Animation 3 and Animation 4 the cart starts at rest, so we have $x = x_0 + 0.5\, at^2$, or $a = 2(x - x_0)/t^2$. The cart has a displacement of 2 m in 1 s in Animation 3 and a displacement of -2 m in 1 s in Animation 4. Therefore, the accelerations are 4 m/s^2 and -4 m/s^2 respectively.

What about Animation 5? The cart has an initial velocity and slows to a stop (it has a positive velocity and a negative acceleration). How can we calculate this cart's acceleration? We cannot do it with the data given (again without cheating by looking at the velocity data). Why? While it is true that we could estimate the initial velocity by $\Delta x / \Delta t$, this method will not always yield good results since this is the average velocity during the time interval and not the instantaneous velocity at $t = 0$ s. The best way to measure this cart's acceleration is to turn on the velocity calculation and use either $v = v_0 + at$ or $v^2 = v_0^2 + 2a(x - x_0)$. This yields an acceleration of about -3.7 m/s^2. Note that if you had used $\Delta x / \Delta t$ for an estimate of the initial velocity you would have gotten 3 m/s, which differs from the actual 3.7 m/s.

If you leave the velocity calculation on for all of the animations, you can now use $v = v_0 + at$ to calculate all of the accelerations.

Illustration 2.5: Motion on a Hill or Ramp

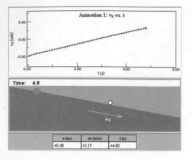

A putted golf ball travels up a hill and then down again (position is given in meters and time is given in seconds). When an object (like a golf ball) travels up or down an inclined ramp or hill, its motion is often characterized by constant, nonzero acceleration. If the incline of the hill is constant, then the motion of the object can also be considered straight-line motion (or one-dimensional motion). It is convenient to analyze the motion of the golf ball by defining the $+x$ axis to be parallel to the hill and directed either upward or downward along the hill as shown in Animation 1.

Here are some characteristics of the motion that you should convince yourself are true:

- In **Animation 1** the $+x$ direction is defined to be down the hill. Therefore, when the ball moves down the hill, it is moving in the $+x$ direction and thus v_x is positive. When the ball moves up the hill, it is moving in the $-x$ direction and thus v_x is negative.

- As the golf ball is traveling up/down the hill, is it slowing down or speeding up? Well, the answer to this question depends on what you mean by slowing down and speeding up. As the ball rolls up the hill its velocity is negative (because of how the x axis is defined) and decreasing in magnitude (a smaller negative number). At the top of the hill, its velocity is zero, and as it travels down the hill, the ball is speeding up. Therefore, its speed decreases, reaches zero, and then increases. How can this be if v_x is always increasing? Speed is the magnitude of velocity (and is always a positive number). As the ball travels up the hill, v_x increases from -5 m/s to zero; yet its speed decreases from 5 m/s to zero. Note that the phrases "speeding up" and "slowing down" refer to how the speed changes, not necessarily to how the velocity changes.

- Is the acceleration of the golf ball increasing, decreasing, or constant? To answer this, look at the slope of the graph at every instant of time. The slope of the velocity vs. time graph (velocity in the x direction) is equal to the acceleration (in the x direction). Does it change or is it the same? Notice that it is constant at all times and is in the positive x direction (as defined by the coordinates).

- Besides using the graph to calculate acceleration, you can also use the velocity data from the data table. Since average acceleration is the change in velocity divided by the time interval, choose any time interval, measure v_{xi} and v_{xf}, and calculate the average acceleration $a_{x\,\text{avg}}$. Since the acceleration is constant, the average and instantaneous accelerations are identical.

- The direction of acceleration can also be found by subtracting velocity vectors pictorially. **Animation 2** shows the black velocity vectors at $t = 0.2$ s and $t = 1.0$ s. To subtract vectors, drag v_i away from its original position (you can drag the little circle on the arrow's tail) and then drag the red vector $-v_i$ into place and use the tip-to-tail method. The direction of the acceleration is in the same direction as the change-in-velocity vector. Now, try **Animation 3**, which shows the velocity vectors at $t = 1.2$ s and $t = 2.0$ s. Compare the change-in-velocity vector for each of the two time intervals. You will find that they are the same. Since the acceleration is constant, the change in velocity is constant for any given time interval.

- The area under a v_x vs. time graph is always the displacement, Δx. You can use the graph to find Δx from $t = 0$ to $t = 3$ s. Use the data table to check your answer by determining the displacement from $x - x_0$. What is the displacement from $t = 0$ to $t = 6$ s? If you answer anything other than 0 m, you should revisit the definition of displacement.

See **Illustration 3.2** for more details on what happens to the acceleration when the angle of the hill is varied.

Illustration 2.6: Free Fall

A ball is dropped from rest near the surface of Earth as shown in **Animation 1**. Its motion is described as *free fall* (position is given in meters and time is given in seconds). If the $+y$ direction is defined to be upward, then the ball's acceleration is constant and has a value of -9.8 m/s^2. If the $+y$ direction is defined to be downward, then the acceleration is $+9.8$ m/s^2.

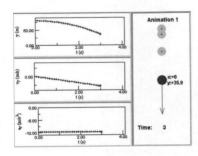

Animation 2 shows the ball thrown upward such that it travels upward, reaches some maximum height, and falls back to your hand at the same height at which is was first thrown. Consider the motion of the ball only while it is in the air and not in your hand (If it's in your hand, it's not considered free fall.). The green vector shown is the velocity vector of the ball, v_y. When the ball is at its peak, what is v_y?

Look at the graphs. At the peak of its motion, the ball goes from upward (a positive velocity) to downward (a negative velocity). It changes velocity smoothly, and the velocity must go through zero at this turnaround point. The acceleration is the change in velocity, which is constant throughout the motion at -9.8 m/s^2. You can measure the velocity at two different times by clicking in the velocity vs. time graph to see that $\Delta v / \Delta t$ is constant.

Animation 3 shows the ball thrown downward. Notice that while the ball travels farther (note that it is off screen for most of its motion) and moves much faster than when it is dropped from rest, the slope of the velocity vs. time graph is still -9.8 m/s^2 as shown by the acceleration vs. time graph.

When you get a good-looking graph, right-click on it to clone the graph and resize it for a better view.

Exploration 2.1: Compare Position vs. Time and Velocity vs. Time Graphs

Shown are three different animations, each with three toy monster trucks moving to the right. Two ways to describe the motion of the trucks are position vs. time graphs and velocity vs. time graphs (position is given in centimeters and time is given in seconds).

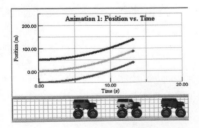

Answer the following questions that focus on the velocity and acceleration of the monster trucks.

a. How does the initial position affect the various graphs?

b. Describe the motion of the trucks by analyzing the position vs. time graphs.

c. Once you have completed (a) and (b), check your answers by analyzing the velocity vs. time graphs.

When you get a good-looking graph, right-click on it to clone the graph and resize it for a better view. This is especially important for viewing points near the origin.

Exploration 2.2: Determine the Correct Graph

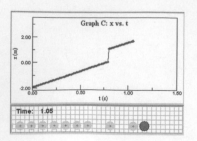

a. View the animation of the red ball by selecting **Ball Only** and describe its motion in words (position is given in meters and time is given in seconds).

b. Now view the three possible position vs. time graphs A, B, and C by clicking the links in the table. Which one is the correct graph? Give at least one reason why each of the other two graphs is incorrect.

c. Now view the three possible velocity vs. time graphs D, E, and F by clicking the links in the table. Which one is the correct graph? Give at least one reason why each of the other two graphs is incorrect.

Exploration 2.3: A Curtain Blocks Your View of a Golf Ball

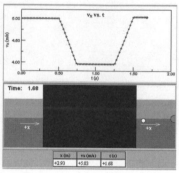

A putted golf ball (not shown to scale) rolls on a green. A black curtain blocks your view of the ball, but otherwise does not interfere with the ball's motion (position is given in meters and time is given in seconds). The animation is a side view of the ball rolling on the green.

Analyze the velocity vs. time graph for the ball and describe the terrain of the green behind the curtain that blocks your view. Give valid reasons for your answer.

After you answer this question, **check the correct answer.**

Exploration 2.4: Set the *x(t)* of a Monster Truck

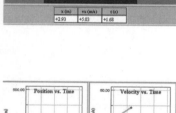

By now you have seen the equation $x = x_0 + v_0 t + 0.5 a t^2$. Perhaps you have even derived it for yourself. But what does it really mean for the motion of objects?

The animation allows you to explore all three terms in the equation: the initial position by changing x_0 from -50 cm to 50 cm, the velocity term by changing v_0 from -15 cm/s to 15 cm/s, and the acceleration term by changing a from -5 cm/s^2 to 5 cm/s^2.

Use the animation to guide your answers to the following questions (position is given in centimeters and time is given in seconds).

a. How does changing the initial position affect the position vs. time graph?

b. How does changing the initial position affect the velocity vs. time graph?

c. How does changing the initial velocity affect the velocity vs. time graph?

d. How does a positive initial velocity vs. a negative initial velocity affect the velocity vs. time graph?

When you get a good-looking graph, right-click on it to clone the graph and resize it for a better view.

Exploration 2.5: Determine *x(t)* and *v(t)* of the Lamborghini

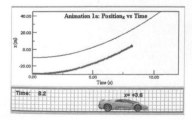

a. Find the position of the toy Lamborghini as a function of time, *x(t)*, for each animation (position is given in centimeters and time is given in seconds). Note that the graph depicts the position as a function of time. Use the "check function" button to see the actual position vs. time graph and use this as a guide for your analysis.

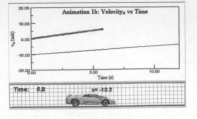

Remember to use the proper syntax such as $-10 + 0.5*t$, $-10 + 0.5*t*t$, *or* $-10 + 0.5 * t^\wedge 2$. *Revisit* **Exploration 1.3** *to refresh your memory.*

b. Find the velocity of the toy Lamborghini as a function of time, $v(t)$, for each animation (position is given in centimeters and time is given in seconds). Use the "check function" button to see the actual velocity vs. time graph and use this as a guide for your analysis. (If you have taken calculus, this exercise should be particularly straightforward.)

Exploration 2.6: Toss the Ball to Barely Touch the Ceiling

To show your coordination, you try to toss a ball straight upward so that it just barely touches the ceiling (position is given in meters and time is given in seconds). What initial velocity is required? In this Exploration the acceleration of the ball is -9.8 m/s^2. Calculate this initial velocity and then test your answer by typing the initial velocity in the text box and clicking the "set velocity and play" button.

Exploration 2.7: Drop Two Balls; One with a Delayed Drop

Two giant tennis balls are released from rest at a certain height. One (the ball on the right) can be dropped after the first ball is dropped. You may change the time delay from 0 to 2.5 s (enter the time delay in the text box and click the "set delay and play" button). The ghost images mark the balls' positions every 0.5 s (position is given in meters and time is given in seconds).

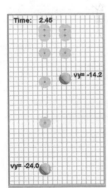

  Choose a 1-s delay (for simplicity) and then answer the following questions.

a. Once the second tennis ball (the ball on the right) is released, does the difference in the speeds increase, decrease, or stay the same?

b. Once the second tennis ball (the ball on the right) is released, does their separation increase, decrease, or stay the same?

c. Is the time interval between the instants at which they hit the ground smaller than, equal to, or larger than the time interval between the instants at which they were released?

Exploration 2.8: Determine the Area Under $a(t)$ and $v(t)$

A 1.0-kg cart on a track experiences several different constant accelerations as shown in the animation (position is given in meters and time is given in seconds). The red dot shows you where position measurements are taken. In addition, the graph of either the acceleration vs. time or the velocity vs. time is shown (use the check box to toggle between the two) along with data in a table. One cell of the table shows the calculation of the area under the curve (the integral $\int a\ dt$ or $\int v\ dt$) as it is plotted in the graph shown.

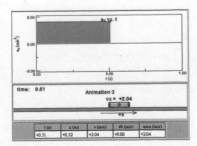

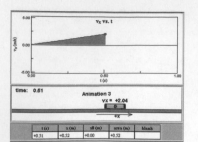

View all five animations and answer the questions below for the acceleration vs. time graph.

a. What is the initial velocity in each animation?

b. What is the final velocity in each animation?

c. What is the difference between the final velocity and the initial velocity $(v - v_0)$ in each animation?

d. What is the total area under the curve calculated during each animation?

e. How are your answers for (c) and (d) related? Does this make sense? Why?

View all five animations and answer the questions below for the velocity vs. time graph (use the check box to view the velocity vs. time graphs):

f. What is the initial position in each animation?

g. What is the final position in each animation?

h. What is the displacement of the cart $(x - x_0)$ in each animation?

i. What is the total area under the curve calculated during each animation?

j. How are your answers for (f) and (g) related? Does this make sense? Why?

Problems

Problem 2.1

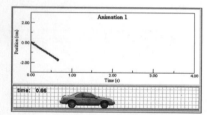

Which of the animations shows the correct position vs. time graph for the toy T-bird (position is given in centimeters and time is given in seconds)? For simplicity, the red dot is shown to indicate where measurements are taken.

Problem 2.2

A hockey puck sliding on ice collides and rebounds from a wall on a hockey rink. A top view is shown in the above animation (position is given in meters and time is given in seconds).

a. For each time interval in the data table below, calculate the displacement, distance traveled, average velocity, and average speed of the puck.

b. For which time interval(s) listed in the table above is the displacement equal to the distance traveled?

c. Is the magnitude of the displacement always equal to the distance traveled?

d. In general, if an object moves in a straight line but does not change direction, will the magnitude of its displacement

Time interval	Displacement (m)	Distance traveled (m)	Average velocity (m/s)	Average speed (m/s)
$t = 1.5$ s to 12.0 s				
$t = 1.5$ s to 6.0 s				
$t = 6.0$ s to 12.0 s				

during any interval equal its distance traveled during the same interval? If the answer is no, which will be greater?

e. In general, if an object moves in a straight line but changes direction at some point, will the magnitude of its displacement during an interval that includes the change in direction equal its distance traveled? If the answer is no, which will be greater?

f. Finally, qualitatively draw the acceleration vs. time graph for the animation.

Problem 2.3

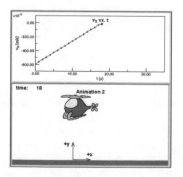

Which helicopter flies according to the velocity vs. time graph shown above (position is given in meters and time is given in seconds)?

Problem 2.4

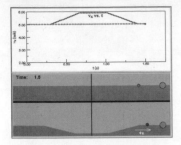

Two balls are putted with the same initial velocity on separate greens. The *x* axis is always defined to be parallel to the ground (even when the blue ball is on the hill). The first ball (red) rolls on a flat surface. The second ball (blue) rolls on a flat surface, then down a hill, onto a flat surface, then up a hill, and onto a flat surface again (position is given in meters and time is given in seconds).

a. Which ball will reach the hole first? Make your prediction and then press "play" to view the animation to see the result.

b. Which ball has a greater average velocity during the animation time?

Problem 2.5

Observe the two animations (position is given in meters and time is given in seconds).

a. Sketch the velocity as a function of time for each animation. Draw both functions on the same graph in order to compare the two animations.

b. Make a second velocity vs. time sketch for Animation 1. On your graph, make sure to have a long enough time interval that the ball's position is between $x = -2$ m and $x = 2$ m **twice**. What is the area under this portion of the velocity curve? Remember the convention that area under a curve above the axis is considered positive and area below the axis is considered negative. How would your answer change if you did this for Animation 2?

Problem 2.6

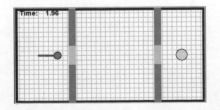

Shown is a red golf ball that you are to putt into the green hole (position is given in centimeters and time is given in

seconds). Two time-dependent obstacles are in your way. Find the initial velocity that will score a hole-in-one. Show all your work leading up to this velocity.

Problem 2.7

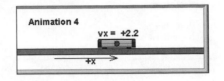

A 1.0-kg cart on a track experiences several different constant accelerations as shown in the animation (position is given in meters and time is given in seconds). The red dot shows you where position measurements are taken. What is the acceleration of the cart in each animation?

Note: *In Animation 1, the click-drag to read coordinates is disabled. In Animations 4-6, the time display is disabled. In Animations 2 and 3, the calculation for the velocity is disabled.*

Problem 2.8

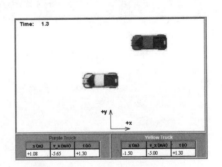

The purple truck is catching up to the yellow truck (position is given in meters and time is given in seconds).

a. If the trucks continue, at what clock reading, *t*, will the purple truck pass the yellow truck?

b. At what position, *x*, does the purple truck pass the yellow truck?

c. On one graph, plot *x* vs. *t* for each truck. Verify your answers for parts (a) and (b).

Problem 2.9

The animation simulates the motion of a helium balloon with the effect of air resistance neglected. A graph of *y* velocity vs. time for the balloon is also shown (position is given in meters and time is given in seconds).

a. Is the *y* velocity increasing, decreasing, or constant?

b. Is the *y* acceleration increasing, decreasing, or constant?

c. What is the acceleration of the balloon?

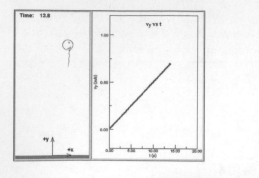

Problem 2.10

At $t = 10$ s, the rope holding cargo on a hot-air balloon is cut and the cargo is in free fall (position is given in meters and time is given in seconds). The table gives you the coordinates of the cargo.

a. What is the velocity of the cargo before the rope is cut?

b. At what instant does the cargo reach its maximum height? Calculate t and compare it to what you measure in the animation (to the nearest 0.05 s).

c. What is the maximum height (relative to the ground) reached by the cargo? Calculate this height and compare it to what you measure in the animation.

d. Calculate the instantaneous velocity of the cargo at the instant (just before) it hits the ground.

Problem 2.11

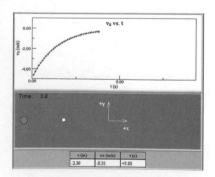

A golf ball is putted on a level, but wet, green. A top view is shown in the animation (position is given in meters and time is given in seconds).

a. Is the velocity of the golf ball constant, increasing, or decreasing during the interval from $t = 0$ to $t = 5$ s?

b. Is the speed of the golf ball constant, increasing, or decreasing during the interval from $t = 0$ to $t = 5$ s?

c. What does the area under the velocity vs. time graph (from zero to the curve) from $t = 0$ to $t = 9.4$ s correspond to?

d. What is the ball's average acceleration between $t = 0$ and $t = 9.4$ s?

e. Is the acceleration of the golf ball constant, increasing, or decreasing during the interval from $t = 0$ to $t = 5$ s?

Problem 2.12

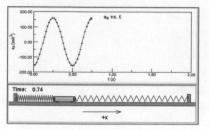

Two springs are attached to the ends of a cart that is on a cart track. If the cart is pulled back and released, it will move back and forth as shown in the animation. The graph shows the acceleration of the cart as a function of time (position is given in meters and time is given in seconds).

a. At what time(s) is the acceleration of the cart equal to zero?

b. Where is the acceleration a maximum? A minimum?

c. What is the area under the graph for the interval from $t = 0.5$ s to $t = 1.0$ s?

Problem 2.13

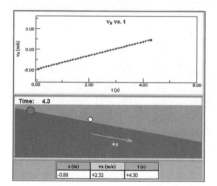

A golf ball is putted up a steep hill on a green. A side view is shown in the animation (position is given in meters and time is given in seconds). The positive x direction is defined to be parallel to the hill and down the hill. What should the minimum initial velocity of the ball be in order to make it into the hole located at $x = -3.6$ m?

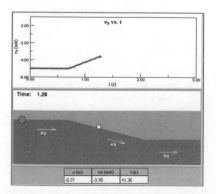

Problem 2.14

Press "play" to see a putted golf ball roll on a green as shown in the animation (position is given in meters and time is given in seconds).

a. What is the acceleration of the ball for each region (the lowest flat surface, the hill, and the highest flat surface)?

b. When the ball is on the hill, would you say that the velocity of the ball is increasing, decreasing, or constant? What about its speed?

c. Suppose you want to putt the ball so that it just barely makes it to the hole at the top of the hill. What should the initial velocity of the ball be (i.e., the velocity of the ball at the bottom of the hill)?

Problem 2.15

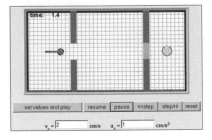

Shown is a red golf ball that you are to putt into the green hole (position is given in centimeters and time is given in seconds). Two time-dependent obstacles are in your way. Find the initial velocity and the acceleration that will score a hole-in-one. Show all your work leading up to this velocity and acceleration.

Problem 2.16

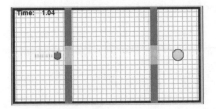

Shown is a red golf ball that you are to putt into the green hole (position is given in centimeters and time is given in seconds). Two time-dependent obstacles are in your way. Find the initial velocity and the acceleration that will score a hole-in-one. Show all your work leading up to this velocity and acceleration.

Problem 2.17

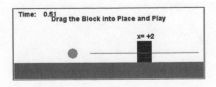

Drag the black rectangle into position (between $x = 0.5$ m and $x = 3.5$ m) before you play the animation (position is given in meters and time is given in seconds). Play the animation in order to measure the time when the ball collides with the rectangle. After the ball collides with the rectangle, you must click "reset" to move the rectangle to a new position and play the animation again.

a. How would you describe the motion of the ball before it collides with the rectangle? How many measurements are needed to confirm your description?

b. Is the x trajectory of the ball best described using the equation for constant velocity or the equation for constant acceleration? Support your answer with actual measurements.

c. Find an equation for the x trajectory of the ball (before it hits the rectangle). What minimum number of measurements is required to determine this trajectory?

Problem 2.18

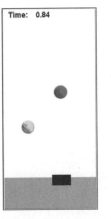

A giant tennis ball launcher shoots a giant red tennis ball straight up into the air. At the instant it leaves the launcher ($t = 0$ s), a giant green tennis ball is dropped from rest (position is given in meters and time is given in seconds).

a. If you want the red ball to return to the launcher at the same instant that the green ball hits the ground, what should the red tennis ball's initial velocity be?

b. With the initial velocity you calculated in part (a), what would be the maximum height of the red ball shot from the launcher?

Make sure to measure the position of each ball consistently.

Problem 2.19

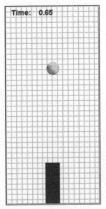

A giant tennis ball launcher shoots a giant tennis ball straight up into the air (position is given in meters and time is given in seconds). Because of the cylinder of the launcher, you can't see where the ball is first launched (assume it is launched at $t = 0$ s). If you double the initial velocity of the ball, what is the new position of the maximum height?

Two-Dimensional Kinematics

Topics include vectors (decomposition and addition), inclined plane, two-dimensional kinematics, projectile motion, and uniform circular motion.

Illustration 3.1: Vector Decomposition

A red vector is shown on a coordinate grid, and several properties of that vector are given in a data table (position is given in meters). How can you represent this vector? There are two ways: component form and magnitude and direction form. Both ways of representing vectors are correct, but in different circumstances one way may be more convenient than the other.

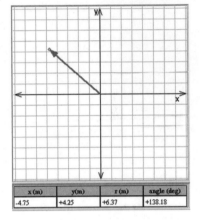

x (m)	y(m)	r (m)	angle (deg)
-4.75	+4.25	+6.37	+138.18

You can drag the head of the vector by click-dragging the small circle at the vector's head.

Magnitude and Direction Form: When you think of a vector, such as the one depicted, you are thinking of the magnitude and direction form. We describe the magnitude as the size of the vector (depicted in the table as *r*, which is always a positive number) and the direction as an angle (also depicted in the table and given in degrees). This angle is measured from the positive *x* axis to the direction that the vector is pointing.

Component Form: When you are solving problems in two dimensions, you often need to decompose a vector into the component form. So how do you do that? Look at the **show components** version of the animation. As you drag around the red vector, the maroon vectors show you the *x* and *y* components of the red vector (these are also shown in the table as *x* and *y*). Try to keep the length of the vector the same and change the angle. How did the components change with angle? As the angle gets smaller, the *x* component of the vector gets larger (it approaches the magnitude of the vector) and the *y* component of the vector gets smaller (it approaches zero). As the angle approaches 90° the *x* component of the vector gets smaller (it approaches zero) and the *y* component of the vector gets larger (it approaches the magnitude of the vector). Mathematically this is described by the statement that

$$x = r\cos(\theta) \quad \text{and} \quad y = r\sin(\theta).$$

Once in component form, we can of course go back to magnitude and direction form by using the relationships

$$r = (x^2 + y^2)^{1/2} \quad \text{and} \quad \theta = \tan^{-1}(y/x).$$

Notice that the magnitude of the vector, here *r*, must be positive as stated above.

Illustration 3.2: Motion on an Incline

Galileo was the first person to realize that a well-polished (a very slippery or frictionless) inclined plane could be used to reduce the effect of gravity. He realized that if you started with a vertical incline (angle of 90°), the scenario was equivalent to free fall. If the incline was horizontal (an angle of 0°), the object would not move at all.

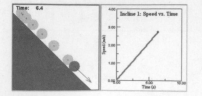

He therefore reasoned that as you decreased this angle from 90° the acceleration would decrease. He was able to measure this acceleration and, thereby, determine the acceleration due to gravity. Mathematically, this amounts to the realization that as a function of the incline's angle

$$g_{\text{eff}} = g \sin(\theta),$$

where g_{eff} is the acceleration down the incline. See Illustration 2.5 and Chapter 4 for more details.

By varying the type of object sliding down the slippery (frictionless) incline, he was able to show that all objects accelerate at the same rate. Try the experiment for yourself (time is given in seconds and distance is given in meters) with the above three animations.

Galileo started his objects from rest on an incline. What did he find from his experiments? Galileo's conclusion was that during successive equal-time intervals the objects' successive displacements increased as odd integers: 1, 3, 5, 7, ... What does that really mean? Consider the chart below, which converts Galileo's data into data we can more easily understand (data shown for an incline whose angle yields an acceleration of 2 m/s^2):

Elapsed time (s)	Displacement during the time interval (m)	Total displacement (m)
1	1	1
2	3	4
3	5	9
4	7	16

The third column is constructed by adding up all of the previous displacements that occurred in each time interval to get the net or total displacement that occurred so far. What is the relationship between displacement and time? The displacement is related to the square of the time elapsed. Does that idea look familiar? It should. We found in Chapter 2 that $x = x_0 + v_0 t + 0.5at^2$ or that, for no initial velocity, $x = x_0 + 0.5at^2$ or that Δx is proportional to t^2.

Illustration 3.3: The Direction of Velocity and Acceleration Vectors

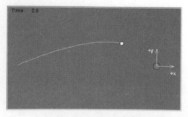

A putted golf ball rolls across the green as shown in the animation (position is given in meters and time is given in seconds). The animation represents the top view of the motion of the ball.

What is the direction of the velocity of the golf ball at any instant of time? **View the velocity vector** to check your answer (notice that a black line tangent to the object's path is also drawn). The direction of the velocity vector is determined by a fairly simple rule: It is always tangent to the path and in the direction of motion. The "direction of motion" is basically the direction of the object's displacement during a very small time interval. Since the displacement divided by this very small time interval approaches the instantaneous velocity (see **Illustration 2.3**), the instantaneous velocity must point in the direction of motion. This also directly follows from the definition of instantaneous velocity as the derivative of the position vector with respect to time.

What about the acceleration vector? It points in the direction of the change in velocity during any small time interval. This is again a result of the definition of acceleration. However, it also follows an interesting rule.

The acceleration vector can be resolved into two components, a component tangent to the path (called the tangential acceleration) and a component perpendicular to the path (called the radial component). **View the velocity and acceleration vectors**. The radial component of the acceleration is related to the change in direction of the velocity vector and points along the radius of curvature. The tangential component of the acceleration is related to the change in the magnitude of the velocity vector. In other words, it is related to the change in the speed. If the object is slowing down, then the tangential component of the acceleration is opposite to the velocity. If the object is speeding up, then the tangential component of the acceleration is in the same direction as the velocity.

Click here to view the velocity vector (blue), acceleration vector (orange), and acceleration components (yellow for the tangential component and red for the radial component).

Illustration 3.4: Projectile Motion

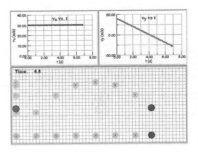

A purple ball undergoes projectile motion as shown in the animation (position is given in meters and time is given in seconds). The blue and red objects illustrate the *x* and *y* components of the ball's motion. Ghost images are placed on the screen every second. To understand projectile motion, you must first understand the ball's motion in the *x* and *y* directions separately (any multidimensional motion can be resolved into components).

Consider the *x* direction. Notice that the *x* coordinate of the projectile (purple) is identical to the *x* coordinate of the blue object at every instant. What do you notice about the spacing between blue images? You should notice that the displacement between successive images is constant. So what does this tell you about the *x* velocity of the projectile? What does it tell you about the *x* acceleration of the projectile? This should tell you that the object moves with a constant velocity in this direction (which is also depicted on the left graph).

Now consider the *y* direction. Notice that the *y* coordinate of the projectile (purple) is identical to the *y* coordinate of the red object at every instant. What do you notice about the spacing between successive images for the red object? You should notice that the displacement between successive images gets smaller as the object rises and gets larger as the object falls. This means that it has a downward acceleration. By studying the right-hand graph, we can also see that the *y* acceleration is constant.

A particularly important point to understand for the motion of a projectile is what happens at the peak. What is the velocity of the projectile at the peak? This is a tricky question because you have a good idea that the *y* velocity is zero. However, does this mean that the velocity is zero? Remember that velocity has two components, v_x and v_y. At the peak, v_x is not zero. Therefore, the velocity at the peak is not zero. **Click here** to view the velocity and acceleration vectors.

Illustration 3.5: Uniform Circular Motion and Acceleration

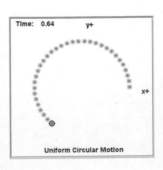

Uniform Circular Motion

Uniform circular motion is an interesting mixture of one- and two-dimensional concepts. For uniform circular motion the speed of the object must be constant. This is the uniform in uniform circular motion. **Show the ball undergoing uniform circular motion**. So is an object moving in a circle with a constant speed accelerating? Yes! Why? The velocity is changing with time. Watch the animation (position is shown in meters and time is shown in seconds). The animation depicts an object moving in a circle at a constant speed. To determine the acceleration, we need to consider the change in velocity for a change in time.

Since the speed does not change in time, what does change in time? It is the direction of the velocity that changes with time. **Draw two velocity vectors corresponding to two different times** to convince yourself that the direction of the velocity changes with time. Recall that velocity has both a direction (which always points tangent to the path, the so-called tangential direction) and a magnitude, and either or both can change with time. Where does the change in velocity point? **Calculate the acceleration.** The acceleration points toward the center of the circle. This direction—toward the center of the circle—is called the centripetal or center-seeking direction. It is often also called the radial direction, since the radius points from the center of the circle out to the object (the net acceleration points in the opposite direction).

Therefore, for uniform circular motion, the acceleration always points toward the center of the circle, *no matter the cause.* This is despite the fact that the velocity and the acceleration point in changing directions as time goes on. However, we get around this apparent difficulty in describing direction by defining the centripetal or radial direction and the tangential direction (the direction tangent to the circle). These directions change, but the velocity is always tangent to the circle and the acceleration is always pointing toward the center of the circle. The following animation **shows velocity and acceleration** as the object undergoes uniform circular motion.

Illustration 3.6: Circular and Noncircular Motion

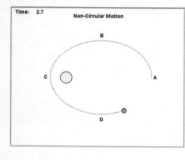

A planet (green) orbits a star (yellow) as shown in the two animations. One animation depicts the **uniform circular motion** of a planet and the other one depicts the **noncircular motion** of a planet (position is given in 10^3 km and time is given in years). This illustration will compare the two motions by focusing on the velocity and the acceleration of the planet in each of the animations.

Start the **uniform circular motion** animation and watch the planet's motion. How would you describe the motion of the planet (consider velocity and acceleration)? The speed of the planet is certainly a constant since the motion of the planet is uniform. But using our usual xy coordinates, the velocity certainly changes with time. Recall that the term velocity refers to both the magnitude and direction. However, if we use the radial and tangential directions to describe the motion of the planet, the velocity can be described as tangential and the acceleration can be described as being directed along the radius (the negative of the radial direction). **Click here** to view the velocity vector (blue) and the black line tangent to the path. **Click here** to view the acceleration vector (red), too. Notice that the acceleration vector points toward the star at the center of the circle.

Start the **noncircular motion** animation and watch the planet's motion. How would you now describe the motion of the planet (consider velocity and acceleration)? The speed of the planet is certainly no longer a constant since the motion of the planet is no longer uniform. Again using our usual xy coordinates, the velocity certainly changes with time since now both the direction and the magnitude change. However, if we use the radial and tangential directions to the path of the planet, the velocity can be described as tangential and the acceleration can be described as being directed along the radius. **Click here** to view the velocity vector (blue) and **click here** to view the acceleration vector (red), too. Notice that the velocity and the acceleration are no longer perpendicular for most of the orbit of the planet.

Notice that between points A and C the planet is speeding up, and between points C and A the planet is slowing down. This means that at points A and C the tangential component of acceleration is zero. It turns out that for a planet orbiting a star (if there are no other planets or stars nearby) the acceleration of the planet is directed exactly toward the star whether the motion of the planet is uniform or not.

Exploration 3.1: Addition of Displacement Vectors

Suppose that you use a radar system to track an airplane (the red circle) and the airplane travels according to the animation shown.

a. Draw a vector for the displacement of the airplane from $t = 0$ s to $t = 8$ s. To do this, click the "Draw Vector" button. When a vector appears, drag it to the position of the airplane at $t = 0$ s. Then play the animation, stop it at $t = 8$ s, and adjust the tip of the vector until it is at this position.

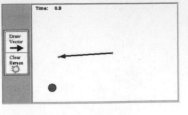

b. Now draw a displacement vector for the airplane from $t = 8$ s to $t = 16$ s. Use the same procedure as before. Be sure to click the "Draw Vector" button so that you can have a new vector to work with. You should see both the first displacement vector and the second displacement vector.

c. Now draw a displacement vector for the airplane from $t = 0$ s to $t = 16$ s. Use the same procedure as before. What do you notice? To add vectors like this, you can connect the vectors from tail to head. The result, called the resultant vector, is the vector drawn from the tail of the first vector to the head of the last vector.

d. **Click here** to view the correct answer. How does your result compare to the correct answer?

Exploration 3.2: Run the Gauntlet, Controlling *x*, *v*, and *a*

Drag the tip of the arrow to control the position, velocity, or acceleration of the object depending on which animation you choose.
Use the animation to answer the following questions (position is given in meters and time is given in seconds).

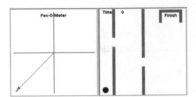

a. Can you navigate to the goal on the right? We call this challenge Running the Gauntlet.

b. Which controller (the position, velocity, or acceleration) is harder to use? Why?

Exploration 3.3: Acceleration of a Golf Ball That Rims the Hole

A putted golf ball "rims" the hole as shown in the animation (position is given in centimeters and time is given in seconds). Velocity vectors for the ball at the instant just before it hits the hole and the instant just after it hits the hole are shown. Note that the ball's speed does not change upon hitting the edge of the hole; this would not occur for an actual golf ball that rims the hole.

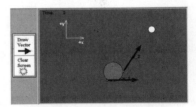

Suppose we want to find the average acceleration of the golf ball at some instant when it is in contact with the hole.

a. Draw the change-in-velocity vector using the velocity vectors shown. Click "Draw Vector" to add a vector to the animation and click "Clear Screen" to erase all drawn vectors.

b. What is the magnitude and direction of the change in velocity during this interval?

c. What is the average acceleration during this interval?

d. For the animation shown, at what instant do you think the instantaneous acceleration will equal the average acceleration of the golf ball during the time interval from 0.9 s to 1.2 s?

e. Click here to view the acceleration vector. If your change-in-velocity vector is still drawn on the screen, then you can stop the animation at the point where the acceleration vector and change-in-velocity vector match up. Did this occur at the instant you predicted?

Exploration 3.4: Space Probe with Constant Acceleration

When you studied projectile motion, you learned that for projectile motion the *x* acceleration is zero and constant (which results in a constant *x* velocity) and the *y* acceleration

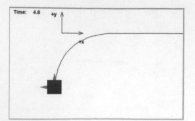

is constant and downward toward Earth with a magnitude of 9.8 m/s^2. What mathematical curve describes the shape of the path of the projectile? Its shape is a parabola. It turns out that the shape of the path of any object that has constant acceleration and an initial velocity that is in a different direction than the acceleration is a parabola.

In the animation shown (position is given in meters and time is given in seconds), a space probe has engines that can fire on all four sides. Two of the engines engage at $t = 2$ s. The acceleration is constant and zero before the engines engage, and it is constant (but not equal to zero) after the engines engage.

a. What is the direction of the x component of the acceleration after the engines engage?

b. What is the y velocity before the engines engage?

c. After the engines engage, how is the y velocity different?

d. Now **click here** to view the velocity and acceleration vectors. Do they match what you predicted?

Exploration 3.5: Uphill and Downhill Projectile Motion

A projectile is launched at $t = 0$ s (position is given in meters and time is given in seconds). You may change the projectile's launch angle and initial speed and the height of the hill by using the text boxes and clicking the "set values and play" button.

For $h = 0$ m, vary the projectile's launch angle and initial speed and consider the following questions.

a. For a given initial speed, what launch angle will provide the maximum range of the projectile?

b. For the value of launch angle in (a), what is the value of the initial speed that will hit the target?

c. What other value(s) of the projectile's launch angle and initial speed will enable the projectile to hit the target?

d. Are these values unique?

e. What is the general relationship between launch angle and initial speed?

For $h = 10$ m, vary the projectile's launch angle and initial speed and consider the following questions.

f. For a given initial speed, what launch angle will provide the maximum horizontal displacement of the projectile?

g. What value(s) of the projectile's launch angle and initial speed will enable the projectile to hit the target?

h. Are these values unique?

i. Are these values the same as in (c)?

For $h = -10$ m, vary the projectile's launch angle and initial speed and consider the following questions.

j. For a given initial speed, what launch angle will provide the maximum horizontal displacement of the projectile?

k. What value(s) of the projectile's launch angle and initial speed will enable the projectile to hit the target?

l. Are these values unique?

m. Are these values the same as in (c) and (g)?

Exploration 3.6: Uniform Circular Motion

A point (red) on a rotating wheel is shown in the animation (position is given in meters and time is given in seconds).

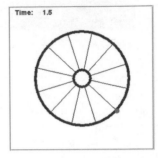

a. Note that the speed of the red point is constant. Is its velocity constant?

b. Click here to view the velocity vector. After viewing the vector rethink your answer: Is the velocity of the red point constant?

c. What is the direction of the red point's acceleration vector? **Click here** to view the acceleration and velocity vectors.

d. How does the speed of the red point compare to the speed of another point, say a green one, which is at only half the radius of the red point? **Click here** to view both points. For clarity the green point is shown on the opposite side from the red one.

e. Why is the speed of the green point less than the speed of the red point?

f. How does the magnitude of the acceleration of the red point compare to the magnitude of the acceleration of the green point? **Click here** to view both points and their velocity and acceleration vectors.

Problems

Problem 3.1

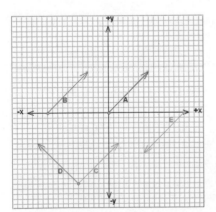

Five vectors are shown on the coordinate grid (position is given in meters). You can change the position of a vector by click-dragging at the base of the vector. Click **restart** to return the vectors to their original positions.

a. Rank the x components of the five vectors shown (smallest to largest).

b. Rank the y components of the five vectors shown (smallest to largest).

c. What are the components of the vector that results when Vector B is added to Vector D?

Problem 3.2

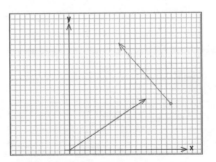

Two vectors are shown on the coordinate grid (position is given in meters).

a. What are the x and y components of the blue vector?

b. What are the x and y components of the red vector?

Now drag the circle at the tail of the red vector so that it is on top of the blue vector's head. The vector sum is now a vector that reaches from the tail of the first (blue) vector to the head of the second (red) vector.

c. What are the components of this vector sum?

d. How do they relate to the components of the original (blue and red) vectors?

Problem 3.3

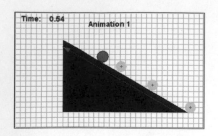

The animations represent the motion of a ball on various surfaces (position is given in meters and time is given in seconds). The "ghosts" are placed at equal time intervals. Such a picture is often called a motion diagram.

Answer the following questions using the coordinate system specified in each animation by the **red arrow**. Please indicate ties by ( ). *For example a suitable response could be: 1,(2,3),4,5,6.* For parts (a), (b), and (c), use the ghost images to qualitatively rank the quantities.

a. Rank each case from highest to lowest displacement.

b. Rank each case from highest to lowest final velocity.

c. Rank each case from highest to lowest acceleration (assume constant acceleration).

Now use the usual x and y coordinates that you can access by click-dragging in the animation.

d. Calculate the displacement vector for each animation.

e. Calculate the acceleration vector for each animation (assume that in Animation 3 and Animation 6 the ball starts at rest and that in Animation 1 and Animation 4 the ball ends at rest).

Problem 3.4

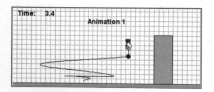

A bowling ball is lifted from rest onto a shelf by an external agent (position is given in meters and time is given in seconds). For each quantity below, rank the animations (numbered 1 through 4) from least to greatest. Ties in ( ) please. *For example, a suitable response could be: 1,(2,3),4.*

Quantity	Ranking
magnitude of displacement	
magnitude of average velocity	

Problem 3.5

A flying helicopter is shown in the animation (position is given in meters and time is given in seconds).

a. Sketch a graph of x position vs. *time* for the helicopter.

b. Sketch a graph of y position vs. *time* for the helicopter.

c. What is the x velocity of the helicopter at any instant?

d. What is the y velocity of the helicopter at any instant?

e. What is the speed of the helicopter at any instant?

Problem 3.6

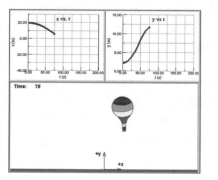

A hot-air balloon travels as shown in the animation (position is given in meters and time is given in seconds). The x and y positions of the hot-air balloon's basket are shown in the graphs.

a. During approximately what time interval is the magnitude of the y velocity increasing?

b. During approximately what time interval is the magnitude of the y velocity decreasing?

c. At approximately what instant of time does the y acceleration change from positive to negative?

d. What is the y velocity from $t = 87$ s until $t = 200$ s?

e. What is the y acceleration from $t = 87$ s until $t = 200$ s?

f. What is the x velocity from $t = 87$ s until $t = 200$ s?

g. What is the x acceleration from $t = 87$ s until $t = 200$ s?

h. What is the x displacement from $t = 0$ s until $t = 200$ s?

i. What is the y displacement from $t = 0$ s until $t = 200$ s?

Problem 3.7

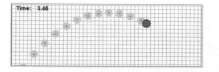

A projectile is launched as shown in the animation (position is given in meters and time is given in seconds). Where does the ball reach its minimum speed, and what is its speed when it gets there?

Problem 3.8

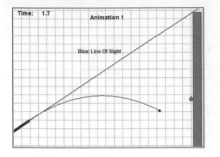

A hunter (off screen) aims his rifle at an apple in a tree as shown in the animation (position is given in meters and time is given in seconds). At the instant the bullet leaves the rifle, the apple starts falling from rest. Which of the above animations correctly depicts the hunter's aim that hits the apple?
Note: All three show the apple being hit, but only one animation depicts correct physics.

Problem 3.9

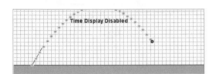

A projectile is launched with an initial speed of 20 m/s as shown in the animation (position is given in meters). The time display is suppressed, but you can still click–drag to get coordinates. A line is also shown that represents the initial direction of the velocity.

a. What is the launch angle?
b. What are v_{0x} and v_{0y}?
c. What is the maximum height the projectile will reach?
d. How long does it take the projectile to reach that height?
e. What is the total time that the projectile is in the air?

Problem 3.10

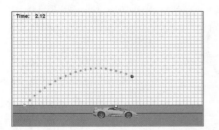

A projectile is launched when the yellow Lamborghini (not shown to scale) goes by at $t = 0$ s (position is given in meters and time is given in seconds). You may change the projectile's launch angle and initial speed by using the text boxes and clicking "set values and play." Find the relationship between v_0 and θ such that the projectile will always hit the car. When you determine the relationship, make sure to test it with a few values of v_0 and θ.

Problem 3.11

A red ball slides off a table as shown in the animation (position is given in meters and time is given in seconds). Ignore air friction. If the ball collides with the other table such that v_y remains the same and v_x changes sign upon collision, where will the red ball land?

Problem 3.12

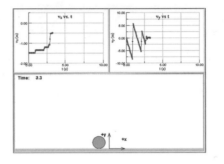

A bouncing basketball is shown in the animation (position is given in meters and time is given in seconds). While the basketball is in the air, its motion is characterized by projectile motion.

a. What is the average y acceleration of the ball during the interval of the first bounce, that is from $t = 0.85$ s to $t = 0.95$ s?
b. What is the average x acceleration of the ball during the interval of the first bounce, that is from $t = 0.85$ s to $t = 0.95$ s?
c. What is the magnitude of the acceleration of the ball during this interval?
d. While the ball is in the air (between the bounces), is the x velocity increasing, decreasing, or constant? What is the x acceleration while the ball is in the air?
e. As the ball rolls to a stop, between $t = 3.0$ s and $t = 8.0$ s, what is the x acceleration of the ball? What is the y acceleration of the ball during this interval?

Problem 3.13

The animation shows a putted golf ball as it travels toward the hole (position is given in meters and time is given in seconds).

a. Is the acceleration of the golf ball between $t = 0$ and $t = 4.2$ s constant, increasing, or decreasing?
b. What is the average acceleration of the golf ball during this time interval?

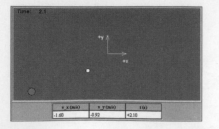

c. Calculate the x displacement of the golf ball from $t = 0$ to $t = 4.2$ s and show that it is the same as what you measure on the animation.

d. Calculate the y displacement of the golf ball from $t = 0$ to $t = 4.2$ s and show that it is the same as what you measure on the animation.

e. What is the magnitude of the displacement of the golf ball from $t = 0$ to $t = 4.2$ s?

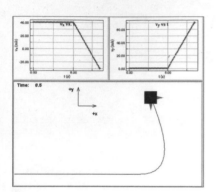

Problem 3.14

In the animation shown, a space probe has engines that can fire on all four sides (position is given in meters and time is given in seconds). Two of the engines engage at $t = 5$ s. The acceleration is constant and zero before the engines engage, and it is constant (but not equal to zero) after the engines engage.

a. What is the initial velocity of the probe just before the engines fire?

b. What is the acceleration of the probe after the engines fire?

c. Assuming the engines continue to fire in the same way, what will be the position and velocity of the probe at $t = 25$ s?

d. At what instant is $v_x = 0$? At what instant is $v_y = 0$?

Problem 3.15

An object travels along a circular path as shown in the animation (position is given in meters and time is given in seconds).

a. At $t = 2$ s, what is the direction of the velocity of the object?

b. At $t = 2$ s, what is the approximate direction of the acceleration of the object? You do not need to give an exact direction, just an approximate direction based on what you know about the direction of the radial component and

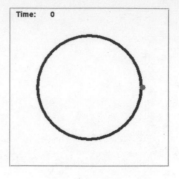

the direction of the tangential component.

c. At $t = 4$ s, what is the direction of the velocity? If it is zero, indicate so.

d. At $t = 4$ s, what is the approximate direction of the acceleration? If it is zero, indicate so.

e. At $t = 6$ s, what is the direction of the velocity?

f. At $t = 6$ s, what is the approximate direction of the acceleration?

Problem 3.16

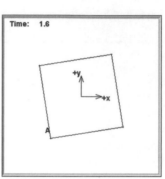

A square rotates as shown in the animation (position is given in meters and time is given in seconds). A corner of the square is labeled A. For all of the following questions consider the motion of point A from $t = 0.5$ s to $t = 2.5$ s.

a. What is the displacement of point A during this time interval?

b. What is its distance traveled during this interval?

c. What is its average velocity during this interval?

d. What is its average speed during this interval?

Problem 3.17

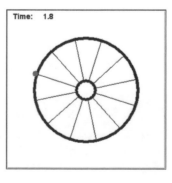

A point (red) on a rotating wheel is shown in the animation (position given in meters and time given in seconds).

a. What is the period of the wheel (the time it takes the red point to complete one revolution)?

b. What is the speed of the red point?

c. What is the magnitude of the acceleration of the red point?

d. At $t = 5.0$ s, what is the direction of the velocity vector and what is the direction of the acceleration vector for the red point?

Newton's Laws

Topics include Newton's law and forces (weight, normal force, and tension).

Illustration 4.1: Newton's First Law and Reference Frames

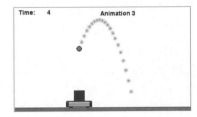

On first glance it may seem like Newton's first law (an object at rest remains at rest and an object in motion remains in motion unless acted on by a net force) is contained within Newton's second law. This is actually not the case. The first law is also a statement regarding reference frames. This is the information NOT contained in the second law. Sometimes the first law is also called the law of inertia. It defines a certain set of reference frames in which the first law holds, and these reference frames are therefore called inertial frames of reference. Put another way, Newton's first law states that if the net force on an object is zero, it is possible to find at least one reference frame in which that object is stationary. There are many frames in which the object is moving with a constant velocity.

A ball popper on a cart (not shown to scale) is shown moving on a track in three different animations (position is given in meters and time is given in seconds). In each animation the ball is ejected straight up by the popper mechanism at $t = 1$ s.

Let us first consider **Animation1**. In this animation the cart is stationary. But is it really? We know that we cannot tell if we are stationary or moving at a constant velocity (in other words in an inertial reference frame). Recall that if we are moving relative to Earth at a constant velocity we are in an inertial reference frame. So how can we tell if we are moving? How about the cart? We cannot tell if there is motion as long as the relative motion with respect to Earth can be described by a constant velocity. In Animation 1 the cart *could* be stationary. In this case, we expect—and actually see—that the ball lands back in the popper. However, if the cart was moving relative to Earth and we were moving along with the cart, the motion of the ball and the cart would look exactly the same!

What would the motion of this ball and cart look like if the cart moved relative to our reference frame (or if we moved relative to its reference frame)? Animations 2 and 3 show the motion from different reference frames. What do these animations look like? Both animations resemble projectile motion. The motion of the ball looks like motion in a plane as opposed to motion on a line. Does the ball still land in the popper? Would you expect this? Sure. There is nothing out of the ordinary going on here. Since there are no forces in the x direction, the motion of the ball (and cart) should be described by constant velocity in that direction. Therefore the ball and the cart have the same constant horizontal velocity.

For more on reference frames and relative motion, see Chapter 9.

Illustration 4.2: Free-Body Diagrams

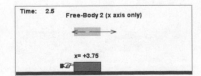

In Illustration 4.2 an 8-kg block is pushed across the floor (position is given in centimeters and time is given in seconds). Along with the motion, several possible free-body diagrams are drawn for both the x and y components of the force. Note that only one of the possible free-body diagrams is correct for each component.

Look at the motion of the block by pressing "play." How do we analyze the motion of the block using forces? Well, the first thing we do is draw a picture that shows only the object and the direction of the forces. The picture we draw is called a free-body diagram. First we will analyze the forces in the x direction and then the forces in the y direction.

Consider the forces in the x direction (Free-Body x). What forces act? How big are they? How do we know? Click each of the four x-direction free-body diagrams. Which one do you think is correct? Usually we know all of the forces that act, but here we just know of the push that is shown in **Free-Body 1x**. Is that the only force acting in the x direction? Newton's second law says that a net force acting on an object means that the object must be accelerating (the object's velocity changes). Does the block's velocity change? No (You can tell either by looking at the block's motion or by calculating the velocity and showing that it does not change.); therefore there must be another force acting, that of friction that opposes the motion. This eliminates **Free-Body 1x** and **Free-Body 3x** because they show only one force. The second force not only opposes motion, but in this animation it is exactly the same size as the push. This means that **Free-Body 2x** is not correct either. Therefore, **Free-Body 4x** depicts the correct free-body diagram for the forces that act in the x direction. (The form of the frictional force will be considered in detail in Chapter 5.)

Now consider the forces in the y direction (Free-Body y). What forces act? How big are they? How do we know? Click each of the four y-direction free-body diagrams. Which one do you think is correct? Usually we know all of the forces that act, but here we just know of the force of gravity that is shown in **Free-Body 1y**. Is that the only force acting in the y direction? Since the block does not accelerate in the y direction, there must be another force acting. This eliminates **Free-Body 1y** and, **Free-Body 2y** because they show only one force. The force that is missing is the so-called normal force (the force of the table acting on the block) that opposes gravity. The normal force not only opposes motion in the y direction but here is exactly the same size as the gravitational force, the object's weight. This means that **Free-Body 3y** cannot be correct either. Therefore, **Free-Body 4y** depicts the correct free-body diagram for the forces that act in the y direction. Note that the normal force is not always equal to the weight. If there is an acceleration in the y direction or if the block is on an incline, the normal is not equal to the weight.

Note that in order to solve the complete problem we would draw all of the forces on one free-body diagram. We have done the analysis here by breaking up the motion into components. Given what we have said above, what does the complete free-body diagram look like? The **complete free-body diagram** animation shows the combination of the forces in the x direction and the forces in the y direction.

Illustration 4.3: Newton's Second Law and Force

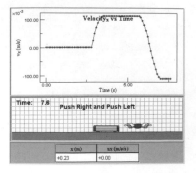

Although most physicists would agree that the concept of force is not as fundamental as the concept of a conservation law, it is still considered central to the study of physics. A force is a push, a pull, or any other interaction, exerted by one object on another object. We know from experience that a push or a pull often causes an object to move. This allows us to quantify the definition of force in terms of a quantity that was defined previously: acceleration.

If an object's mass remains constant, the magnitude of a force exerted on an object is proportional to the time rate of change of the velocity (i.e., acceleration). Specifically, $\Sigma\, F = m\, a$.

Use this definition as you consider the results of Illustration 4.3 (position is given in meters and time is given in seconds). Set the mass in the text box before you select the graph type: velocity or acceleration.

The two-handed image ("handy") interacts with the 1.0-kg cart in the animation if the image is near the left-hand or right-hand end of the cart. The arrow below the cart shows the direction and strength of the force exerted on the cart. You will have to move the image to keep it behind the cart since the interaction changes direction if the image passes through the center of the cart. Start the animation and explore it for a few minutes. Reset the animation if the cart goes off the end of the track.

Now select velocity (and then acceleration). Drag the handy image to the left of the cart and try to apply the force for as brief a period of time as you can. This will result in a force applied to the cart only for a short period of time and then no force will act. What do the resulting velocity and acceleration graphs look like? The velocity graph should show an increasing velocity for the instant handy is acting on the cart; then it should have a slope of zero. The velocity only changes when the force is acting. The acceleration graph should give a spike during the application of the force and be zero otherwise. Repeat the same process when the image is to the right of the cart. What changes? Because force is a vector, the applied force is now in the negative x direction. Therefore, the velocity and the acceleration are now both negative as well.

Now select velocity (and then acceleration). Drag the handy image to the left of the cart and then keep dragging it to the right as the cart moves. This will result in a constant force applied to the cart. What do the resulting velocity and acceleration graphs look like? The velocity graph should have a constant slope upward while the acceleration graph should give a constant acceleration during the application of the force. Repeat the process when the image is to the right of the cart. What changes? Because force is a vector, the applied force is now in the negative x direction. Therefore the velocity and the acceleration are now both negative as well.

What changes on the velocity and acceleration graphs will occur if the mass of the cart is doubled or decreased by a factor of two? Try it and find out. Since acceleration is equal to the force over the mass, an increase in mass means a smaller acceleration, and a decrease in mass means a larger acceleration.

Illustration 4.4: Mass on an Incline

A mass is on a frictionless incline as shown in the animation (position is given in meters and time is given in seconds). You may adjust m, the mass of the block (100 grams $< m <$ 500 grams), and θ, the angle of the incline ($10° < \theta < 45°$), and view how these changes affect the motion of the block.

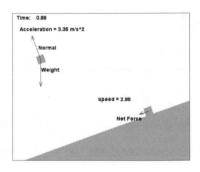

One of the first things to stress about this type of problem is that, for a suitable set of coordinates, while it is a two-dimensional problem, the motion of the block is one dimensional. Since the motion of the block is down the incline, let's choose that direction for the x axis. Since coordinate axes are perpendicular, let's also call the direction normal to the incline, the y axis. This does two things for us: The net force (and therefore the acceleration) is now on axis (the x axis) and we do not need to decompose the normal force. Check the box and click the "register values and play" button to see the free-body diagram for the block and the net force acting on the block.

What force determines the acceleration of the block? It is the part of the gravitational force that is down the incline ($mg \sin \theta$). Therefore, the other component of the gravitational force ($mg \cos \theta$) must be equal to the normal force since we do not see the block flying off of the incline. The acceleration of the block is g sin θ down the incline.

Now try changing the mass of the block. How do you think the block's acceleration will change as you change the mass?

Now change the angle of the incline. How do you think the angle of the incline affects the acceleration of the block? In the animation you are limited to $10° > \theta < 45°$.

Can you predict, from either the formula or the animation, what will happen to the normal force and the acceleration when $\theta = 0°$ and $\theta = 90°$?

Illustration 4.5: Pull Your Wagons

Two toy wagons, attached by a lightweight rope (of negligible mass), are pulled with a constant force using another lightweight rope (again of negligible mass) as shown in the animation (position is given in centimeters and time is given in seconds). The mass of the red wagon is 2.0 kg and the mass of the blue wagon is 1.2 kg. What is the force of the hand on the rope, and what is the tension in the rope joining the two wagons? To answer these questions, you must apply Newton's second law. However, when applying Newton's second law, you must first define the system that you are considering. Let's answer each question separately.

What is the force of the hand on the rope? Begin by defining the system to which we will apply Newton's second law. Since we want to determine the force of the hand on the rope, we start by choosing the rope to be our system. What forces act on the rope? It may help to draw a free-body diagram. **View a free-body diagram of the rope along with the animation**.

Note that there are two forces on the rope, the force of the hand on the rope in the $+x$ direction and the force of the red wagon on the rope in the $-x$ direction. These two forces are equal in magnitude; therefore, the net force on the rope is zero. But how can it be zero if the rope's acceleration is NOT zero? Since the mass of the rope is negligible, we set the mass of the rope equal to zero, and because of Newton's second law, the net force on the rope is zero. Of course, the rope's mass, in reality, is not zero; however, it's close enough to zero that we can say that it's approximately zero. Ultimately, this means that the tension in the rope is constant.

What is the force of the red wagon on the rope? What system should we now consider? We have two choices: (1) Consider the red wagon as the system or (2) consider the blue wagon, the red wagon, and the rope between them as the system. Either choice can lead you to the answer, but choice (2) is the most direct and best choice in order to solve the problem most quickly.

Consider the two wagons and rope as the system as depicted in **this animation**. The gray box represents the system. Now, draw a free-body diagram for the system and then **view the animation again** in order to check your answer. Once you have drawn the free-body diagram and identified the forces, you can apply Newton's second law, determine the force of the red wagon on the rope, and solve for the force of the hand on the rope.

That answers the first question. Now for the second question: What is the tension in the rope joining the two wagons? You can answer this question by following a similar procedure. Identify your system, draw a free-body diagram, and apply Newton's second law.

Illustration 4.6: Newton's Third Law, Contact Forces

Illustration 4.6 shows graphs of position, velocity, and acceleration vs. time for a 2-kg red block (not shown to scale) pushed by a 12-N force on a frictionless horizontal

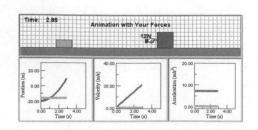

surface (position is given in meters and time is given in seconds). The red block is in contact with (and therefore pushes on) the green 1-kg block (also not shown to scale). Click here to **show and play the physical situation**. Note that on the position vs. time graph each block's trajectory is shown in a color-coded $x(t)$ function, while in the velocity and acceleration vs. time graphs, a single $v(t)$ or $a(t)$ is shown (the blocks move together and therefore must have the same velocity and acceleration). The blocks may not move together when you set the contact forces.

Now it is up to you to determine what contact forces are required to make the motion of the blocks physical. When you are ready, select the "set values and play" button with the default forces. What happens? The red block "moves through" the green one because the forces are not correct. The red block has the 12-N force acting on it and the green block has no forces acting on it. Of course each object's weight and normal force act in the vertical direction, but they cancel for each object. Here we are just considering the horizontal forces that could give a net force.

Try some values for the forces and check to see if you can get the same motion of the blocks and the same graphs as the physical situation.

Were you able to get the motion correct? Let us now go about it systematically instead of by exploring (or guessing). **Show and play the physical situation with both masses as one system**. If we look at things this way we have one object of mass 3 kg and a net force of 12 N, which means an acceleration of 4 m/s² (this is borne out by the acceleration graph).

What next? We could analyze the forces acting on the first mass, but let's analyze the second mass since it has only the first mass pushing on it. Because it has an acceleration of 4 m/s² and a mass of 1 kg, it must experience a force of 4 N from the push of the red mass. What about the red mass? Newton's third law says it must experience an equal and opposite force, here a force of −4 N. Try these values out (−4 N for the force on the red block and 4 N for the force on the green block) to see if you believe what Newton's third law says the forces should be.

Exploration 4.1: Vectors for a Box on an Incline

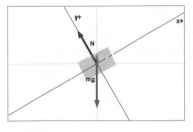

Exploration 4.1 represents a free-body diagram for a 20-N block on a 30° frictionless incline (the length of the vectors is given in newtons). The light gray lines represent the traditional xy axis, and the black lines represent the coordinates along the incline. The blue vector represents the normal force; the green vector represents the weight. You may move the tails of the blue and green vectors to add them and use the red vector to represent their resultant vector by dragging the red vector's tip.

a. Determine the resultant force from the diagram.

b. Determine the acceleration of the block.

Exploration 4.2: Change the Two Forces Applied

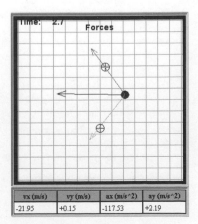

Drag either of the crosshair cursors or the ball (position is given in centimeters and time is given in seconds). The cursors each exert a constant force on the black ball (either attraction or repulsion) if they are **within 10 cm** of the ball. When the ball hits a wall, the wall exerts a force on the ball causing it to recoil. The green and red arrows display the forces due to each cursor, and the blue arrow represents the net force.

For attraction and repulsion, drag the black ball around to see the net force. When you get the ball and the cursors into an orientation you like, click the "play" button to see the effect of the forces on the ball. Briefly explain how and why the ball moves according to the forces applied.

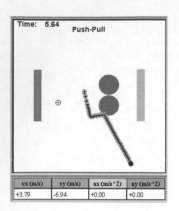

Exploration 4.3: Change the Force Applied to Get to the Goal

Drag the crosshair cursor close to the black ball (position is given in meters and time is given in seconds). Notice that the cursor exerts a force, that is, a push or a pull, on the ball depending on the force you select. The ball will bounce off the purple spheres and will bounce off the soft walls around the animation. The animation will end if the ball hits either rectangle. The blue arrow represents the net force.

a. Try to get the ball to hit the green rectangle and not the red rectangle.

b. Given an applied force, how does the ball move?

c. Does the ball always move the way you expect? Why or why not?

Exploration 4.4: Set the Force on a Hockey Puck

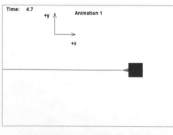

A 250-gram hockey puck is acted upon by a single force. It is free to slide on the ice (position is given in meters and time is given in seconds) in any direction. You can set the force vector by changing its magnitude ($0 \text{ N} < F < 10 \text{ N}$) and direction. The force vector is shown in the animation as a red arrow. You also can set the initial velocity components ($-15 \text{ m/s} < v < 15 \text{ m/s}$).

a. When the initial velocity is zero, in what direction does the ball travel for a given force?

b. When the initial velocity is not zero, in what direction does the ball travel for a given force? Hint: The best way to do this is to pick a nonzero v_{0x} or v_{0y}, not both. Also turn on the ghosts.

c. Try $F = 5 \text{ N}$, $\theta = 270°$, $v_{0x} = 7 \text{ m/s}$, and $v_{0y} = 15 \text{ m/s}$. Does this motion look familiar? Turn on the ghosts to help with the answer.

Exploration 4.5: Space Probe with Multiple Engines

A space probe is designed with four engines that can fire in the $+x$, $-x$, $+y$, and $-y$ directions, respectively (position is given in meters and time is given in seconds). For each of the situations below, first predict the motion of the space probe. Your prediction should be a detailed description of the motion of the probe. Only after you make a prediction, check it by viewing the animation. An example is shown in the first row of the table.

Situation	Your prediction	Animation
The space probe has a constant velocity in the $+x$ direction when suddenly an engine exerts a force on the probe in the $+x$ direction.	The probe will have an acceleration in the $+x$ direction. Therefore, since it is already traveling in that direction when the engine fires, it will speed up and will continue moving in the $+x$ direction.	**Animation 1**
The space probe has a constant velocity in the $+x$ direction when suddenly an engine exerts a force on the probe in the $-x$ direction.		**Animation 2**
The space probe has a constant velocity in the $+x$ direction when suddenly an engine exerts a force on the probe in the $+y$ direction.		**Animation 3**
The space probe has a constant velocity in the $+x$ direction when suddenly an engine exerts a force on the probe in the $-y$ direction.		**Animation 4**

Situation	Your prediction	Animation
The space probe has a constant velocity in the +x direction when suddenly an engine exerts a force on the probe in the −y direction and another engine exerts a force in the −x direction.		**Animation 5**
The space probe has a constant velocity in the +x direction when suddenly an engine exerts a force on the probe in the +y direction and another engine exerts a force in the +x direction.		**Animation 6**
The space probe has a constant velocity in the +x direction when suddenly all four engines fire simultaneously.		**Animation 7**

Exploration 4.6: Putted Golf Ball Breaks Toward the Hole

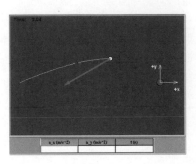

A putted golf ball rolls toward the hole on a green. The animation shows a top view of the ball on the green. The acceleration vector (orange) of the ball is shown on the animation, and the components of the ball's acceleration are displayed in the data table.

The net force on the golf ball is in the same direction as the acceleration of the golf ball, according to Newton's second law. This means that if you know the mass of the golf ball and the acceleration of the golf ball, you can calculate the net force on the golf ball.

a. Is the net force on the golf ball in the animation constant during the interval from $t = 0$ to $t = 4.8$ s?

b. If not, does its magnitude and/or direction change?

c. If the mass of a golf ball is 0.046 kg, what is the net force on the golf ball at $t = 1.0$ s?

d. For practice, calculate the net force on the golf ball at $t = 2.0$ s, $t = 3.0$ s and $t = 4.0$ s as well.

Exploration 4.7: Atwood's Machine

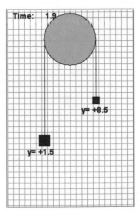

A 10-kg mass, M, is attached via a massless pulley to another variable mass m (position is given in meters and time is given in seconds). You can test the limits of the formula for the acceleration of the Atwood's machine (not shown to scale) by changing the ratio of the masses above.

a. Draw a free-body diagram for each mass.

b. Solve for the acceleration of m in terms of g, M, and m.

c. Which, if any, of the following statements regarding the motion of the masses are true?

- when M = m then: $a = g$.
- when M = m then: $a = 0$.
- when M $\gg$ m then: $a = g$.
- when M $\gg$ m then: $a = 0$.
- when M < m then: $a = 0$.
- when M < m then: $a = g$.
- when M < m then: $a < 0$.

Verify your answer(s) to (c) by using the animation and your answer for (b).

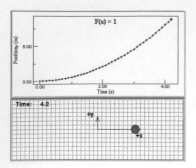

Exploration 4.8: Enter a Formula for the Force Applied

This Exploration allows you to choose initial conditions and forces and then view how that force affects the red ball. You can right-click on the graph to make a copy at any time. If you check the "strip chart" mode box, the top graph will show data for a time interval that you set. Note that the animation will end when the position of the ball exceeds +/−100 m from the origin.

Remember to use the proper syntax, such as: $-10+0.5 * t$, $-10+0.5 * t * t$, and $-10+0.5 * t\hat{} 2$. Revisit **Exploration 1.3** to refresh your memory.

Differential equations can be difficult to solve analytically. One way around this is to use a numerical method to generate a solution at discrete time steps. The above animation does just that by advancing the position of the red ball from its initial value at time t_0 to a new value at $t_1 = t_0 + dt$. This process can be repeated over and over to approximate the solution as a function of time.

Clearly there are pitfalls in the above procedure. If the time step is too large (1 year for example) interesting phenomena can be missed. This is clearly not an informative dataset if something interesting happens during the time interval. On the other hand, if the time step is too small (1 nanosecond for example) the computer may take a very long time to plot a representative set of points so that you can see the motion of the ball.

For each of the following forces, first describe the force (magnitude and direction) and then predict the motion of the ball. How close were you? Don't forget to determine how the initial position and velocity affect the motion of the ball for each of the forces.

a. $F_x(x, t) = 1$.
b. $F_x(x, t) = -1$.
c. $F_x(x, t) = 1 * \text{step}(3\text{-}t)$. This function is a constant until $t = 3$ s when it turns off.
d. $F_x(x, t) = x$.
e. $F_x(x, t) = -x$.
f. $F_x(x, t) = \cos(x)$.
g. $F_x(x, t) = \cos(t)$.

Problems

Problem 4.1

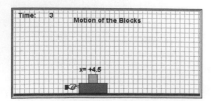

A red block is pushed and moves as shown in the animation. In addition, a green block sits on the red block and moves as well.

a. Which free-body diagram is correct? Give reasons why the other three diagrams are incorrect.

b. How would your answer to (a) change if the blocks did not move?

Problem 4.2

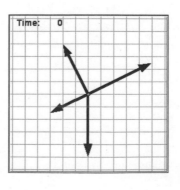

A free-body diagram for a 20,000-kg airplane at some instant is shown in the animation (grid size is given in 40,000 newtons). Generally, all of the external forces on an airplane can be resolved into four components called weight, lift, thrust, and drag. You can move a vector around by click-dragging at its tail.

a. What is the net force on the airplane at this instant?

b. What is the acceleration of the airplane at this instant?

c. What can you definitely say about the velocity of the airplane at this instant?

d. Suppose a classmate in your study group proclaims that "the net force on the airplane is zero and therefore the airplane must be on the ground and at rest." What is the error in your classmate's statement?

Problem 4.3

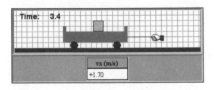

A 100-kg wagon with a 20-kg block on its frictionless bed is pulled to the right with a constant force (position is given in meters and time is given in seconds). Does the animation obey Newton's laws? Support your answer.

Problem 4.4

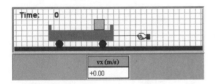

A 100-kg wagon with a 20-kg block on its frictionless bed is pulled to the right with an unknown force (position is given in meters and time is given in seconds). Sketch a plot of the force exerted by the hand on the cart as a function of time.

Problem 4.5

A 0.010-kg buoy is dropped into a lake as shown in the animation. Before it hits the water, it is in free fall.

a. Before the buoy hits the water, what is the net force on the buoy?

b. At $t = 1.2$ s, what is the net force on the buoy?

c. At $t = 1.2$ s, what is the force of the water on the buoy?

d. At $t = 4.5$ s, what is the net force on the buoy? Approximately, what is the velocity of the buoy at this instant? Can the velocity of an object be zero even though the net force on the object is not zero?

e. At $t = 4.5$ s, what is the force of the water on the buoy?

f. At $t = 11.0$ s, what is the net force on the buoy?

g. At $t = 11.0$ s, what is the force of the water on the buoy?

h. Describe the velocity of the buoy at this instant ($t = 11.0$ s); is it increasing, decreasing, or constant?

Problem 4.6

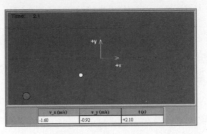

The animation shows a putted golf ball of mass 0.050 kg as it rolls toward the hole. The putter hit the ball before $t = 0$ s and is no longer in contact with the ball (position is given in meters and time is given in seconds).

a. What is the net force on the golf ball during the interval from $t = 0$ to $t = 4.2$ s?

b. What is the force of the putter on the golf ball during this interval?

Problem 4.7

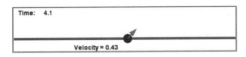

A 20-kg ball has a hole with a rod passing through. The rod exerts a force as needed that constrains the ball to move along the rod. An applied force is now added (the "pulling" force) so the ball is pulled as shown (position is given in meters and time is given in seconds). The force vector is shown as a red arrow, and the force makes an angle θ with the horizontal. The velocity is given in meters/second. You may adjust the angle and/or the magnitude of the pulling force ($F < 7$ N).

a. How does the acceleration change as you vary the "pulling" force for a constant angle?

b. How does the acceleration change as you vary the angle for a constant pulling force?

c. Combine your answers above to obtain a general mathematical formula for the acceleration of the ball due to an arbitrary applied force.

d. Determine the general mathematical formula for the normal force the rod exerts on the ball when an arbitrary force is applied to the ball.

Problem 4.8

A 50-kg box is riding in an elevator that accelerates upward or downward at a constant rate (position is given in meters and time is given in seconds). The box rests on a digital scale that records its apparent weight in newtons. The green arrow represents the instantaneous velocity of the elevator and its contents. Adjust the value of the acceleration

$(-9.8 \text{ m/s}^2 \le a \le 9.8 \text{ m/s}^2)$ and see how it affects the apparent weight.

a. What type of force is recorded on the scale?

b. Draw a free-body diagram for the box when its acceleration is 4.9 m/s², 0 m/s², and −4.9 m/s².

c. Write a formula for the scale reading as a function of the acceleration of the elevator, the mass of the box, and g.

d. Determine the value of the elevator's acceleration that would make the force that the scale exerts on the box vanish. In other words, how can the box become apparently "weightless?"

Problem 4.9

A 10-kg mass is attached via a massless string over a massless pulley to a hand (position is given in meters and time is given in seconds). The masses in each animation are identical.

a. Rank the animations according to the acceleration of the mass, from greatest to least (positive is up).

b. Rank the animations according to the tension in the string, from greatest to least (positive is up).

Indicate ties by placing the animation numbers in () please. For example, a suitable response could be: 1,2,(3,4),5,6.

c. Calculate the acceleration of the mass in each animation.

d. Calculate the tension of the string in each animation.

Problem 4.10

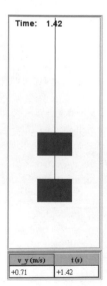

Two boxes, each of mass 2.0 kg, are connected by a lightweight rope. The boxes are hoisted upward with a constant acceleration as shown in the animation (position is given in meters and time is given in seconds).

a. Draw a free-body diagram for each box.

b. What is the tension in the top rope?

c. What is the tension in the rope connecting the two boxes?

Consider an alternative situation in which the system has the same acceleration but the rope between the boxes is not lightweight. It is a steel cable with a total mass of 1.0 kg.

d. What would be the force of the steel cable on the top box?

e. What would be the force of the steel cable on the bottom box?

Problem 4.11

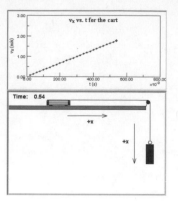

A 1.0-kg cart (not shown to scale) on a low-friction track is connected to a string and a hanging object as shown in the animation. Neglect any effects of the pulley on the motion of the system (position is given in meters and time is given in seconds).

a. What is the tension in the string?

b. What is the mass of the hanging object?

Note that the coordinates for each object (the positive x direction) are already chosen for you.

Problem 4.12

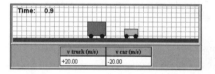

A large 2000-kg truck and a small compact car collide head-on as shown in the animation (position is given in meters and time is given in seconds). Assume the collision takes place in 0.05 seconds.

a. Describe the force on each vehicle before, during, and after the collision. Be sure to estimate the magnitudes of these forces and give their directions.

b. Which vehicle, the car or the truck, experiences the greater force during the collision?

Problem 4.13

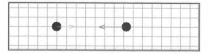

Newton's third law states that whenever two objects interact, they exert equal and opposite forces on each other (position is given in meters and time is given in seconds). The balls in the animations can be dragged around. As you do so, notice how the sizes of the arrows change. Each arrow represents the force on an object (the length of the arrow indicates the magnitude of the force). Which animation, if any, obeys Newton's third law?

Newton's Laws 2

Topics include Newton's second law, forces (static and kinetic friction, air resistance, springs), and uniform circular motion.

Illustration 5.1: Static and Kinetic Friction

How does friction change our analysis? Well, friction is due to the fact that objects, even smooth ones, are actually very bumpy on a microscopic level. Because of this, atoms from each surface are in such close contact that they form chemical bonds with each other. These bonds need to be broken in order for motion to occur.

The direction of the frictional force is opposite to the direction of motion, and the magnitude of the frictional force is proportional to the magnitude of the normal force, N. There are two kinds of frictional forces: static and kinetic friction.

Static friction is the force when there is no relative motion between surfaces in contact. Conversely, kinetic friction is the force when there is relative motion between surfaces in contact. We represent the coefficient of friction by the Greek letter μ, and use subscripts corresponding to static and kinetic friction, respectively: μ_s and μ_k. It is always the case that $\mu_s > \mu_k$.

Consider what happens when you pull on a stationary block up to the point where there is still no motion. Set the mass to 100 kg and vary $F_{applied}$. The frictional force f_s matches $F_{applied}$ up to the frictional force's maximum value, $f_{s;\,max} = \mu_s N = 392$ N. After that the frictional force dramatically decreases, the block accelerates, and the frictional force becomes the kinetic friction of motion, $f_k = \mu_k N$.

So what are μ_s and μ_k? Well, given that there is no motion until when $F_{applied} = 392$ N, $f_{s;\,max}$ is approximately 392 N. Therefore, given that the normal force is 980 N, $\mu_s = 0.4$. Now there is an acceleration when $F_{applied} = 392$ N. Since the change in $v = 9.8$ m/s in 5 seconds, $a = 1.96$ m/s^2. Given this, $ma = F_{applied} - f_k = F_{applied} - \mu_k N$. Therefore, after some algebra, $\mu_k = 0.2$.

Illustration 5.2: Uniform Circular Motion: F_c and a_c

Uniform circular motion is an interesting mixture of one- and two-dimensional concepts. During uniform circular motion, the speed of the object must be constant. This is the uniform in uniform circular motion. So is an object moving in a circle with a constant speed accelerating? Yes! Why? The velocity is changing with time. Watch the animation (position is shown in meters and time is shown in seconds). The animation depicts an object moving in a circle at a constant speed. To determine the acceleration we need to consider the change in velocity for a change in time.

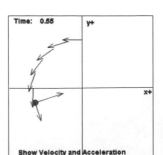

Since the speed does not change in time, what does change in time? It is the direction that changes with time. **Draw two velocity vectors** to convince yourself that the direction of the velocity changes with time. Recall that velocity has a direction (which always points tangent to the path, the so-called tangential direction) and a magnitude, and either or both can change with time. In what direction does the change in velocity point? **Calculate the acceleration.** It points toward the center of the circle. Since the object is accelerating, this motion must be due to a force (or a set of forces, a net force) that points solely toward the center of the circle. (Note: if the motion is nonuniform circular motion, the net force can point in another direction.) This direction—toward the center of the circle—is called the centripetal or center-seeking direction. It is often also called the radial direction, since the radius points from the center of the circle out to the object (the net force points in the opposite direction).

Therefore, for uniform circular motion, the acceleration always points toward the center of the circle. This is despite the fact that the velocity and the acceleration point in changing directions as time goes on. However, we get around this apparent difficulty in describing direction by defining the centripetal or radial direction and the tangential direction (the direction tangent to the circle). These directions change, but the velocity is always tangent to the circle, and the net force is always pointing toward the center of the circle. The following animation **shows velocity and acceleration** as the object undergoes uniform circular motion.

Illustration 5.3: The Ferris Wheel

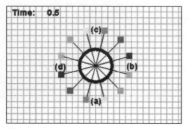

So how does our analysis of Newton's laws change when an object is moving in uniform circular motion? There are really only two things to remember.

First, the net force is always toward the center of the circle for uniform circular motion. This net force is responsible for the acceleration toward the center of the circle, the centripetal acceleration we saw in **Illustration 5.2**.

Second, because the centripetal acceleration is a positive number, v^2/r, it can never be negative. So unlike linear motion in which you have a choice of where to place the coordinate axes (to make life easier or more difficult), the choice here is critical. Your choice of coordinates must have one axis with its positive direction pointing toward the center of the circle.

In the animation, a Ferris wheel rotates at constant speed as shown (position is shown in meters and time is shown in minutes). Each square represents a chair on the Ferris wheel.

Consider a rider at point (a). What does the free-body diagram for a chair on the Ferris wheel look like at this point? To answer this question we must determine the applied forces that act on a rider when the rider is at point (a). At point (a) there are the normal force and the weight acting in opposite directions. Are the forces the same size or different? They must be different and the normal force must be bigger. Why? We know that the net force must point toward the center of the circle and that the net force is $ma = m\,v^2/r$ for uniform circular motion.

What is the acceleration of the rider when the rider is at point (a)? As stated above we know the acceleration must be v^2/r, where $v = 2\pi r/T$, where T is the period of one revolution.

What about the answers to these questions when the rider is at points (b), (c), and (d)? Well, the forces may be different or point in different directions, but the results are the same. The net force must be toward the center of the circle and be $m\,v^2/r$.

Illustration 5.4: Springs and Hooke's law

Springs are interesting objects that for a range of stretching and compression follow Hooke's law. Hooke's law states that the force that the spring exerts is $F = -k\,\Delta x$, where k is the spring constant and Δx is the displacement of the spring from its equilibrium position. In this Illustration the spring can be stretched by click-dragging the blue ball (position is given in centimeters and force is given in newtons). Slowly drag the spring back and forth and watch the graph.

Where is the equilibrium position of the spring? Given that Δx is measured from the equilibrium of the spring, look for the position where $F = 0$ N. This occurs at $x = 30$ cm. This is the equilibrium position.

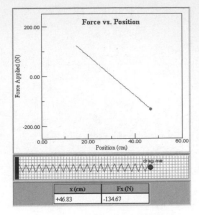

What is the spring constant of the spring? It is not the force shown in the table divided by the position shown in the table. Why not? Recall that the "Δx" in the spring force equation is measured from equilibrium. Therefore, at maximum extension, $x = 20$ cm and the force is -160 N. Therefore, $k = 800$ N/m. Given that the negative of the slope of the line on the graph should also be k, we can measure the spring constant by finding the slope and we get the same result.

The fact that spring forces are variable with position means that while we can determine the force, we cannot (given what we currently know) determine the velocity and position vs. time for an object attached to a stretched spring. Why? The force is not constant (it varies with position), and therefore the acceleration is not constant. This means we cannot use the kinematic equations for constant acceleration. What can we do? We can use concepts that you will learn about in Chapters 6 and 7.

Illustration 5.5: Air Friction

In this Illustration we compare the motion of a red projectile launched upward to that of an identical green projectile launched upward but subjected to the force of air friction. To make the motion easier to see, we have given both projectiles a slight horizontal velocity and do not consider frictional effects in this direction either. In addition, we show the free-body diagrams for each projectile (the force of gravity is drawn with a fatter vector so it is easier to see).

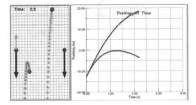

Watch the **Position Graph** animation and look at the free-body diagram. First, what is the direction of the force of air friction? It opposes motion, just as static and kinetic friction do. Consider the **Velocity Graph** animation. If we look at the motion on the way up, the velocity is positive, and therefore the force of friction opposes the motion and is downward, hence $|a_y| > g$ on the way up. At the top of the arc, the velocity is zero, and hence $|a_y| = g$. On the descent, the velocity is downward, and the force of air friction is therefore upward and hence $|a_y| < g$. Therefore, $|a_y|$ is greater on the way up! This is borne out by the **Acceleration Graph**. At some point, the frictional force has exactly the same size as the force of gravity. When this occurs there is no longer a net force, and the acceleration of the projectile is zero. The velocity corresponding to this situation is called the terminal velocity.

These animations are valid at low speeds. We can experimentally determine that the force of air friction is proportional to the velocity at low speeds, with $\mathbf{R} = -b\,\mathbf{v}$, where $\mathbf{R}$ is the resistive or drag force and b is a constant that depends on the properties of the air and the size and shape of the object. One benefit of this model is that the mathematics is a little easier to handle than for the high-speed case.

For massive small objects at high speeds (not depicted, but you can look at **Exploration 5.6** to view this model) we can experimentally determine that the force of air friction is proportional to the velocity squared. The magnitude of the drag force can be represented as $R = 1/2\, D\rho\, Av^2$, where ρ is the density of air (mass/volume), A is the cross sectional area of the object, v is the magnitude of the velocity, and D is the drag coefficient (0.2–2.0). Sometimes the drag force is written as bv^2 with the assignment that $b = 1/2\, D\rho A$. We can solve for the velocity as a function of time, but it is harder. We must be careful in this model if we have two-dimensional motion, since the x and y motions are no longer independent.

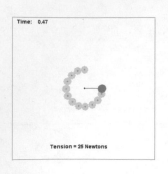

Exploration 5.1: Circular Motion

A puck travels in a circular path on a frictionless table, propelled by a string pulling from the center of the circle (position is shown in meters and time is shown in seconds). You may adjust the mass (10 g $< m < 500$ g), the speed (1 m/s $< v < 50$ m/s), and/or the radius (0.5 m $< r < 3.5$ m). The tension is displayed on the screen.

How does the tension in the string depend upon the mass, the speed of the block, and the radius of the circle?

a. If you only vary the mass, how does the tension change?

b. If you only vary the velocity, how does the tension change?

c. If you only vary the radius, how does the tension change?

Exploration 5.2: Force an Object Around a Circle

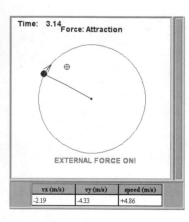

In this Exploration you are looking down at a black ball on a table top. Drag the crosshair cursor (position is given in meters and time is given in seconds) to **within 5 m** of the 0.2-kg ball. The cursor will then exert a constant force on the black ball. You may choose either an attractive or a repulsive force. In addition, the black ball is constrained to move in a circle by a very long wire. The blue arrow represents the net force acting on the mass, while the bar graph displays its speed in meters/second.

For both attraction and repulsion, drag the cursor around to see how the net force varies.

a. At the beginning of the animation (before you move the cursor), in what direction does the net force point?

b. With this force, does the ball move?

c. What type of applied force makes the ball acquire a tangential velocity?

d. Describe the direction of the force that makes the ball acquire the maximum tangential velocity for the force applied.

e. When the ball has a tangential velocity, in which direction does the net force point when the cursor is nearby? In what direction does the acceleration point?

f. With the object moving, drag the cursor far away from the ball. In what direction is the net force now? What is the direction of the acceleration? Why?

Exploration 5.3: Spring Force

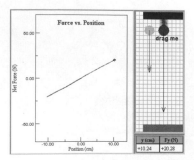

The spring-ball system shown in the animation can be stretched by click-dragging the dark blue ball (position is given in meters and time is given in seconds). The black arrow attached to the ball shows the net, i.e., total, force on the ball. The pale blue ball on the left is the *free-body diagram* for the dark blue ball. The red and green arrows attached to the pale blue ball show the spring and gravitational forces, respectively. The acceleration due to gravity is 9.8 m/s^2 in this animation.

a. Find the mechanical equilibrium for this system when the spring constant is 1.0 N/m, 2.0 N/m, 3.0 N/m, and 4.0 N/m.

b. Use your equilibrium measurements to find the mass of the ball. Hint: What forces act on the ball?

c. Use your equilibrium measurements to find the natural length of the spring, that is, the length of the spring without an attached mass.

Exploration 5.4: Circular Motion and a Spring Force

A 1-kg mass is attached to the end of a spring of spring constant $k = 10$ N/m and natural length $l_0 = 5$ m (position is shown in meters and time is shown in seconds). You are to set the spring in motion by setting its initial position $(x_0, 0)$ and its initial velocity $(0, v_{0y})$.

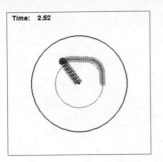

a. Find the v_{0y} needed for circular motion at a radius of 10 m (the red circle).

b. Determine the period of such a motion.

Exploration 5.5: Enter a Formula for the Force

This Exploration allows you to choose initial conditions and forces *with damping*, and then view how that force affects the red ball. You can right-click on the graph to make a copy at any time. If you check the "strip chart" mode box, the top graph will show data for a time interval that you set. Note that the animation will end when the position of the ball exceeds +/−100 m from the origin.

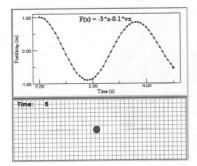

Remember to use the proper syntax such as $-10+0.5 * t$, $-10+0.5 * t * t$, and $-10+0.5 * t^\wedge 2$. Revisit **Exploration 1.3** to refresh your memory.

For each of the following forces, first describe the force (magnitude and direction) and then predict the motion of the ball. How close were you? Don't forget to determine how the initial position and velocity affect the motion of the ball for each force.

a. $F_x(x, vx, t) = 1-0.05 * vx$.
b. $F_x(x, vx, t) = 1-0.5 * vx$.
c. $F_x(x, vx, t) = 1-vx$.
d. $F_x(x, vx, t) = -9.8-vx$.
e. $F_x(x, vx, t) = x-vx$.
f. $F_x(x, vx, t) = \cos(x)-vx$.
g. $F_x(x, vx, t) = \cos(t)-vx$.

Exploration 5.6: Air Friction

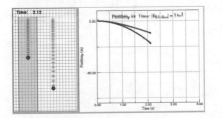

Two identical balls are dropped. The one on the left is in a resistive medium represented by varying shades of blue. The resistive force is represented as bv^n, where b is a constant between 0 and 2 and n is an integer between 0 and 2 (note that as you vary n, the units of b also change).

Select values for b and n, and then click on a graph link to show the motion and that particular graph. *When you get a good-looking graph, right-click on it to clone the graph and resize it for a better view.*

a. How does your choice of n (0, 1, 2) affect the unit of b?

b. For $b = 1$, how does your choice of n (0, 1, 2) affect the position vs. time graph?

c. For $b = 1$, how does your choice of n (0, 1, 2) affect the velocity vs. time graph?

d. For $b = 1$, how does your choice of n (0, 1, 2) affect the acceleration vs. time graph?

e. For $b = 1$, how does your choice of n (0, 1, 2) affect the terminal velocity?

Exploration 5.7: Enter a Formula, F_x and F_y, for the Force

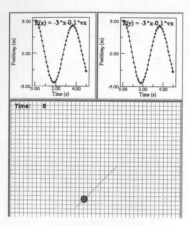

This Exploration allows you to choose initial conditions and forces and then view how that force affects the red ball. You can right-click on the graph to make a copy at any time. If you check the "strip chart" mode box, the top graph will show data for a time interval that you set. Note that the animation will end when the position of the ball exceeds $+/-100$ m from the origin.

Remember to use the proper syntax such as $-10+0.5 * t$, $-10+0.5 * t * t$, and $-10+0.5 * t\hat{}2$. Revisit **Exploration 1.3** to refresh your memory.

For each of the following forces, first describe the force (magnitude and direction) and then predict the motion of the ball. How close were you? Don't forget to determine how the initial position and velocity affect the motion of the ball for each force.

F_x	F_y	x_0	x_0	v_{0x}	v_{0x}
1	1	0	0	0	0
1	-1	0	0	0	0
$-x$	$-2 * y$	10	10	0	0
$-0.5 * vx$	$-9.8-0.5 * vy$	-20	0	20	20

Problems

Problem 5.1

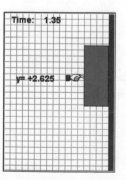

Consider a 2-kg physics textbook (not drawn to scale) pressed against a wall, which has a coefficient of static friction of $\mu_s = 0.3$ and a coefficient of kinetic friction of $\mu_k = 0.2$ as shown in the animation (position is given in meters and time is given in seconds).

a. Draw a free-body diagram for the book, showing all the forces that act.

b. What is the net force on the book? Include both magnitude and direction in your answer.

c. What is the minimum force of the push represented by the hand?

d. What happens to the motion of the book and to the forces if the push is greater than your answer to (c)?

Problem 5.2

Consider a 2-kg physics textbook (not drawn to scale) pressed against a wall, which has a coefficient of kinetic friction of $\mu_k = 0.4$ as shown in the animation (position is given in meters and time is given in seconds).

a. Draw a free-body diagram for the book, showing all the forces that act.

b. What is the net force on the book? Include both magnitude and direction in your answer.

c. What is the force of the push represented by the hand?

Problem 5.3

Consider a 2-kg physics textbook (not drawn to scale) pressed against a wall, which has a coefficient of kinetic friction of $\mu_k = 0.4$ as shown in the animation (position is given in meters and time is given in seconds).

a. Draw a free-body diagram for the book, showing all the forces that act.

b. What is the net force on the book? Include both magnitude and direction in your answer.

c. What is the force of the push represented by the hand?

Problem 5.4

A woman pushes on a block with an unknown force as shown in the animation (position is given in meters and time is given in seconds). At $t = 2$ seconds she doubles the force applied to the block. Determine the coefficient of kinetic friction between the block and the table.

Problem 5.5

A 4-kg block sits on an 8-kg block that is pushed across the floor as shown (position is given in meters and time is given in seconds).

a. Which free-body diagram is correct? Give reasons why the other three diagrams are incorrect.

b. Draw the correct free-body diagram and label the force that causes each interaction.

Be sure to include the force of friction if two surfaces are rubbing against each other. Remember that the length of an arrow is proportional to the values of the quantity being represented, and its length does not represent the actual size of the quantity.

Problem 5.6

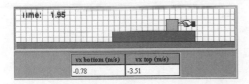

A 10.0-kg block sits on a 20-kg block as shown (position is given in meters and time is given in seconds). There is friction between the top and the bottom block, but the surface between the bottom block and the table is frictionless.

a. Draw free-body diagrams for both blocks.

b. Find the net force on each block.

c. Find the force of the push.

Problem 5.7

A 12-kg box sits on a rough (meaning that there is friction) 26.56° ramp as shown in the animation (position is given in meters and time is given in seconds).

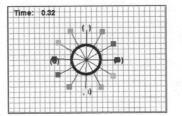

a. Draw a free-body diagram for the box, showing all the forces that act.

b. What is the net force on the box? Include both magnitude and direction in your answer.

c. What is the value for the coefficient of kinetic friction?

Problem 5.8

In the animation, a Ferris wheel rotates at constant speed as shown (position is given in meters and time is given in minutes). Each square represents a chair on the Ferris wheel.

a. Draw the free-body diagram for a chair on the Ferris wheel when it is at the points (a), (b), (c), and (d).

b. What forces act on a rider when the rider is at points (a), (b), (c), and (d)?

c. What is the acceleration of the rider when the rider is at points (a), (b), (c), and (d)?

d. If the rider has a mass of 100 kg, what is the size and direction of the net force on the rider at points (a), (b), (c), and (d)?

Problem 5.9

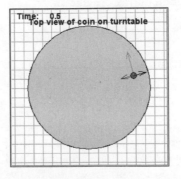

A mass sits on a turntable as shown (position is given in meters and time is given in seconds).

a. What force provides the centripetal acceleration?

b. Which vector represents the net force on the object?

Problem 5.10

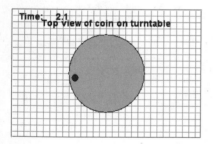

A puck resting on an air hockey table is attached to a string and given an initial tangential push such that it travels in a circle at constant speed (position is given in meters and time is given in seconds).

a. What is the magnitude of the acceleration of the puck?

b. Draw the free-body diagram for the puck.

c. What is the tension in the string if the puck has a mass of 0.1 kg?

Problem 5.11

A 5-gram coin is on a rotating turntable as shown (position is given in meters and time is given in seconds).

a. What is the coin's acceleration during the animation?

b. Draw a free-body diagram for the coin.

c. Determine the minimum value of μ_s for this motion to occur.

Problem 5.12

The spring can be stretched by click-dragging the blue ball as shown in the animation (position is given in centimeters and time is given in seconds). Slowly drag the spring back and forth out of the equilibrium position and answer the following questions.

a. Over what range of compression and stretching is Hooke's law valid?

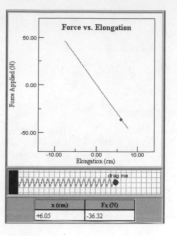

b. Find the elastic limit of the spring.

c. Determine the spring constant of the spring.

Problem 5.13

A 200-gram brick falls onto a platform as shown in the animation (position is given in meters and time is given in seconds). The animation stops when the brick is in equilibrium. Determine the spring constant of the spring.

Problem 5.14

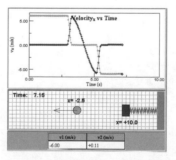

A ball on a frictionless table is fired at a block that is connected to a very light spring as shown (position is given in meters and time is given in seconds).

a. For each animation, draw the force vs. time graph for the orange ball.

b. Suppose you wanted to determine $x(t)$ for the orange ball. Are Newton's laws an effective way to determine $x(t)$ for the orange ball? Why or why not?

Work

Topics include work, kinetic energy, dot products, constant forces, and variable forces.

Illustration 6.1: Dot Products

We talk about work as the amount of force in the direction of an object's displacement, Δx, multiplied by the displacement, Δx. No displacement, no work. Work is positive if $\mathbf{F}$ and Δx point in the same direction and negative if $\mathbf{F}$ and Δx point in opposite directions. This statement is fine if the force and displacement lie on the same axis (in other words lie on a line). What happens if they do not lie on the same axis? When working in two dimensions, the force and the displacement can point in any direction. So how much of a given force is in the direction of the displacement? (We could consider the equivalent description of the amount of the displacement in the direction of the force.)

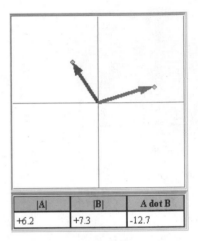

|A|	|B|	A dot B
+6.2	+7.3	-12.7

To answer this question we must use the mathematical construction of the scalar or dot product. The dot product is defined as the scalar product of two vectors: $\mathbf{A}$ dot $\mathbf{B} = \mathbf{A} \cdot \mathbf{B} = A\,B\cos(\theta)$, where θ is the angle between the two vectors and A and B are the magnitudes of the vectors $\mathbf{A}$ and $\mathbf{B}$, respectively.

Drag the tip of either arrow (position is given in meters). The **red arrow is A** and the **green arrow is B**. The magnitude of each arrow is shown and the dot product is calculated. When is the dot product zero? The dot product is zero when the vectors are perpendicular. For any two vectors, the magnitude of the dot product is a maximum when the vectors point in the same (or opposite) directions and a minimum when the vectors are perpendicular. Note also that the assignment of which vector is $\mathbf{A}$ and which vector is $\mathbf{B}$ does not matter.

So the dot product has the right properties to help us mathematically describe WORK. In general, for a constant force,

$$WORK = \mathbf{F} \cdot \Delta x = F\,\Delta x\cos(\theta),$$

where $\mathbf{F}$ is the constant force and Δx is the displacement. F and Δx are the magnitude of the vectors, respectively. You may have heard or seen "$WORK = Fd$," which is not always correct. That statement ignores the vector properties of $\mathbf{F}$ and Δx and can lead you into thinking that the definition of WORK is FORCE times DISTANCE, which it is not.

Illustration 6.2: Constant Forces

This Illustration shows a block on a table subjected to an applied force and a frictional force. The mass of the block, the force applied, and the coefficient of kinetic friction can be controlled by typing values for these variables into the text boxes and pressing the "set values and play" button. The coefficient of static friction is fixed at $\mu_s = 0.4$.

Consider the animation with a 100-kg block and vary the force applied from 0 to 391 N. What happens in the animation? The block does not move. What is the work done on the block due to the applied force? What is the work done on the block due

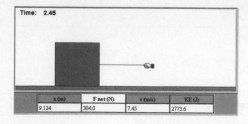

to the force of static friction? What is the work done on the block due to the normal force? What is the work done on the block due to the force of gravity? The work done by each of these forces is zero. How do you know? First there is no displacement. If there is no displacement there can be no work. In addition, the normal force and the force of static friction can never do work. The normal force cannot do work because if there is a displacement it would always be perpendicular to the normal force, hence no work. The force of static friction can also never do work. When there is static friction there can never be a displacement. After all, static friction implies the block is static and thus not moving. Since there is no work done on the block, the block's kinetic energy cannot change.

Now consider the animation with a 100-kg block and an applied force of 446 N. What happens in the animation? The block moves and, in fact, it accelerates. What is the work done on the block due to the applied force? What is the work done by the force of kinetic friction? What is the work done on the block due to the normal force? What is the work done on the block due to the force of gravity? The net work done by the sum of these forces is now not zero. How do you know? There is a change in kinetic energy. This can only happen when there is work done on the block. The force of gravity does not do any work on the block because the force is perpendicular to the displacement in this animation. The normal force, as said above, can never do work. The work done by the applied force will be positive.

The force of kinetic friction reduces the kinetic energy of the block by $|F_{\text{k friction}} \Delta x|$ because the frictional force and the block's displacement are in opposite directions. Kinetic friction will always oppose motion, so it will always reduce kinetic energy. Note that we do not say the work done on the block by kinetic friction. This phrase is not correct. The work done by friction on the block is the energy the block loses, and this is not equal to $-F_{\text{k friction}} \Delta x$. Some of the kinetic energy dissipated by friction, $|F_{\text{k friction}} \Delta x|$, is transferred to the table as thermal energy (the table heats up), while some of it remains with the block as thermal energy (the block heats up). Therefore, $-F_{\text{k friction}} \Delta x$ is not the work done on the block by friction; $-F_{\text{k friction}} \Delta x$ is the total work done by friction (done on the block and the table) and is the amount that the kinetic energy is reduced.

The change in kinetic energy of the block is the net force $F_{\text{applied}} - F_{\text{k friction}}$ times the displacement (this is because the net force points in the direction of the displacement). Therefore, in the table, F_{net} times $x = KE$ (because x is the displacement since the block starts at $x = 0$ m and the final KE is the change in KE since the block starts with no kinetic energy).

Illustration 6.3: Force and Displacement

Two forces that we often use as examples when talking about work are the force of gravity and the elastic force of springs. We know from earlier chapters that the character of the two forces is different. The gravitational force on an object is always mg, while the spring force is dependent on how much the spring is stretched or compressed from equilibrium. As a consequence, the form for the work done by each force will be different.

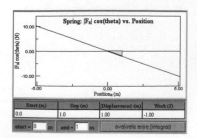

In general, for a constant force, $WORK = \mathbf{F} \cdot \Delta \mathbf{x} = F \Delta x \cos(\theta)$, where $\mathbf{F}$ is the constant force and Δx is the displacement. F and Δx are the magnitude of the vectors, respectively.

The graph shows $F \cos(\theta)$ vs. distance for a 1-kg object near the surface of Earth (position is given in meters). By checking the box you make the graph represent the $F \cos(\theta)$ vs. distance graph for a mass on a spring with $k = 2$ N/m (the equilibrium point of the spring is conveniently set at $x = 0$ m). You can enter in values for the starting and stopping points for the calculation of the work and then click the "evaluate area (integral)" button to calculate the work.

Begin by looking at the $F \cos(\theta)$ vs. distance graph for gravity (F_y vs. y). Gravity is a constant force (near the surface of Earth). Therefore, the magnitude of work done by gravity will be $|mg \, \Delta y|$. So, consider a ball at $y = 0$ m that drops to $y = -2$ m. Is the work done by gravity positive or negative? Use the graph to calculate the work. It is indeed positive (a negative force in the y direction and a negative displacement in the y direction means $\cos(\theta) = 1$). This is because the force is in the same direction as the displacement. What about lifting an object up from $y = -2$ m to $y = 0$ m? The work done is negative since the force is in the opposite direction from the displacement [$\cos(\theta) = -1$]. We can use $|F \, \Delta y|$ because the force does not vary over the displacement. But what if the force does vary, as in the case of a spring?

Check the check box to see the graph representing a spring force. Enter in values for the starting and stopping points for the calculation of the work and then click the "evaluate area (integral)" button. Enter in $x = 0$ m for the starting point and $x = 4$ m for the ending point, representing the stretching of a spring. Is the magnitude of the work done $|F \, \Delta x|$? Why or why not? The magnitude of the work is not $|F \, \Delta x|$. In the case of the spring, the magnitude of the work is $0.5 \, k \, x^2$, which is the area under the force function (it is also the integral of $F \, dx$). Note also that the work is negative: The force and the displacement are in the opposite direction [$\cos(\theta) = -1$].

Enter in $x = 4$ m for the starting point and $x = 0$ m for the ending point. What happens to the sign of the work done by the spring now?

Illustration 6.4: Springs

The fact that the spring force varies with position means that while we can determine the force, we cannot determine the velocity of an object attached to a stretched spring using kinematic equations for constant acceleration. Why? The force is not constant (it varies with position) and therefore the acceleration is not constant. What can we do? We can use the work-energy theorem.

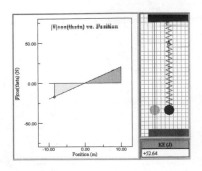

The spring-ball system shown in the animation can be stretched by click-dragging the 1-kg dark blue ball (position is given in meters and time is given in seconds). The black arrow attached to the ball shows the net, i.e., total, force on the ball. The pale blue ball on the left is the *free-body diagram* for the dark blue ball. The red and green arrows attached to the pale blue ball show the spring and gravitational forces, respectively. The acceleration due to gravity is 9.8 m/s^2 in this animation.

Hooke's law states that the force that the spring exerts is $F = -k \, x$, where k is the spring constant and x is measured from the equilibrium position of the spring. In this Illustration the initial position of the spring and the spring constant can be changed by using the text boxes.

So how do we determine the work done by the spring? We need to calculate the integral of $F \cos(\theta) \, \Delta x$, where F and Δx are the magnitude of the force vector and the displacement vector, respectively. We must calculate the integral because the force is not a constant.

Consider $k = 2$ N/m and $y = 5$ m and run the animation. Initially the spring is compressed and the net force points down and the infinitesimal displacement points down; therefore, $\cos(\theta) = 1$. We determine that the work done by the spring is initially positive, yielding an increasing kinetic energy. After passing through equilibrium ($y = 0.1$ m for this k), however, the net force is now upward, while the infinitesimal displacement still points down. Thus $\cos(\theta) = -1$, and the work is negative. Therefore, the kinetic energy decreases until the spring is maximally extended and $v = 0$ m/s. The process then reverses with a positive amount of work until the mass again passes through equilibrium and the spring does a negative amount of work until the mass is at rest at its starting position $y = 5$ m. The process repeats indefinitely if there are no resistive forces.

Illustration 6.5: Circular Motion

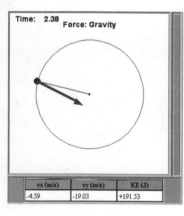

A 1-kg black ball is constrained to move in a circle as shown in the animation (position is given in meters, time is given in seconds, and energy on the bar graph is given in joules). In the **no external force** animation the wire is horizontal on a frictionless tabletop and the force of the wire is the only force that acts. In the **gravity** animation the wire is vertical and the ball is subjected to gravity (downward as usual) as well as the force of the wire. You may set the initial velocity and then choose either configuration. The blue arrow represents the net force acting on the mass, while the bar graph displays its kinetic energy in joules.

For **no external force**, select various initial velocities and then **set v and play: no external force**. In what direction does the net force point? Here the only force acting is the force of the wire pulling on the ball to make it go in a circle. The direction of this force is always toward the center of the circle (a centripetal force). With this force, does the black ball's speed change? No. The ball's velocity changes, but its speed does not. The work-energy theorem tells us that since there is no work done by the force of the wire (its force is perpendicular to the ball's displacement) there can be no change in the ball's kinetic energy. In general, any centripetal force cannot ever do work.

For **gravity**, select various initial velocities and then **set v and play: gravity**. In what direction does the net force point now? Well, this is a bit more complicated. There is a force toward the center of the circle as there was before (again due to the wire), but now there is also the force of gravity downward. Therefore, the net force does not point toward the center of the circle any more. With this force, does the black ball's speed change? Yes. While the part of the force due to the wire cannot do any work, the part of the force that is due to gravity can and does do work.

Exploration 6.1: An Operational Definition of Work

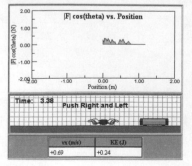

This Exploration allows you to discover how work causes changes in kinetic energy.

Drag "handy" to the front and/or the back of the cart to impart a force (position is given in meters and time is given in seconds). Look at the force $\cos(\theta)$ vs. position graph as well as the final velocity.

a. What can you say about the relationship between the force applied and the work done?

b. How does the application of the force change the kinetic energy?

c. What happens when the mass changes?

Make sure to push from both sides when the cart is both stationary and moving to the left and to the right.

Exploration 6.2: The Two-Block Push

Two blocks are pushed by identical forces (position is given in meters and time is given in seconds), each block starting at rest at the first vertical rectangle (start). The top block is twice the mass of the bottom block, $m_1 = 2m_2$. The graphs and tables are initially blank. **Animation without Graphs and Tables**.

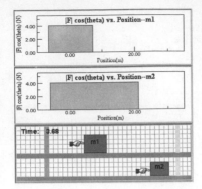

a. Which object has the greater kinetic energy when it reaches the second vertical rectangle (finish)? Why?

b. Once you have answered the above question, click **Animation with Graphs and Tables** to see if you were correct. Consider both the graphs and the tables.

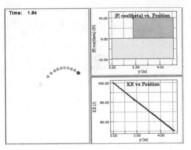

If you were incorrect in your answer, can you figure out why you answered incorrectly? What is the correct rationale you should have used to answer this question? Use the graphs and tables where appropriate.

Exploration 6.3: The Gravitational Force and Work

A 1-kg ball is subjected to the force of gravity as shown in the animation (position is given in meters and time is given in seconds). The ball starts at $x = 0$ m and $y = 0$ m. You can vary the ball's initial velocity and view how this affects the motion of the ball and the ball's kinetic energy. Also shown are the graphs of force $\cos(\theta)$ vs. position and kinetic energy vs. position.

With $v_{0x} = 0$ m/s and $v_{0y} = 0$ m/s,

a. In what direction does the net force on the ball point?

b. How would you describe the kinetic energy vs. position graph?

With $v_{0x} = 10$ m/s and $v_{0y} = 0$ m/s

c. What is the ball's minimum kinetic energy?

With $v_{0x} = 10$ m/s and $v_{0y} = 10$ m/s,

d. How would you describe the kinetic energy vs. position graph? Be explicit about what happens to the kinetic energy.

e. What is the condition for the work done by gravity to be zero?

f. What is the ball's minimum kinetic energy?

With $v_{0x} = -10$ m/s and $v_{0y} = 10$ m/s,

g. How would you describe the kinetic energy vs. position graph? Be explicit about what happens to the kinetic energy.

h. What is the condition for the work done by gravity to be zero?

i. What is the ball's minimum kinetic energy?

Exploration 6.4: Change the Direction of the Force Applied

A 20-kg ball has a hole with a rod passing through. The rod exerts a force as needed that constrains the ball to move along the rod. An applied force is now added (the "pulling" force) so the ball is pulled as shown (position is given in meters and time is given in seconds). The force vector is shown as a red arrow, and the force makes an angle θ with the horizontal. The velocity is given in meters/second. You may adjust the angle and/or the magnitude of the pulling force ($F < 7$ N).

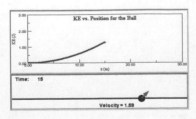

a. How does the work done by the pulling force change as you vary the pulling force for a constant angle?

b. How does the work done by the pulling force change as you vary the angle for a constant pulling force?

c. Combine your answers above to obtain a general mathematical formula for the work done on the ball due to an arbitrary pulling force.

d. Determine the general mathematical formula for the work done by the force the rod exerts on the ball when an arbitrary force is applied to the ball.

Exploration 6.5: Circular Motion and Work

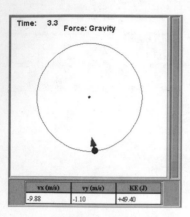

A 1-kg black ball is constrained to move in a circle as shown in the animation (position is given in meters, time is given in seconds, and energy on the bar graph is given in joules). In the animation the wire is vertical and the ball is subjected to gravity (downward as usual), as well as the force of the wire. You may set the initial velocity and then play the animation. The blue arrow represents the net force acting on the mass, while the bar graph displays its kinetic energy in joules.

a. Set the speed fast enough to get the ball over the top. Then restart and examine forces at positions near, say, $-45°$ and $45°$ (hanging straight down is $-90°$). Label your forces as F_g (gravity), F_{wire}, and F_{net}.

b. Given your force diagrams, there are positions where the speed of the ball is changing more rapidly than others. Take each of the positions you considered and rank them from highest tangential acceleration to lowest.

c. Assume that the ball can get to $y = 10$ m. How much kinetic energy does the ball lose in going from $y = -10$ m to $y = 10$ m? Is this independent of v_{0x} initial?

d. What is the work done by gravity when the ball goes from $y = -10$ m to $y = -10$ m?

e. Determine the minimum speed that the ball must have to go over the top. Once you have an answer, check it using the animation.

Exploration 6.6: Forces, Path Integrals, and Work

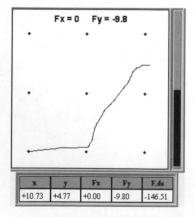

Move your cursor into the animation, then click-drag the crosshair cursor with the mouse. The bar graph on the right displays the work done by the force along the path. For your reference, there are circles every 10 m that form a coordinate grid (position is given in meters and the result of the integral is given on the bar graph in joules). Use the "reset integral" button to re-zero the work calculation between paths.

For each force, answer the following questions:

a. Starting at the origin (the center, $x = 0$ m and $y = 0$ m) and moving to $x = 0$ m and $y = 10$ m, what is the work done by the force?

b. Starting at $x = 0$ m and $y = 10$ m and moving to $x = 0$ m and $y = 0$ m, what is the work done by the force?

c. Starting at the origin (the center, $x = 0$ m and $y = 0$ m) and moving to $x = 0$ m and $y = -10$ m, what is the work done by the force?

d. Starting at the origin (the center, $x = 0$ m and $y = 0$ m) and moving to $x = 10$ m and $y = 0$ m, what is the work done by the force?

e. Starting at the origin (the center, $x = 0$ m and $y = 0$ m) and moving to $x = -10$ m and $y = 0$ m, what is the work done by the force?

f. Starting at the origin (the center, $x = 0$ m and $y = 0$ m), choosing your own path around the animation and ending back at the origin, what is the work done by the force?

When you are through, feel free to experiment with forces of your own choosing.

Problems

Problem 6.1

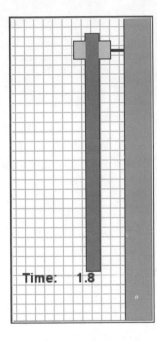

A 2-kg hammer strikes a 1.5-gram nail at $t = 1.8$ s as shown in the animation (position is given in centimeters and time is given in seconds).

a. Determine the work done on the hammer by the nail.

b. Use your calculation in (a) to determine the average force exerted on the nail by the hammer.

Problem 6.2

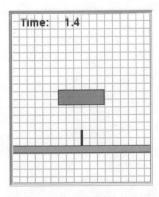

A 1.5-kg brick falls a given height onto a 15-gram spike as shown in the animation (position is given in meters and time is given in seconds).

a. Determine the work done on the brick by the nail.

b. Use your calculation in (a) to determine the average force exerted on the nail by the brick.

Problem 6.3

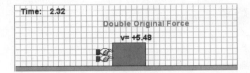

A woman pushes on a 2.5-kg block with an unknown force as shown in the animation (position is given in meters and time is given in seconds). At $t = 2$ seconds she doubles the force applied to the block.

a. Determine the total work done on the block and table in the first two seconds of the animation.

b. Determine the total work done on the block and table in the final two seconds of the animation.

Problem 6.4

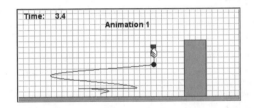

A bowling ball is lifted from rest onto a shelf by an external agent (position is given in meters and time is given in seconds). The bowling ball starts at rest and ends up at rest when the animation ends. For each quantity below, rank the animations (numbered 1 through 4) from least to greatest.

   Indicate ties by placing the animation numbers in () please. For example, a suitable response could be: 1,2,(3,4),5,6.

Quantity	Ranking
Work done on the bowling ball by gravity	
Work done on the bowling ball by the external agent	
Total work done on the bowling ball	

Problem 6.5

A 5.0-kg block (called Block 1) is lifted from rest by an external agent, then returned to its original position as shown in the animation (position is given in meters and time is given in seconds). An identical block (called Block 2) is pushed along the surface with a force of 10 N. As with Block 1, it is returned to its original position at the end of the animation. Both blocks start and end at rest.

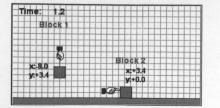

a. Determine the work done by gravity on Block 1 during the animation.

b. Determine the work done by gravity on Block 2 during the animation.

c. Determine the work done by the normal force on Block 2 during the animation.

d. Determine the total work done by friction (done on Block 2 and the table) during the animation.

Problem 6.6

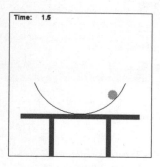

A mass of 2 kg is in a rather large bowl and moves as depicted in the animation (position is given in meters and time is given in seconds). There is no friction between the mass and the bowl, so the mass slides along the surface of the bowl (it does not roll at all). Determine the velocity of the mass at the bottom of the bowl.

Problem 6.7

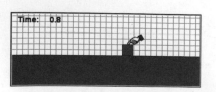

A 12-kg box is pushed at constant speed (the box is already moving at this constant speed at $t = 0$ s and continues to do so even after the animation ends) as shown in the animation (position is given in meters and time is given in seconds). The hand pushes on the box at an angle of 60° from the **vertical**. Note that there are four forces acting on the box: gravity, the force of the hand, the normal force, and friction.

During the animation,

a. Is the work done on the box by the external force (hand) positive, negative, or zero?

b. Is the work done on the box by the normal force positive, negative, or zero?

c. Is the work done on the box by gravity positive, negative, or zero?

d. Is the work done by friction (done on the box and the table) positive, negative, or zero?

e. Is the total work done on the box positive, negative, or zero?

Problem 6.8

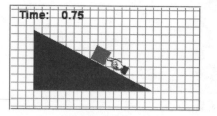

A 12-kg box slides up a 26.56° frictionless ramp at a constant speed as shown in the animation (position is given in meters and time is given in seconds). Note that both gravity and the hand do work on the box.

a. What is the work done on the box by the external force (hand) during the animation?

b. What is the work done on the box by gravity during the animation?

c. What is the total work done on the box during the animation?

Problem 6.9

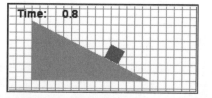

A 12-kg box slides down a rough 26.56° ramp at constant speed (it is already traveling at this constant speed at $t = 0$ s and continues to do so even after the animation ends) as shown in the animation (position is given in meters and time is given in seconds). Note that both gravity and friction do work on the box.

a. What is the work done by friction (done on the box and the ramp) during the animation?

b. What is the work done on the box by gravity during the animation?

c. What is the total work done on the box during the animation?

Problem 6.10

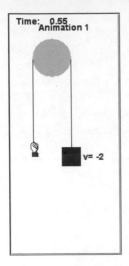

A 10-kg mass is attached via a massless string over a massless pulley to a hand (position is given in meters and time is given in seconds). The masses in each animation are identical.

a. Rank the animations according to the work done on the mass by gravity, from greatest to least.

b. Rank the animations according to the work done on the mass by the tension in the string, from greatest to least.

c. Rank the animations according to the total amount of work done on the mass, from greatest to least.

Indicate ties by placing the animation numbers in () please. For example, a suitable response could be 1,2,(3,4),5,6.

d. Calculate the work done on the mass by gravity during each of the animations.

e. Calculate the work done on the mass by the tension in the string during each of the animations.

f. Calculate the total amount of work done on the mass during each of the animations.

Note: For this Problem, the mouse-down for coordinates has been disabled.

Problem 6.11

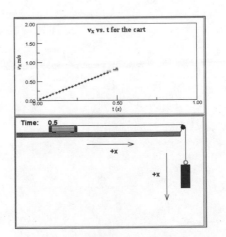

A 2.5-kg cart on a low-friction track is connected to a string and a 0.5-kg hanging mass as shown in the animation. Neglect any effects of the pulley on the motion of the system (position is given in meters and time is given in seconds). During the animation,

a. What is the work done on the hanging mass due to the tension in the string?

b. What is the work done on the hanging mass due to gravity?

c. What is the work done on the cart due to the tension in the string?

d. What is the work done on the cart due to gravity?

e. What is the work done on the cart due to the normal force?

f. What is the total amount of work done on the two-object system?

g. What is the final kinetic energy of the two-object system?

Note that the coordinates for each object (the positive *x* direction) are already chosen for you.

Problem 6.12

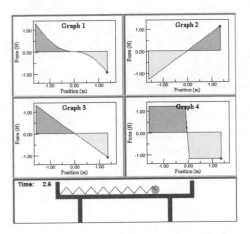

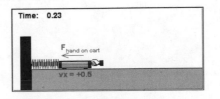

A ball on an air track is attached to a compressed spring (at $x = 0$ m the spring is unstretched) as shown in the animation (position is given in meters and time is given in seconds). Which area properly represents the work done by the spring during the animation (assume $v = 0$ m/s at the beginning and end of the animation)?

Problem 6.13

A cart sits on a track. A compressible spring is connected to the cart and to a barrier at the end of the track. At $t = 0$ s, the spring is compressed 0.5 m from its unstretched position, and you have to push on the cart to keep it in equilibrium. Then, by applying a varying force, you allow the spring to relax and then cause it to stretch while maintaining equilibrium during the entire process. The spring constant is 50 N/m. The frictional force of the track on the cart is negligible. Treat the cart as a point particle (position is given in meters and time is given in seconds).

a. What is the work done by the force of your hand on the cart during the interval between $t = 0$ and when the spring is fully stretched?

b. What is the work done by the spring on the cart during this same interval?

c. What is the total work done on the cart during this interval?

d. What must the force of your hand on the cart be to keep the cart in equilibrium when the spring is fully compressed?

e. What must the force of your hand on the cart be to keep the cart in equilibrium when the spring is fully stretched?

f. Why is the work done by your hand on the cart not equal to the product of this force component [calculated in part (e)] and the displacement of the cart?

Problem 6.14

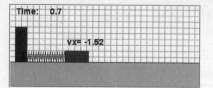

A 0.50-kg cart resting on an air track oscillates as shown in the animation (position is given in meters and time is given in seconds). What is the spring constant of the spring?

Energy

Topics include kinetic energy, potential energy, conservative and nonconservative forces, external forces, and collisions.

Illustration 7.1: Choice of System

The animation represents a ball sliding on a curved wire (position is given in meters, time is given in seconds, and energy on the bar graph is given in joules) subject to the forces of gravity, the normal force, and friction. Note that the wire does not depict the potential energy function of the ball (see **Illustration 7.3** for the Illustration on Potential Energy Diagrams). There are also three bar graphs that accompany the animation. They represent the kinetic energy (**orange**), gravitational potential energy (**blue**), and the energy dissipated due to friction (**red**). The two animations represent two different systems in which to analyze the motion via energy.

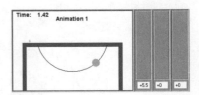

First play **Animation 1**. Note that in this animation there is no potential energy due to gravity and no energy dissipated due to friction. How can this be? Well, in this case we have chosen the system to be just the ball. **Animation 1: show system**. As a consequence, the system is not isolated because the ball experiences an *external force* due to gravity in addition to the *external* dissipative force of friction. Gravity does positive and then negative work on the ball, changing the ball's kinetic energy. In addition, the force of friction dissipates energy by doing a negative amount of work on the ball.

Now play **Animation 2**. What is going on here? What is the system now? Here there is potential energy due to gravity as well as energy dissipated due to friction. The system includes Earth and the room, and therefore the total energy must include gravitational potential energy and the frictional energy. **Animation 2: show system**. Given that we have defined a system that includes Earth and the room, the total energy (found by adding up all three bar graphs) should stay constant.

Illustration 7.2: Representations of Energy

There are many different ways to represent motion (as we have already seen). The same is true for energy. For example we can represent an object's kinetic energy using a graph of kinetic energy vs. time, a bar chart of kinetic energy that changes with time, or as a value in a table that changes with time (position is given in meters, time is given in seconds, and energy on the bar graph is given in joules). All three representations give us the same information, just in different forms. So why might we want to use a different representation? Well, it depends on what concept we are trying to illustrate and which representation gets to the heart of that concept. The collision occurs at a distance due to magnets on the front of each cart that allow the carts to collide without touching.

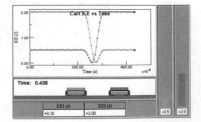

The graph shows us an instant-by-instant accounting of the kinetic energy of each cart. This is important if we wish to analyze every detail of the kinetic energy of the carts involved in the collision. Usually we are interested in whether or not energy is conserved in a given collision. In this case a graph gives this information, but it also gives us much more information. Notice that during the collision (kinetic) energy appears to be missing! This energy must be accounted for, so where is it? It is temporarily stored by the magnets attached to the carts. If there was a spring in between the carts, the energy would have been temporarily stored there instead. This energy is then transferred back to the carts by the end of the collision. This is why we compare the kinetic energies **before** the collision to those **after** the collision and often do not attempt to analyze the details of the collision itself.

Another way to answer the question of energy conservation, therefore, is with the bar chart (it is color coded) or with a table (the values are labeled). We simply compare the values—either the size of the bars or the values from the table—from **before** and **after** the collision. Are they the same? If yes, the energy, specifically the kinetic energy, of the two-cart system is conserved.

Illustration 7.3: Potential Energy Diagrams

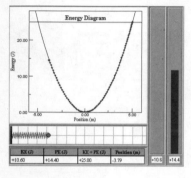

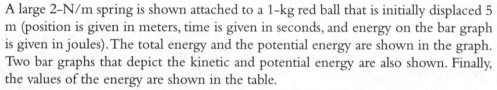

A large 2-N/m spring is shown attached to a 1-kg red ball that is initially displaced 5 m (position is given in meters, time is given in seconds, and energy on the bar graph is given in joules). The total energy and the potential energy are shown in the graph. Two bar graphs that depict the kinetic and potential energy are also shown. Finally, the values of the energy are shown in the table.

The potential energy diagram is an important diagram because it depicts the potential energy function, often just called the potential. This terminology is unfortunate since it can lead to confusion with the electric potential. The potential energy function is plotted vs. position, and therefore it tells you the potential energy of an object if you know its position. The potential energy function for a mass on a spring is just $PE(x) = 0.5 k x^2$. Here $PE(x) = x^2$. Note that, depending on your text, you may have seen the potential energy function represented as either $V(x)$ or $U(x)$. We use the book-independent version $PE(x)$. In addition to the potential energy function, a horizontal teal line represents the total energy of the system.

Because of the form of the above potential energy function, it is easy to get confused as to what it is actually showing and what it represents. If you have not done so already, run the animation. The red dot on the potential energy curve does NOT represent the actual motion of a particle on a bowl or roller coaster. In other words, it does NOT represent the two-dimensional motion of an object. It represents the one-dimensional motion of an object, here the one-dimensional motion of a mass attached to a spring. The motion of the red mass is limited to between the turning points represented by where the total energy is equal to the potential energy.

Now **also show the kinetic energy on the graph too**. Watch the kinetic energy and potential energy change as the mass moves and the spring ceases to be stretched and then gets compressed. Notice that the potential energy added to the kinetic energy always adds up to the total energy. Therefore, if you know the total energy and the potential energy function, you know the kinetic energy of the object at any position in its motion.

Clearly, the force depicted is a spring force. How can we be sure? Well, there is a relationship between the force and the potential energy function. This relationship is expressed as $F_x = -d\,(PE)/dx$. Therefore, since $PE(x) = x^2$, $F_x = -2x$, which tells us that $k = 2$ N/m (as stated in the first line of the Illustration).

Illustration 7.4: External Forces and Energy

When we talk about energy we tend to focus on the change in kinetic energy and change in potential energy, where the change in potential energy is the negative of the work done by conservative forces. But what happens with nonconservative and external forces? (Note that some books lump external forces with nonconservative forces.) Well, these are the forces that cause the total energy of the system to change.

In other words, without a nonconservative force or an external force, the total energy would never change. This is what we mean by the statement of energy conservation, $\Delta KE + \Delta PE = 0$.

If there are nonconservative or external forces, the total energy will change. When we say nonconservative forces we usually are thinking about kinetic friction. Kinetic friction is a special force that always decreases the total energy of the system (the amount of work that it does on an object is always negative). If friction exists in a system, and you wait long enough, all of the energy will dissipate. What about external forces? Do they add or take energy from the system? Well, it depends.

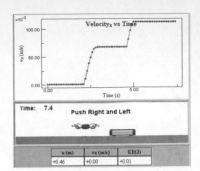

Consider the cart in the animation. The cart interacts with the two-handed image if the image is near the left-hand or right-hand end of the cart (position is given in meters and time is given in seconds). The arrow below the cart shows the direction and strength of the external force applied. Reset the animation if the cart goes off the end of the track.

Move the cart around and look at each of the graphs. Now focus on the **|F| cos(theta) vs. position** graph, which tells you about the work done by the external force (the hand). Is it always positive or is it always negative? It can be positive or negative depending on the circumstances. If the work done by the force is positive, the energy of the system (the cart and Earth) increases. Since the potential energy of the cart remains fixed (since it does not change height), all of this energy is seen as kinetic energy. If the work done by the force is negative, the energy of the system decreases. Again, the change in energy is seen as kinetic energy.

Illustration 7.5: A Block on an Incline

A block is on an incline and slides without friction. Partway down the incline, it hits a spring as shown (position is given in meters, time is given in seconds, and energy on the bar graph is given in joules). You can add the protractor by checking the box. Also shown are the force vectors, one for each force (the **red** ones) and one for the total force acting (the **blue** one). The energy of the system is shown in the three bar graphs on the right: kinetic energy (**orange**), gravitational potential energy (**blue**), and elastic potential energy (**green**).

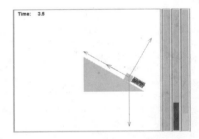

Let's begin analyzing this situation as we would have in Chapters 3 and 4. First, we need to define a convenient set of axes. A convenient set of axes has one axis along the incline and the other axis perpendicular to the incline. This choice allows us to have one direction where there is no acceleration (the direction perpendicular to the incline) and one direction where there is an acceleration (parallel to the incline). There is also another reason for this choice of axes. It allows us to decompose only one force instead of two. We have to decompose the gravitational force into a component along the incline and one perpendicular to the incline. How do we deal with the spring force? Well, the honest answer is that while we can analyze the forces to determine the acceleration, it is not tremendously useful since the spring force is not constant.

Run the animation and look at the normal force and the gravitational force vs. the spring force. The spring is not compressed initially, then it compresses, and then it uncompresses. During this time the net force on the block changes dramatically. Look at the blue net force vector. As a consequence, the acceleration of the block changes dramatically as well. (Note that the net force still points parallel to the incline; its size is what changes dramatically.)

Since the forces change over the course of the motion of the block, the acceleration of the block is not constant throughout the motion of the block. Newton's laws and kinematics clearly fall short in analyzing the motion here. What to do? Use energy! At the starting point of the motion of the block, it has no kinetic energy, and no elastic (spring) potential energy, but it does have gravitational potential energy. As the block moves down the incline some of the gravitational potential energy is converted to kinetic energy. When the block hits the spring, the kinetic energy and the gravitational potential energy get converted to elastic (spring) potential energy.

Watch the animation and describe how all of the potential energy due to the compressed spring gets converted to other types of energy.

Exploration 7.1: Push a Cart Around

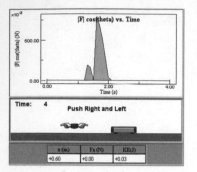

The cart in the animation interacts with the two-handed image if the image is near the left-hand or right-hand end of the cart (position is given in meters and time is given in seconds). Move the image from side to side and observe the resulting graph. The arrow below the cart shows the direction and strength of the force. Restart the animation if the cart goes off screen.

Define the system to be just the cart and answer the following questions assuming that you are moving the cart around with the "handy" image.

a. Is the energy of the system constant? If not, where is it coming from?

b. Does the energy always decrease if the image is to the right of the cart? Does it increase if the image is to the left of the cart?

Exploration 7.2: Choice of Zero for Potential Energy

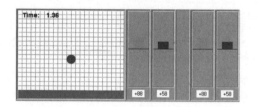

The animation depicts a ball being dropped from $y = 15$ m onto the ground at $y = 0$ m (position is given in meters and time is given in seconds). For this animation we will assume that the ball undergoes a very hard collision with the ground, which also conserves energy. Also shown are two pairs of bar graphs representing the different types of energy associated with the ball: the kinetic energy (**orange**) and the gravitational potential energy (**blue**). The bar graphs on the left show the kinetic energy and the potential energy as measured from $y_{ref} = 0$ m. The bar graphs on the right show the kinetic energy and the potential energy with a varying zero potential energy point. You can vary the zero point from -15 m $< y_{ref} < 15$ m by changing the value in the text box and clicking the "set value and play" button.

Change the zero point for the potential energy from zero to a variety of positive values and a variety of negative values. Answer the following questions about the animation.

a. For zero points that are less than zero, does the gravitational potential energy shift up or down?

b. Is all of this energy accessible to the ball? In other words, can it all be converted to kinetic energy?

c. For zero points that are greater than zero, does the gravitational potential energy shift up or down?

d. For $y_{ref} = -15$ m, how much potential energy does the ball start out with? How much does it have when it hits the ground? What is the change in potential energy?

e. For $y_{ref} = 15$ m, how much potential energy does the ball start out with? How much does it have when it hits the ground? What is the change in potential energy?

f. How do your answers for (d) and (e) compare? Why?

Exploration 7.3: Elastic Collision

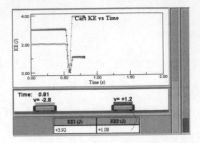

The initial velocities of the two carts in the above animation can be changed by entering new values into the text fields (position is given in meters, time is given in seconds, and energy on the bar graph is given in joules). As the carts approach one another, they begin to repel due to the magnets carried by each of them, thereby changing their velocities. The two color-coded bar graphs on the right show the instantaneous kinetic energy of the carts.

When you get a good-looking graph, right-click on it to clone the graph and resize it for a better view.

a. Run the animation using 2 m/s and −2 m/s for the velocities of the left and right carts, respectively. What is the change in kinetic energy of the left cart? The right cart? What is the total change in energy?

b. Simulate collisions using other values of equal but opposite velocities. How does this effect each cart's change in kinetic energy? The change in the total energy?

c. Stop the animation just as the collision is about to take place and step forward in time so that the animation is paused during the collision process. What happens to the total energy during the collision process?

d. Does the last result imply that the two-cart system is not isolated?

e. Run the animation using 1 m/s and −2 m/s for the velocities of the left and right carts, respectively. What is the change in total kinetic energy produced by the collision?

Exploration 7.4: A Ball Hits a Mass Attached to a Spring

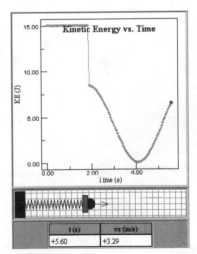

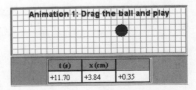

Whenever objects interact, energy is likely to be converted from one form to another and/or dissipated (position is given in meters and time is given in seconds). Consider two models of a ball hitting a 0.4-kg rectangle attached to a massless spring. After the collision the masses stick together and oscillate. **Animation 1** represents an ideal spring and frictionless conditions, while **Animation 2** represents a more realistic spring, and friction takes its inevitable toll on the system (only the kinetic energy of the ball is shown in the graph). Consider a system made up of the mass, the rectangle, and the massless spring as you answer the following questions. The potential energy of the spring is zero when the spring is uncompressed and, since the spring is massless, it has no kinetic energy.

a. What is the mass of the black ball?

b. What is the initial energy of the system?

Answer the following questions for each animation.

c. What is the energy of the system immediately after the collision?

d. Draw energy diagrams for the three objects that make up the system at the following times: $t = 0$ s, $t = 1.90$ s, $t = 4.10$ s, $t = 6.30$ s, and $t = 8.55$ s.

e. For **Animation 2** only: Approximately how long does it take for 80% of the initial energy to be dissipated?

Exploration 7.5: Drag the Ball to Determine PE(x)

Potential energy is energy associated with the configuration of an object or a system (position is given in meters and time is given in seconds). Since potential energy can be converted to kinetic energy, an operational way to determine potential energy is to let the system evolve from an unknown configuration to a known configuration and measure the kinetic energy. You can use this technique to measure and plot potential energy functions, $PE(x)$.

Plot the potential energy as a function of position for both animations. Note that these interactions may or may not be physical interactions.

Procedure: Reset will initialize the system to a known potential energy. This initial configuration has been marked with a small red dot. Assume this configuration has zero potential energy, $PE_0 = 0$, and the object has a mass of 1 kg. Use the mouse to move the object to a new position and release it. The object will have zero initial velocity when it is released. If the object returns to the original position you can record the velocity and calculate the kinetic energy. This kinetic energy must have come from the potential energy at the new position if the interaction is conservative.

Note: Animation will stop after 100 s.

Exploration 7.6: Different Interactions

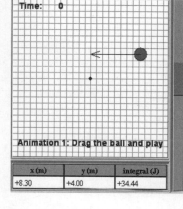

The animations show a red ball that you can drag with the mouse (position is given in meters, time is given in seconds, and energy on the bar graph is given in joules). The bar graph shows the negative of the force on the ball integrated over the displacement from the origin. This is the negative of the work done on the ball to get it to this position. This integral, when the force is conservative, is also the potential energy associated with the ball when it is at this position. Also shown is a table with a calculation of position and the negative of the work.

a. Briefly describe the force in each animation.

b. Which of the forces is conservative? Why?

c. For the conservative forces, draw the potential energy function.

Exploration 7.7: Exploring Potential Energy Functions

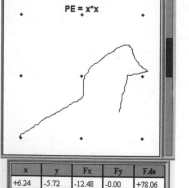

Select a possible potential energy function. Drag the crosshair cursor with the mouse. The bar graph on the right displays the work done along path by the force that corresponds to the given potential energy function. For your reference, there are circles every 10 m that form a coordinate grid (position is given in meters and the result of the integral given on the bar graph is in joules).

a. Describe each potential energy function in words.

b. How does the work relate to the change in potential energy along a certain path?

c. What happens when you drag the cursor through a closed path (a path that begins and ends at the same point)?

d. What is the force that is responsible for each potential energy function? Write the force in the x and y direction as a function of x and y, $F_x(x, y)$ and $F_y(x, y)$.

When you are finished with this Exploration, feel free to enter your own potential energy function.

Problems

Problem 7.1

A mass of 2 kg is in a rather large bowl and moves as depicted in the animation (position is given in meters and time is given in seconds). There is no friction between the mass and the bowl so it slides along the surface of the bowl (it does not roll at all). Determine the velocity of the mass at the bottom of the bowl.

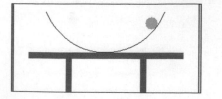

Problem 7.2

A 12-kg box slides up a 26.56° frictionless ramp at constant speed as shown in the animation (position is given in meters and time is given in seconds). Note that the hand does work on the box.

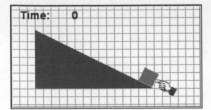

a. What is the work done on the box by the external force (hand) during the animation?

b. What is the change in gravitational potential energy of the box during the animation?

c. What is the change in kinetic energy of the box during the animation?

Problem 7.3

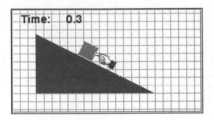

A 12-kg box slides down a 26.56° frictionless ramp at constant speed as shown in the animation (position is given in meters and time is given in seconds). Note that the hand does work on the box.

a. What is the work done on the box by the external force (hand) during the animation?

b. What is the change in gravitational potential energy of the box during the animation?

c. What is the change in kinetic energy of the box during the animation?

Problem 7.4

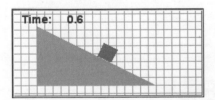

A 12-kg box slides down a rough 26.56° ramp at constant speed (it is already traveling at this constant speed at $t = 0$ s and continues to do so even after the animation ends) as shown in the animation (position is given in meters and time is given in seconds). Note that friction does work on the box.

a. What is the work done by friction (done on the box and the ramp) during the animation?

b. What is the change in gravitational potential energy of the box during the animation?

c. What is the change in kinetic energy of the box during the animation?

Problem 7.5

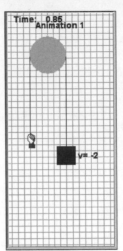

A 10-kg mass is attached via a massless string over a massless pulley to a hand (position is given in meters and time is given in seconds). The masses in each animation are identical.

a. Rank the animations according to the change in gravitational potential energy of the mass, from greatest to least.

b. Rank the animations according to the work done on the mass by the tension in the string, from greatest to least.

c. Rank the animations according to the change in kinetic energy of the mass, from greatest to least.

Indicate ties by placing the animation numbers in () please. For example, a suitable response could be 1,2,(3,4),5,6.

d. Calculate the change in gravitational potential energy of the mass during each of the animations.

e. Calculate the work done on the mass by the tension in the string during each of the animations.

f. Calculate the change in kinetic energy of the mass during each of the animations.

Problem 7.6

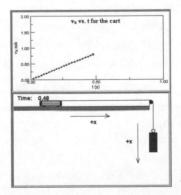

A 2.5-kg cart on a low-friction track is connected to a string and then to a 0.5-kg hanging mass as shown in the animation. Neglect any effects of the massless pulley on the motion of the system (position is given in meters and time is given in seconds).

During the animation,

a. What is the work done on the hanging mass due to the tension in the string?

b. What is the change in gravitational potential energy of the hanging mass?

c. What is the work done on the cart due to the tension in the string?

d. What is the change in gravitational potential energy of the cart?

e. What is the work done on the cart due to the normal force?

f. What is the total work done by the tension on the two-object system?

g. What is the change in potential energy of the two-object system?

h. What is the change in kinetic energy of the two-object system?

Note that the coordinates for each object (the positive x direction) are already chosen for you.

Problem 7.7

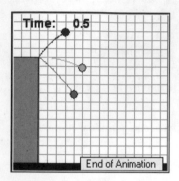

Three balls are thrown off the top of a building, all with the same speed but with different launch angles (position is given in meters and time is given in seconds). The components of the initial velocities are given.

- The blue ball has an initial velocity of (6 m/s, 8 m/s).
- The green ball has an initial velocity of (10 m/s, 0 m/s).
- The red ball has an initial velocity of (8 m/s, −6 m/s).

a. Rank the three balls according to which one hits the ground first.

b. Rank the three balls according to which one has the greatest speed the instant before impact with the ground.

c. Now calculate te speed of each of the balls the instant before impact with the ground.

Problem 7.8

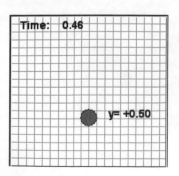

A ball is dropped on a hard floor as shown in the animation (position is given in meters and time is given in seconds). Assume that the acceleration due to gravity is 9.8 m/s^2.

a. What is the speed of the ball the instant before it hits the ground?

b. How much energy (in % of original energy) is lost in the collision with the floor?

c. What is the coefficient of restitution for the ball?

The coefficient of restitution, for the collision where one object does not move, is the ratio $|v_f| / |v_i|$.

Problem 7.9

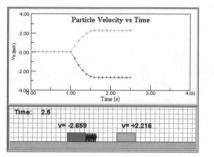

Two carts are in close proximity. A massless spring is attached to the end of the red cart and is compressed. The massless spring is released such that the two carts are "pushed" apart as shown in the animation (position is given in meters and time is given in seconds). The mass of the green cart is 1.5 kg. Consider a system made up of the two carts and the massless spring.

a. What is the velocity of the center of mass of the carts after the massless spring is released (assume that since the spring is massless it cannot have a kinetic energy)?

b. What is the mass of the red cart?

c. What is the change in kinetic energy of the system due to the release of the spring?

d. What was the change in potential energy of the spring?

Problem 7.10

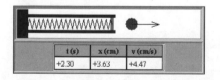

A spring gun is loaded with a 500-gram projectile (position is given in centimeters and time is given in seconds). The spring is massless and therefore has no kinetic energy.

a. How much potential energy is converted to kinetic energy in the spring gun?

b. How much potential energy has been converted to kinetic energy when the ball is at the following positions: −5 cm, −4 cm, −3 cm, −2 cm, −1 cm, and 0 cm?

c. Plot the potential energy of the spring as a function of distance.

Problem 7.11

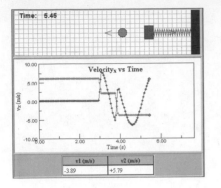

v1 (m/s)	v2 (m/s)
-3.89	+5.79

A 1.0-kg projectile bounces off of an object ($m = 1$ kg) attached to a massless spring as shown (position is given in meters and time is given in seconds). The table entries, v1 and v2, show the velocities of the projectile and the target, respectively. Assume that the collision is elastic.

a. There are five important time intervals during the animation. What are they? Briefly describe what is happening during these intervals.

b. Draw a graph of energy vs. time for the kinetic energy of the projectile, the kinetic energy of the target, and the potential energy of the spring.

When you get a good-looking graph, right-click on it to clone the graph and resize it for a better view.

Problem 7.12

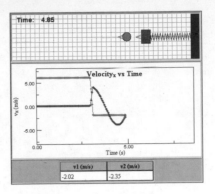

v1 (m/s)	v2 (m/s)
-2.02	-2.35

A 0.5-kg projectile bounces off of an object ($m = 1$ kg) attached to a massless spring as shown (position is given in meters and time is given in seconds). The table entries, v1 and v2, show the velocities of the projectile and the target, respectively. Assume that the collision is elastic.

a. Draw a graph of energy vs. time for the kinetic energy of the projectile, the kinetic energy of the target, and the potential energy of the spring.

b. What percent of the initial energy was transferred to the target-spring system during the collision?

When you get a good-looking graph, right-click on it to clone the graph and resize it for a better view.

Momentum

Topics include momentum, impulse, conservation of momentum, one-dimensional and two-dimensional collisions, and center of mass.

Illustration 8.1: Force and Impulse

So what do we mean by a force? Newton considered a net force as something that caused a time rate of change of momentum, $\Delta\mathbf{p}/\Delta t$ or $d\mathbf{p}/dt$. However, C.D. Broad (*Scientific Thought*, 1923) wrote, "It seems clear to me that no one ever does mean or ever has meant by 'force', rate of change of momentum." So if Newton's statement seems odd it is because you are used to a special—and famous—case of Newton's general statement of the second law, that of $\Sigma\,\mathbf{F}_{net} = m\mathbf{a}$.

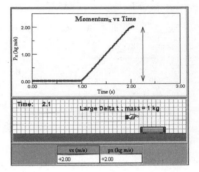

Consider the force applied over a **small Δt**. Notice the change in momentum (position is given in meters and time is given in seconds). The arrow represents the change in momentum. Initially the mass of the cart is 1 kg. Change the mass to 2 kg. Does the change in momentum differ? No! But what does change is the final velocity; it is half of the velocity when the mass was 1 kg. The same force results in the same change in momentum in the same time interval.

Another way to represent this is in terms of the integral (the area) under a force vs. time graph. Check the box to see this graph. This area is called the impulse, which is a fancy name for $\Delta\mathbf{p}$. What can you say about the impulse received by the cart, independent of its mass? Check the second box to find out. Again, it should be, and is, the same.

Consider the animation with the **large Δt**. The difference between the animations is that in **large Δt** the force acts for a longer time and therefore the force causes a larger change in momentum. Again, the arrow represents the change in momentum, which is larger than the **small Δt** case.

Illustration 8.2: The Difference between Impulse and Work

In **Illustration 8.1**, we learned a change in momentum was due to a net applied force. What about kinetic energy? Well, it is also due to a net applied force, but in a different way. Recall that we talk about work as the amount of force in the direction of an object's displacement multiplied by the displacement. No displacement, no work. Work is positive if $\mathbf{F}$ and $\Delta\mathbf{x}$ (or $d\mathbf{x}$) point in the same direction and negative if $\mathbf{F}$ and $\Delta\mathbf{x}$ (or $d\mathbf{x}$) point in opposite directions.

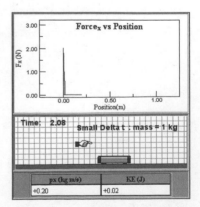

Consider the force applied by the hand over a **small Δt** (this happens automatically at $t = 1$ s). Notice the change in momentum (position is given in meters and time is given in seconds). Initially the mass of the cart is 1 kg. Change the mass to 2 kg. Does the change in momentum differ? No! But what does change is the final velocity; it is half of the velocity when the mass was 1 kg. The same force results in the same change in momentum in the same time interval.

So what happens to the kinetic energy? Does it remain the same upon a change in mass? No. Why not? Recall that the work that will be equal to the change in kinetic energy is related to the displacement the cart undergoes when the force is applied.

Due to the larger mass, the cart does not accelerate as much and therefore does not move as far, so its kinetic energy is less.

Another way to represent this is in terms of the integral (the area) under a force $\cos(\theta)$ vs. distance graph. This area is called the work that is the object's ΔKE. What can you say about the work received by the cart when its mass changes? Check the second box to find out. Again, it should be, and is, different.

What happens when instead of applying a **small** Δt, you apply a **large** Δt? There is a larger impulse because Δt is larger. There is also a larger change in the kinetic energy since Δx is larger as well.

Illustration 8.3: Hard and Soft Collisions and the Third Law

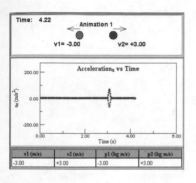

This Illustration models collisions between two particles (position is given in meters and time is given in seconds). Both animations assume identical particles and both animations start with the particles having the same velocities. What is different about the animations is the interaction between the two particles. The interaction in **Animation 1** can be characterized as hard since the acceleration is very large and the interaction is very short range. It is also called a contact interaction because the force turns on when the two particles are in contact.

In **Animation 2** the interaction can be characterized as soft. Vary the masses of the two particles and be sure to pay attention to the scale of the acceleration graph in each animation. In addition, watch the relative acceleration of the particles as you vary the masses. Notice that the accelerations are different when the masses are different.

What can you say about the force experienced by each particle in each animation? The character of the forces is different: One is soft and one is hard. Nonetheless, the forces are always equal and opposite. To see this you must take each object's acceleration and multiply it by its mass. This is exactly the statement of Newton's third law, the law of reciprocity of forces. In this case, there are no net external forces acting on the two particles, so the change in momentum of the two-particle system is zero. In other words, momentum is conserved. Using equations, we would say that since $\Sigma \mathbf{F}_{net} = \Delta \mathbf{p}/\Delta t$ or $\Sigma \mathbf{F}_{net} = d\mathbf{p}/dt$, if the net force on a system is zero, then $\Delta \mathbf{p}/\Delta t = 0$ or $d\mathbf{p}/dt = 0$, which means that the change in momentum over time must be zero. Hence the sum of the two impulses experienced by the balls must be zero. If one particle's momentum goes up, the other particle's momentum must go down by exactly the same amount. Check it out by looking at the tables.

Two-dimensional models show a dramatic difference between hard and soft collisions. (See Problem 8.12 for two-dimensional collisions.) Hard collisions tend not to have much of an effect on incident particles except for the occasional particle that suffers a head-on impact. Soft collisions, on the other hand, produce minor deflections on a large number of particles. The experimental observation of alpha particles being deflected backwards from gold foil led Ernest Rutherford to predict that atoms have a small hard core, the nucleus.

Illustration 8.4: Relative Velocity in Collisions

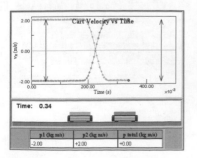

In this set of collisions there are no net external forces acting on the two carts. Enter new values for the velocity of each cart and the mass of the right-moving (orange) cart. Then click the "set values and play" button to register your values and play the animation (position is given in meters and time is given in seconds). We have set limits on the values you can choose:

$$0.5 \text{ kg} < m_1 < 4 \text{ kg}, \quad 0 \text{ m/s} < v_1 < 4 \text{ m/s}, \quad \text{and}$$
$$-4 \text{ m/s} < v_2 < 0 \text{ m/s}.$$

The table gives an instantaneous reading of each cart's momentum as well as the total momentum in the two-cart system. In addition, when you select the check box, arrows representing the magnitude of the relative velocities before and after the collision between the two carts are also shown.

Because the net force on the system of two carts is zero, the change in momentum of the two-particle system is zero. In other words, momentum is conserved. Using equations, we would say that since $\Sigma \mathbf{F}_{net} = \Delta \mathbf{p}/\Delta t$ or $\Sigma \mathbf{F}_{net} = d\mathbf{p}/dt$, if the net force on a system is zero, then $\Delta \mathbf{p}/\Delta t = 0$ or $d\mathbf{p}/dt = 0$, which means that the change in momentum over time must be zero. Hence the sum of the two impulses experienced by the carts must be zero. If one particle's momentum goes up, the other particle's momentum must go down by exactly the same amount.

In elastic collisions, the concept of the **relative velocity** is an important one in analyzing the collision. The relative velocity is defined as $\mathbf{v}_1 - \mathbf{v}_2$ (it could also be defined as $\mathbf{v}_2 - \mathbf{v}_1$ since the choice of 1 and 2 is arbitrary).

Turn on the relative velocity arrows and vary the velocity of each cart and the mass of the right-moving (orange) cart. Determine the relationship between the relative velocity before the collision and the relative velocity after the collision. What did you find? It turns out that the magnitude of the relative velocity before and after an elastic collision is the same. However, the sign of the relative velocity changes from before to after the collision: $(\mathbf{v}_1 - \mathbf{v}_2)_i = -(\mathbf{v}_1 - \mathbf{v}_2)_f$. This relationship can be verified by using the conservation of energy and conservation of momentum equations and a bit of algebra.

Consider an elastic collision where $v_1 = 1$ m/s and $v_2 = -4$ m/s. Clearly the relative velocity before the collision is 5 m/s. What must it be after the collision? -5 m/s. Try it and find out if this is true. Does it matter if you change the mass of the orange cart?

Illustration 8.5: The Zero-Momentum Frame

Is physics different when viewed in different reference frames? Well, it can certainly look different. Consider the collision in the animation as seen initially in the reference frame of Earth (the relative velocity between this frame and Earth's stationary frame is zero). Here both the red ball and the blue ball have the same mass equal to 1 kg. Note that energy and momentum are conserved in the collision with KE = 2 J and $p_x = 2$ kg·m/s before and after the collision.

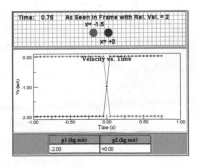

Change v from zero to 2 m/s (position is given in meters and time is given in seconds). How does the collision change? The red ball is now initially stationary, and the blue ball is moving to the left at 2 m/s. Note that in the original collision with $v = 0$ m/s, the red ball was initially moving to the right and the blue ball was initially stationary. In the new frame the momentum of the two-ball system is different. However, the kinetic energy happens to be the same and energy and momentum are conserved.

Now try $v = -2$ m/s. Are energy and momentum still conserved? Even though the values of the kinetic energy and momentum change, the laws of conservation of energy and conservation of momentum still hold.

Now try $v = 1$ m/s. What is the new momentum for the two-ball system? This frame of reference is appropriately called the zero-momentum frame. In this frame the sum of the momentum of all objects in the system is zero. This frame is also called the center-of-mass frame. The center of mass is a coordinate that is a mass-weighted average of the positions of the objects that make up the system. In a two-object system the center of mass is always somewhere in between the two objects. Since the center of mass is a mass-weighted average, the center of mass will always be closer to the object that is more massive. In the case of this animation, where both balls have the same mass, the center of mass is always at the midpoint between the two masses. This point does not move in the zero-momentum frame, but does move in other frames.

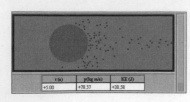

Illustration 8.6: Microscopic View of a Collision

In the animation a red 80-kg ball with an initial kinetic energy of 360 J is trapped inside a box with rigid walls containing a cylinder constructed of 80 small 1-kg spheres (position is given in meters, time is given in seconds, and energy shown on the bar graph is given in joules). The ball crashes into the cylinder and breaks it apart. The bar graph displays the kinetic energy of the red ball. The table displays the time, momentum, and kinetic energy of the red ball.

This animation is meant to simulate a collision between two solid objects, one of which is stationary. The stationary object is a loose collection of smaller objects and approximates a larger solid object. This is only an approximation since this object should stay together, not break apart. When you do collision experiments in the lab the objects colliding do not usually deform this much! Nevertheless, we can learn a lot from this animation. As the red ball hits the blue object, the blue object deforms, absorbing some of the red ball's kinetic energy and momentum. If the blue object were indeed solid, the deformed object—the entire object—would move to the right. We can imagine this by considering the average motion of the small blue balls that make up the larger solid object. We note that the general motion of these blue balls is to the right. Where does all of the initial kinetic energy go? It goes into the kinetic energy of the small blue balls.

Illustration 8.7: Center of Mass and Gravity

The first image shows two blocks of equal mass (position is given in meters). The center of mass of the system is shown by a red dot, and its position is calculated for you. Drag the green block on the right. What do you notice about the location of the center of mass as you change the right block's position?

Now suppose the two blocks have unequal mass as shown in **Animation 2**. Is the center of mass at the center of the system? By viewing the location of the center of mass, you know which block is more massive. So which one is more massive?

How can we calculate the ratio of the mass of the blue block to the mass of the red block? If you only have two blocks, then the ratio of the distances of each block from the center of mass is related to the ratio of their masses. Therefore, by measuring the distance from each block to the center of mass, you can calculate the ratio of their masses. The location of the center of mass is given by $X_{cm} = (x_1 m_1 + x_2 m_2)/(m_1 + m_2)$ for a one-dimensional two-object system.

A concept similar to that of the center of mass is that of the center of gravity. In fact the two are often used interchangeably. The center of mass is defined above; the center of gravity is defined as the point on a system where gravity can be considered to act. The center of gravity takes into account the fact that the force of gravity—and therefore the acceleration due to gravity—is different for different heights above the surface of Earth. For this Illustration, the center of mass is equivalent to the center of gravity. Only if the system is really large might the acceleration due to gravity be different at different parts of the system. This would cause the center of gravity to differ from the center of mass.

Illustration 8.8: Moving Objects and Center of Mass

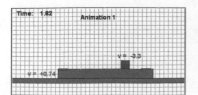

A green block, 1.00 kg, sits on a red block, 4.46 kg, as shown in the animation (position is given in meters and time is given in seconds). All surfaces are frictionless except for the gray patches on the red block. Given the self-propelled motion of the green block in **Animation 1**, are momentum and energy conserved in the animation? If not, why not?

Well, momentum is conserved because there are no external forces. The momentum of the system was zero before the green block moved, is zero when the blocks move, and is again zero when the blocks are stationary. From the point of view of the center of mass, $V_{cm} = 0$ m/s and therefore $P_{cm} = 0$ kg·m/s.

We can see this by considering what happens to the center of mass during **Animation 2**. The center of mass of the system is $X_{cm} = (m_1 x_1 + m_2 x_2)/(m_1 + m_2)$ and is represented by the black dot. Note that the center of mass of the system does not move relative to the ground but does move to the right relative to the right edge of the red block as the red block moves to the right. In fact we can look at the center of mass for each object by replacing each block by a dot as well as shown in **Animation 3**.

What about energy? As is always the case, whether energy is conserved depends on how you define your system. Looking just at the center of mass, since $V_{cm} = 0$ m/s, energy is conserved. However, if we look at the blocks individually, energy (in the sense of mechanical energy) is not conserved. Energy stored in the individual elements of the system (presumably the green block's potential energy) is turned into kinetic energy of both blocks and then is dissipated by friction.

Exploration 8.1: Understanding Conservation Laws

Observe the animation to see if you can discover any conservation laws. You should determine whether your laws hold for the left half of the animation, the right half of the animation, or the entire animation. Properties that you might want to consider are number of particles, color, and sum of the particle speeds.

The animation will run for 100 seconds.

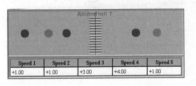

Exploration 8.2: An Elastic Collision

The animation shows an elastic collision between two masses (position is given in centimeters and time is given in seconds).

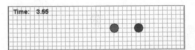

a. Set the initial velocity of the blue ball to zero. For the three conditions of the relative masses of the blue and red balls shown in the table, PREDICT what value (or values) of the initial velocity of the red ball will result in . . .

- both balls moving to the right after the collision.
- the red ball stopping after hitting the blue ball.
- the red ball moving to the left and the blue ball moving to the right after the collision.

Enter the range of initial velocity values for the red ball that results in . . .	. . . both balls moving to the right after the collision.	. . . the red ball stopping after colliding with the blue ball.	. . . the red ball moving to the left and the blue ball moving to the right after the collision.
$m_{red} = m_{blue}$			
$m_{red} = 2 * m_{blue}$			
$m_{red} = 0.5 * m_{blue}$			

AFTER you have made your predictions, test them using the animation. Were you correct? If not, explain.

b. Now set the initial velocity of the blue mass to -20 cm/s, the initial velocity of the red mass to 5 cm/s, and the masses equal. PREDICT the direction each ball will be traveling after impact. AFTER you have made your prediction, try it. Were you correct? If not, explain.

c. Set the initial velocity of the blue mass to -10 cm/s and the red mass to half the mass of the blue ball. PREDICT the velocity the red mass must have in order to completely stop the blue mass when they collide. Now try it. Were you correct? If not, explain.

d. Set the initial velocity of the blue mass to -10 cm/s and the red mass to twice the mass of the blue ball. PREDICT the velocity the red mass must have in order to completely stop the blue mass when they collide. Now try it. Were you correct? If not, explain.

Exploration 8.3: An Inelastic Collision with Unknown Masses

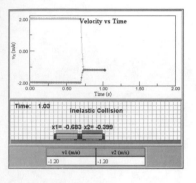

The initial velocities of the two carts in the animation can be changed by entering new values into the text fields (position is given in meters and time is given in seconds). As the carts approach one another they stick together.

Repeat the animation using varying velocities as you answer the following questions. Right-click on the graph to make a copy that can be expanded for better resolution.

a. Run the animation using 2 m/s and -2 m/s for the velocities of the left and right carts, respectively. What is the change in velocity of the left cart? The right cart? What is the ratio of these changes?

b. Simulate collisions using other values of equal but opposite velocities. How does this effect the changes in the velocities? The ratio of the changes?

c. Run the animation using 1 m/s and -2 m/s for the velocities of the left and right carts, respectively. What is the change in velocity of the left cart? The right cart? What is the ratio of these changes?

d. Is the ratio of the changes in the velocities always the same?

e. What is the mass ratio of the carts?

Exploration 8.4: Elastic and Inelastic Collisions and Δp

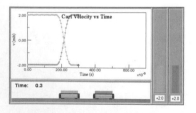

Enter in a new value and click the "set values and play" button to register your values and run the animation (position is given in meters and time is given in seconds). We have set limits on the values you can choose:

$$0.5 \text{ kg} < m_1 < 2 \text{ kg}, \quad 0 \text{ m/s} < v_1 < 4 \text{ m/s}, \quad \text{and}$$
$$-4 \text{ m/s} < v_2 < 0 \text{ m/s}.$$

The bar graph gives an instantaneous reading of each cart's energy and the check box changes the collision type from perfectly elastic to perfectly inelastic.

Answer the following questions for both the elastic and inelastic collisions.

a. Vary the mass and velocities. Is $\Delta \mathbf{p}_1 = -\Delta \mathbf{p}_2$?

b. Why should this be the case?

c. Is the energy of the system constant? If not, where is it going?

Exploration 8.5: Two- and Three-Ball Collisions

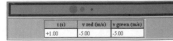

If you drop a rubber ball and it hits the ground at 5 m/s, it bounces back at almost the same speed (position is given in meters and time is given in seconds). But what happens if you drop two balls stacked one upon another? A common lecture demonstration has a professor dropping a light ball and a heavy ball at the same time. The light ball is directly above the heavy ball so that the heavy ball hits the ground first, bounces back, and then hits the light ball which is still on its way down.

This animation uses **two balls** with a mass ratio of 1:10. We consider motion on a horizontal air track so we can ignore the effect of gravity so as to make the physics as clear as possible. The balls move at constant speed to the left before hitting the wall; assume all collisions are elastic.

a. Predict the velocities of the balls after the first set of collisions, that is, when both balls are moving to the right.

b. Predict the velocities if you use three balls with mass ratios of 1:10:100.

c. Now run the animations. Were you correct? If not, explain why.

Note: The animation will run for 100 seconds.

Exploration 8.6: An Explosive Collision

The system's total kinetic energy is increased in a 1,200-J explosion in the animation (position is given in meters and time is given in seconds).

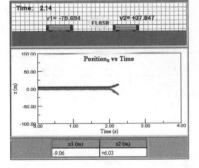

Use a mass ratio of 1:2 for the following questions.

a. Draw energy diagrams for the system before and after the explosion.

b. What percentage of the explosion's energy is converted to kinetic energy?

c. What percentage of the explosion's energy is recoverable as kinetic energy?

d. Is the process shown in the simulation reversible?

Vary the mass of the left cart from 0.1 kg to 1.0 kg for the following questions.

e. Does the larger or the smaller mass receive the most energy?

f. Does the larger or the smaller mass receive the most momentum?

g. Does the ratio of the two masses have any effect on the total resulting kinetic energy?

h. Does the ratio of the two masses have any effect on the recoverable energy?

Exploration 8.7: A Bouncing Ball

The animation represents the seemingly simple example of a ball hitting the ground and bouncing back (position is given in meters and time is given in seconds). The graph can show velocity vs. time or acceleration vs. time and can be zoomed in to see the collision with the ground. Also shown are three bar graphs representing the different types of energy associated with the ball: the kinetic energy (**orange**), the gravitational potential energy (**blue**), and the elastic potential energy (**green**).

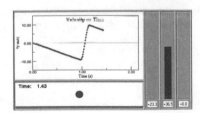

a. There are three important time intervals during the animation. What are they? Briefly describe what is happening during these intervals.

b. Draw energy diagrams, that is, find the values and plot a bar graph for the kinetic energy of the ball.

c. Draw the graph of momentum vs. time. Describe what is happening to the momentum during the three important time intervals. If the momentum of the ball is changing, explain why.

d. Draw the graph of the net force vs. time. Describe what is happening to the net force on the ball during the three important time intervals. If the net force of the ball is changing, explain why.

Problems

Problem 8.1

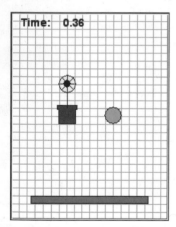

(Time: 2.1)

A 1.5-kg box (position is given in meters and time is given in seconds) slides on ice for 1.5 seconds and then encounters a rough surface.

a. Find the momentum at the start of the simulation.

b. Is the momentum of the box constant during the first 1.5 seconds?

c. Is the momentum of the box constant during the next three seconds?

d. Is momentum conserved during the first 1.5 seconds? Is momentum conserved during the next three seconds?

Problem 8.2

(Time: 0.36)

A flower pot and a basketball collide with a table (position is given in meters and time is given in seconds). Each has exactly the same mass.

a. Which object undergoes the greater change in momentum after colliding with the floor?

b. Which object undergoes the greater change in kinetic energy after colliding with the floor?

c. Is the force of the floor on the flower pot greater or less than the force of the floor on the basketball?

Problem 8.3

Two carts collide with a wall as shown in the animation (position is given in meters and time is given in seconds). Assume the two carts are identical.

(Time: 1)

a. Is kinetic energy constant for either collision?

b. Which cart, top or bottom, undergoes the greater change in kinetic energy due to colliding with the wall?

c. Is this the same cart that undergoes the greater change in momentum?

d. Explain how carts can change their momentum but not their kinetic energy.

Problem 8.4

(Time: 0.64   v= +0   v= +2.6)

Two carts on an air track collide as shown in the animation (position is given in meters and time is given in seconds). If the mass of the red cart is 0.8 kg, what is the mass of the blue cart?

Problem 8.5

(Time: 1.04   v= +7.05)

A large 2500-kg truck (blue) collides with a small car (brown) as shown in the animation (position is given in meters and time is given in seconds). After the collision, the vehicles move at constant velocity. What is the mass of the small car?

Problem 8.6

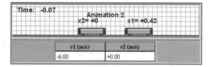

Two identical carts are shown colliding on a frictionless air track (position is given in meters and time is given in seconds). Which animation(s), if any, correctly models the laws of classical physics?

Problem 8.7

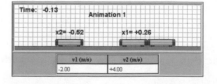

Three identical carts, two of which are attached, are shown colliding on a frictionless air track (position is given in meters and time is given in seconds). Which animation(s), if any, correctly models the laws of classical physics?

Problem 8.8

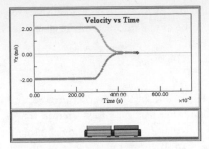

A spring that is attached to the end of a cart is compressed, and the cart is placed next to another cart on a low-friction track. The spring is released such that the two carts are "pushed" apart as shown in the animation (position is given in meters and time is given in seconds). The mass of the green cart is 1.35 kg, and the mass of the orange cart is 0.9 kg.

a. What is the magnitude of the momentum of the green cart after the collision?

b. What is the magnitude of the momentum of the orange cart after the collision?

c. What is the change in momentum of the system due to the release of the spring?

d. What is the change in kinetic energy of the system due to the release of the spring?

Problem 8.9

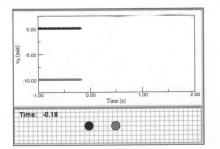

A collision occurs between two pucks on a frictionless surface (position is given in meters and time is given in seconds). Is the collision elastic or inelastic? Note that the masses of the pucks are not necessarily the same.

Problem 8.10

The color-coded graphs show the velocities of the red and black balls, respectively (position is given in meters and time is given in seconds). Would you define the collision shown as elastic, inelastic, totally inelastic, or explosive? Assume both balls have the same mass.

Problem 8.11

Two carts undergo a perfectly inelastic collision (the carts stick together) as shown in the animation. Also shown is a velocity vs. time graph for each cart. You can see the acceleration vs. time

graph by clicking the check box (position is given in meters and time is given in seconds). The two carts have equal speed before the collision. You may vary m_1 from 0.5 kg to 2 kg.

a. For which values of m_1 is the magnitude of the change in momentum, $|\Delta p|$, of the yellow cart greater than, less than, or equal to the magnitude of the change in momentum, $|\Delta p|$, of the bluish cart? Why?

b. For which values of m_1 is the magnitude of the change in acceleration, $|\Delta a_{max}|$, of the yellow cart greater than, less than, or equal to the magnitude of the change in acceleration, $|\Delta a_{max}|$, of the bluish cart? Why?

Problem 8.12

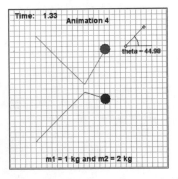

Several two-dimensional collisions between two balls (the green ball is ball 1 and the blue ball is ball 2) are shown (position is given in meters and time is given in seconds). Also shown is a protractor, which you can drag around (by the little circles on its legs) to measure angles.

For each animation

a. Determine the initial momentum of each ball.

b. Determine the final momentum of each ball.

c. Calculate and compare the initial momentum to the final momentum for the two-ball system.

Problem 8.13

Four spheres are shown in the animation. A blue sphere is half as massive as a red one and a purple sphere is twice as massive

as a red one. Where should the purple one be placed in order for the center of gravity to be at the location of the black dot (position is given in meters)?

Problem 8.14

A spring that is attached to the end of a cart is compressed, and the cart is placed next to another cart on a low-friction track. The spring is released such that the two carts are "pushed" apart as shown in the animation (position is given in meters and time is given in seconds).

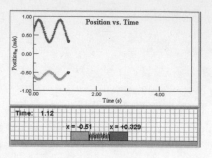

a. Find the ratio of the green mass to the red mass.

b. Find the position of the center of mass.

c. Find the distance from each mass to the center of mass at time $t = 0$ s.

Reference Frames

Topics include relative motion, inertial reference frames, zero-momentum frame, and center of mass.

Illustration 9.1: Newton's First Law and Reference Frames

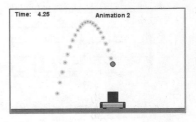

On first glance it may seem like Newton's first law (an object at rest remains at rest and an object in motion remains in motion unless acted on by a net force) is contained within Newton's second law. This is actually not the case. The first law is also a statement regarding reference frames. This is the information NOT contained in the second law. Sometimes the first law is also called the law of inertia. It defines a certain set of reference frames in which the first law holds, and these reference frames are therefore called inertial frames of reference. Put another way, Newton's first law states that if the net force on an object is zero, it is possible to find at least one reference frame in which that object has no acceleration.

A ball popper on a cart (not shown to scale) is shown moving on a track in five different animations (position is given in meters and time is given in seconds). In each animation the ball is ejected straight up by the popper mechanism at $t = 1$ s.

Let us first consider **Animation 1**. In this animation the cart is stationary. But is it really? We know that we cannot tell if we are stationary or moving at a constant velocity (in other words in an inertial reference frame). Recall that if we are moving relative to Earth at a constant velocity, we are in an inertial reference frame. So how can we tell if we are moving? How about the cart? We cannot tell if there is motion as long as the relative motion with respect to Earth can be described by a constant velocity. In Animation 1 the cart **could** be stationary. In this case, we expect—and actually see—that the ball lands back in the popper. However, if the cart was moving relative to Earth, and we were moving along with the cart, the motion of the ball and the cart would look exactly the same!

What would the motion of this ball and cart look like if it moved relative to our reference frame (or if we move relative to its reference frame)? Animations 2, 3, 4, and 5 show the motion from different reference frames.

First look at Animations 2 and 3. What do these animations look like? Both animations resemble projectile motion. The motion of the ball looks like motion in a plane as opposed to motion on a line. Does the ball still land in the popper? Would you expect this? Sure. There is nothing out of the ordinary going on here. Since there are no forces in the x direction, the motion of the ball (and cart) should be described by constant velocity in that direction. Therefore the ball and the cart have the same constant horizontal velocity.

Now look at Animations 4 and 5. What do these animations look like? Neither animation resembles projectile motion. The motion of the ball and cart look like they are being accelerated to either the right or the left (depending on the animation). Notice that the ball still lands in the popper as seen from this reference frame. Why do the ball and cart accelerate? There is nothing that we can see to explain why the ball and cart accelerate. Since Newton's laws must be correct, we must invent a force to describe why the cart and ball are accelerating (a fictitious force, since it really does not exist). Animations 4 and 5 depict motion as seen from noninertial reference frames.

Illustration 9.2: Reference Frames

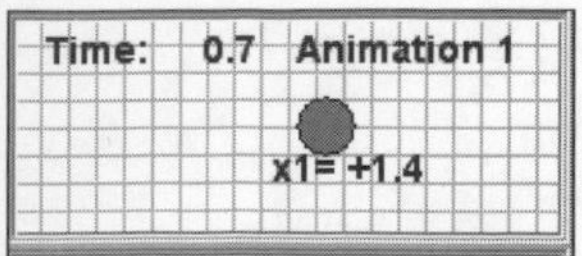

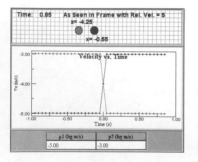

t (s)	x2 (m)
-2.00	-4.00
-1.50	-2.50
-1.00	-1.00
-0.50	+0.50
+0.00	+2.00
+0.50	+3.50

In the two animations, we are given an example of a moving reference frame relative to Earth's (stationary) reference frame. The motion of the orange ball as seen in Earth's reference frame is depicted in the animation by the time, position, and velocity measurements, t, $x1$, and $v1$, respectively (position is given in meters and time is given in seconds). An observer is in another reference frame that is moving with a constant velocity with respect to the surface of Earth. The observer also takes down time, position, and velocity measurements as shown in the table and represented by t, $x2$, and $v2$ respectively. **Animation 1** shows position and **Animation 2** shows velocity.

How do we know that the observer in frame two is moving with respect to Earth's reference frame? At $t = -2$ s, the observer in Earth's frame sees the orange ball at -4 m and moving to the right at a constant velocity of 2 m/s. What does the observer on the other reference frame see? She sees the ball start at the same position, but the ball moves with a different velocity in her frame. She sees it move to the right with a velocity of 3 m/s. Therefore, relative to Earth, our observer in frame 2 is moving with a velocity of 1 m/s.

But in what direction does the observer move? Consider the following question first. What if the observer—in her frame of reference—saw the ball as stationary? We would conclude that the observer was traveling at the same velocity as the ball as seen from the reference frame of Earth. When we move in the direction of the motion of the ball, the ball's relative velocity decreases. Thus, when we move in a direction opposite to the motion of the ball, the ball's relative velocity increases. Therefore the observer is moving to the left, relative to the reference frame of Earth, at 1 m/s!

When a reference frame is moving uniformly (at a constant velocity) with respect to a nonaccelerating (inertial) reference frame, the moving frame of reference is also called an inertial reference frame.

Illustration 9.3: The Zero-Momentum Frame

Is physics different when viewed in different reference frames? Well, it can certainly look different. Consider the collision in the animation as seen initially in the reference frame of Earth (the relative velocity between this frame and Earth's stationary frame is zero). Here both the red ball and the blue ball have the same mass equal to 1 kg. Note that energy and momentum are conserved in the collision with kinetic energy = 2 J and $p_x = 2$ kg·m/s before and after the collision.

Change the velocity from zero to 2 m/s (position is given in meters and time is given in seconds). How does the collision change? The red ball is now initially stationary and the blue ball is moving to the left at 2 m/s. Note that, in the original collision with $v = 0$ m/s, the red ball was initially moving to the right and the blue ball was initially stationary. In the new frame the momentum of the two-ball system is different. However, the kinetic energy happens to be the same and energy and momentum are conserved.

Now try $v = -2$ m/s. Are energy and momentum still conserved? Even though the values of the kinetic energy and momentum change, the laws of conservation of energy and conservation of momentum still hold.

Now try $v = 1$ m/s. What is the new momentum for the two-ball system? This frame of reference is appropriately called the zero-momentum frame. In this frame the momentum of the system is zero. This frame is also called the center-of-mass frame. The center of mass is a coordinate that is a mass-weighted average of the positions of the objects that make up the system. In a two-object system the center of mass is always somewhere in between the two objects. Since the center of mass is a mass-weighted average, the center of mass will always be closer to the object that is more massive. In the case of this animation, where both balls have the same mass, the center of mass is always at the midpoint between the two masses. This point does not move in the zero-momentum frame, but does move in other frames.

Illustration 9.4: Rotating Reference Frames

In an inertial reference frame, momentum and energy are conserved even though observers may disagree on total momentum or total energy. As a consequence, if two observers were in inertial reference frames, the two observers would agree on the forces acting.

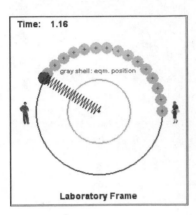

Consider a green mass at the end of a spring (position is shown in meters and time is shown in seconds). If the mass is not moving, the net force on it is zero, and therefore the gray shell represents the **equilibrium position** of the spring.

Now imagine that someone has given the green ball a brief push making it rotate with a constant speed as seen in the **laboratory frame**. We know that the spring must be stretched as shown. Why? Well, we need a force toward the center of the circle because this force is necessary for uniform circular motion. This force is the spring force.

Imagine you (the woman in the animation) are riding on the green mass. From your point of view, **mass's reference frame**, what is the motion of the green ball? It is stationary. In this reference frame the ball does not move. It is not, however, an inertial reference frame because it is accelerating. What happens to the spring in this frame? The spring is stretched from its equilibrium position as before. How would you explain this? From your point of view riding with the rotating mass, since you are not accelerating, the net force on the green ball is zero, and someone or something has stretched the spring by pulling it outward. This force is purely a figment of your imagination. This fictitious force—also called the centrifugal force (no more real because it has a name!)—must be invented by an observer on the green ball if she is going to keep Newton's laws as a fact of Nature. Whenever you are in a rotating reference frame, like a merry-go-round, for example, you are in an accelerating reference frame, and therefore you must invent fictitious forces to make Nature obey Newton's laws.

Exploration 9.1: Compare Momentum in Different Frames

How does the momentum of a particle change when viewed from a different reference frame? The momentum of each ball in the animation is shown in the table (position is given in meters and time is given in seconds). The graph displays color-coded plots of the velocities of the two particles.

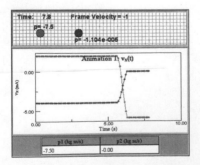

You can view the collision in another inertial reference frame by entering a new value into the frame velocity text box, -10 m/s $< v < 10$ m/s, before you start the animation. Consider the two particles to be an isolated system and answer the following questions using at least two different inertial reference frames for each animation.

a. Does the total momentum depend on your choice of reference frame?

b. Does the change in momentum depend on the reference frame?

c. Is the total momentum conserved in different reference frames?

d. Find the mass and the ratio of the masses of the two balls. Does this result depend on the reference frame?

e. Is there a reference frame in which the total momentum is zero? If so, observe the change in velocity in this reference frame and explain why analysis of the collision is particularly simple in this reference frame.

Exploration 9.2: Compare Energy in Different Frames

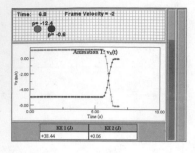

How does the *energy* of a particle change when viewed from a different reference frame? The energy of the two balls in the animation is shown in the table and on the bar graphs to the right. The graph displays the velocity (position is given in meters and time is given in seconds).

You can view the collision in another inertial reference frame by entering a new value into the frame velocity text box, -10 m/s $< v < 10$ m/s, before you start the animation. Consider the two particles to be an isolated system and answer the following questions using at least two different inertial reference frames for each animation.

a. Do the kinetic energies of the individual particles depend on your choice of reference frame?

b. Does the change in total kinetic energy due to the collision depend on the reference frame?

c. Is the total kinetic energy constant in different reference frames? (Be sure to answer for both animations.)

d. Find the mass and the ratio of masses of the two balls. Does this result depend on the reference frame?

e. What is special about the reference frame in which the total momentum is zero? Is the kinetic energy zero in this frame?

Exploration 9.3: Compare Relative Motion in Different Frames

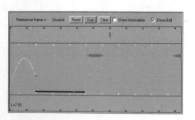

An object may appear to have one motion to one observer and a different motion to a second observer, depending on how the two observers are moving relative to one another. This Exploration lets you view objects from different frames of reference.

There is a river (shown in **green**) in the center of the screen (the yellow dots are stationary with respect to the water) and two **red** boats are moving with respect to the river. There is a rectangular barge that is stationary with respect to the river and changes color based on your reference frame. There is also a person (shown in **blue**) walking on the ground (shown in gray with black dots stationary relative to the ground) near the river.

You can easily change your frame of reference by moving the mouse to different regions and to different objects. For example, if you move the mouse within the river, YOU will become an observer moving with the water. The long thin rectangle (the barge) in the water moves with the water and changes color to represent the reference frame you are in. In addition,

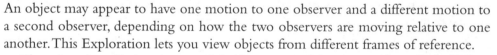

- If you click the "show information" check box, velocity vectors for your frame of reference will be shown with numbers (position is given in meters and speed is given in m/s). Each vector's color corresponds to the color of the object whose reference frame you are in.

- If you click the "show ball" check box, a blue ball is thrown upward from the barge on the river. When you change reference frames you can see how the projectile motion of the ball looks different in different reference frames. Deselect the "show ball" check box to clear the ball's trajectory.

- You can press the mouse button to suspend the animation. If you press with **left** mouse button, the animation will resume when you release it. If you press with **right** mouse button, you need to click it again to resume the animation.
- To change the velocity vectors, it is easier to first suspend the animation. Then click anywhere in the animation to bring up the vectors. Now click near the tip of the arrow and drag it to the left or right.
- There are two numbers near the person. Those are the vertical and horizontal speeds of the person in your frame of reference.
- While the animation is suspended, click near the left leg of that person and drag the mouse up and down. You are changing her vertical speed. Click with the right mouse button to resume the animation. The person will now move toward the river and then swim across it. (The horizontal speed with respect to the ground will increase due to river current).

a. Suspend the animation and then turn on "show information." Change the horizontal and vertical speeds of the person so that she swims straight across the river (as seen from the ground). Once you are satisfied with your choice of velocity, deselect the "show information" check box, click in the animation (things are easier to see this way), and resume the animation.

b. What ground velocity do you have to give her to accomplish this task?

c. Now move the mouse around to change reference frames. Does the person still swim straight across the river in other reference frames? If not, what velocity must you give her (relative to the ground) to accomplish this as seen from the boats? What velocity must you give her (relative to the ground) to accomplish this as seen from the river?

Exploration 9.4: Compare Motion in Accelerating Frames

Is physics different when viewed in different reference frames? The momentum of each ball is shown in the table, and the kinetic energy of each cart is shown in the bar graph in joules (position is given in meters and time is given in seconds). You can change your reference frame using the text box, $-2 \text{ m/s}^2 < a < 2 \text{ m/s}^2$. Answer the following questions.

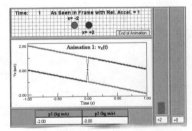

a. Does the total momentum depend on the reference frame?

b. Does the change in momentum depend on the reference frame?

c. Is the total momentum conserved in all reference frames?

d. Find the ratio of the two masses. Is this result the same in all reference frames?

e. Is there a reference frame in which the total momentum is zero?

Exploration 9.5: Two Airplanes with Different Land Speeds

Two airplanes (not shown to scale) travel the same round-trip distance between two cities (position is given in kilometers and time is given in hours). Both airplanes have the same air speed (200 km/hr), but one airplane (the top airplane with the **blue** wingtip) travels faster or slower relative to the ground because it is subject to a headwind and tailwind. *A positive wind velocity means a tailwind on the outbound part of the trip and a headwind on the inbound part of the trip.* The wind velocity can be changed by entering a value $(-199 < v_{\text{wind}} < 199)$ in the text box and registering the value.

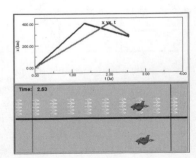

a. *Before* entering a nonzero value in the text box, predict which airplane will reach its destination first if the top (**blue**) airplane is subject to a head/tail wind.

b. Once you have made your prediction, play the animation to see if you were right.

c. If you were incorrect, can you now see why you were incorrect? Explain.

Problems

Problem 9.1

Two objects approach each other as shown in the animation. Before they collide, what is the speed of the green object as measured in the reference frame of the red object (position is given in meters and time is given in seconds)?

Problem 9.2

You are rowing a boat across a river (position is given in meters and time is given in seconds). The river current flows from left to right in the animation. Your goal is to reach your favorite mooring spot on the opposite river bank (shown in red). By rowing, you control the (constant, vertical) velocity of the boat with respect to the water.

a. What is the (horizontal) velocity of the water with respect to the banks?

b. What is the magnitude of the velocity of the boat with respect to the banks?

c. What angle, θ, does the boat's velocity vector make with the lower bank?

Problem 9.3

time	x1	x2
-3.00	-12.00	+15.00
-2.50	-10.00	+11.25
-2.00	-8.00	+8.00
-1.50	-6.00	+5.25
-1.00	-4.00	!3.00
-0.50	-2.00	+1.25
-0.00	-0.00	+0.00
+0.50	+2.00	-0.75
+1.00	+4.00	-1.00

Two space aliens measure the position of an orange star from different space ships. Alien 1 records data as $x1$ and alien 2 records data as $x2$ (position is given in meters and time is given in seconds). You are viewing the same star from an inertial reference frame. Assume that you and the aliens have agreed to use the same distance and time units. You suspect that one of your alien friends is in a noninertial reference frame.

a. Which alien is in an inertial reference frame and which alien is in a noninertial frame?

b. Find the Galilean transformation from your frame into the inertial reference frame.

Problem 9.4

time	x1	v1	x2	v2
-3.00	-15.00	+5.00	+6.00	-2.00
-2.50	-12.50	+5.00	+5.00	-2.00
-2.00	-10.00	+5.00	+4.00	-2.00
-1.50	-7.50	+5.00	+3.00	-2.00
-1.00	-5.00	+5.00	+2.00	-2.00
-0.50	-2.50	+5.00	+1.00	-2.00
-0.00	-0.00	+5.00	+0.00	-2.00

Two physics coworkers measure the position of the object shown from different inertial reference frames (position is given in meters and time is given in seconds). Coworker 1 records data as $x1$ and $v1$ and coworker 2 records data as $x2$ and $v2$. Assume both coworkers have agreed to use meters and seconds to measure distance and time.

a. What is the relative speed of coworker 1 with respect to coworker 2?

b. Find the Galilean transformation that transforms the measurements of coworker 1 into those of coworker 2.

Problem 9.5

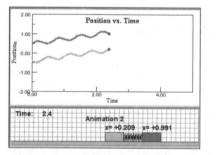

A spring that is attached to the end of a cart is compressed, and the cart is placed next to another cart on a low-friction track. The spring is released such that the two carts are "pushed" apart as shown in the animation (position is given in meters and time is given in seconds). Find the velocity of the center of mass for each animation.

Problem 9.6

Two carts start at rest on similar air tracks as shown in the animation (position is given in meters and time is given in seconds). Write an equation for the x position of the green cart as seen from the orange cart.

Problem 9.7

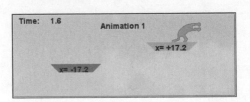

Assume you are sitting on the shore of a lake and observe two boats (not shown to scale) drifting in the lake. Another student is riding in one of the boats (position is given in meters and time is given in seconds).

a. For each of the animations, find the velocity of each boat as seen from the shore.

b. Determine the velocity of the red boat as seen by the student riding in the green boat for each of the animations. This is the relative velocity between the two boats.

You can simulate running along the shore by typing a velocity, -15 m/s $< v < 15$ m/s, into the input field before you select an animation.

c. Does changing the reference frame, i.e., running along the shore, change the relative velocities of the two boats? Try reference frame velocities of $+2$ m/s and -2 m/s.

d. For each of the animations, what velocity should you enter so that the red boat appears stationary?

e. For each of the animations, what velocity should you enter so that the green boat appears stationary?

Problem 9.8

A physics student measures the position and velocity of the ball shown in the animation and reports the results to you in the table shown. You are viewing the same experiment from another reference frame (position is given in meters and time is given in seconds). Your values are shown next to the ball in the animation. Assume that your reference frame is an inertial

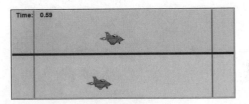

time	x'	v'
-3.00	+0.00	-0.00
-2.50	+0.00	-0.00
-2.00	+0.00	-0.00
-1.50	+0.00	+0.00
-1.00	+0.00	+0.00
-0.50	+0.00	+0.00
-0.00	+0.00	+0.00
+0.50	+0.00	+0.00

reference frame and that both you and the other student use the same units for time and distance.

Do the following for each of the animations:

a. Determine if the other student is in an inertial reference frame.

b. Determine the transformation that transforms the x and v data from your frame to her frame.

Problem 9.9

Two airplanes (not shown to scale) travel the same round-trip distance between two cities (time is given in hours). Both airplanes have the same air speed, but one airplane (the top airplane with the blue wingtip) travels faster or slower relative to the ground because it is subject to a headwind and a tailwind. *A positive wind velocity means a tailwind on the outbound part of the trip and a headwind on the inbound part of the trip.* What is the ratio of the wind speed to the air speed for the top airplane?

CHAPTER

10

Rotations about a Fixed Axis

Topics include linear and angular kinematics, fixed axis rotations, torque, energy, and angular momentum.

Illustration 10.1: Coordinates for Circular Motion

How would you describe the motion of the object shown (position is given in meters and time is given in seconds)? The object is moving in a circle about $x = 0$ m and $y = 0$ m, but the object's x and y coordinates vary with time. They vary in a special way such that x and y are always between -1 m and 1 m. To see this, look at **Animation 2** and watch the x and y values change in the table. This is called the component form. We can also describe the motion in terms of the vector form. In this case, the radius vector, **r**, always has a magnitude of 1 m, but it changes direction. Look at **Animation 3**. We describe the direction of this vector in terms of the angle it makes with the positive x axis. Therefore the angle—when measured in degrees— varies from 0 to 360. It is often convenient to give the angle in a unit different than degrees. We call this unit a radian. The radian unit is defined as 2π radians $= 360°$. Notice that both units are defined in terms of one full revolution. To see the angle given in radians look at **Animation 4**.

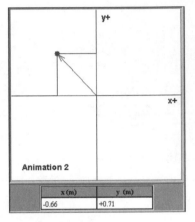

Animation 2

x (m)	y (m)
-0.66	+0.71

So why use radians? Well, it turns out that there is a really nice relationship between angle in radians (θ), the radius (r), and the arc of the circle (s). This geometric relationship states that: $\theta = s/r$. Why is this useful? It allows us to treat circular motion like one-dimensional motion. The arc is the linear distance traveled, which is $s = vt$ when the motion is uniform. This means that $\theta = (v/r)\,t$, since $s = r\theta$. We call v/r by the name omega, ω, and it is the angular velocity. Therefore, $\theta = \omega t$, for motion with a constant angular velocity. When there is a constant angular acceleration, we call it by the name alpha, α, and it is related to the tangential acceleration by a_t/r. So when we are using radians we can use our one-dimensional kinematics formulas with $x \rightarrow \theta$, $v \rightarrow \omega$, and $a \rightarrow \alpha$.

Illustration 10.2: Motion about a Fixed Axis

Many objects rotate (spin) about a fixed axis. Shown is a wheel of radius 5 cm rotating about a fixed axis at a constant rate (position is given in centimeters and time is given in seconds).

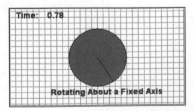

Consider the various points on the surface of the rotating wheel. By watching the line rotate, we can see that the wheel is rotating at a constant rate. In fact, if we watch a point on the surface of the wheel (a radius of 5 cm), we would say it has a constant speed. What about a point halfway out (a radius of 2.5 cm)? It also has a constant speed. But how does this speed compare to the speed of a point on the surface of the wheel?

We can determine this by first considering a quantity that is not related to the radius, the angular speed, ω. The angular speed is the angular displacement divided by the time interval (in this Illustration, since there is no acceleration, the average speed and the instantaneous speed are the same). So what is the wheel's angular speed? From one revolution of the wheel, the angular displacement is 2π and the time interval (called the period, T) is 5 seconds. Therefore the angular speed is the angular displacement over the time interval ($\omega = 2\pi/T$) 0.4π radians/s = 1.256 radians.

How do we relate the angular speed to the linear (tangential) speed of a point on the wheel? First consider the speed of a point on the surface of the wheel. It is again easiest to measure the speed by considering one rotation of the wheel. In this case, the distance traveled by that point is $2\pi r$, and therefore the average (and in this case instantaneous) tangential speed is $2\pi r/T = 2\pi$ cm/s = 6.28 cm/s. The relationship between the angular speed and the tangential speed must be $\omega = v/r$. (Recall that we found above that $\omega = 2\pi/T$.)

This works because there is a relationship between the angular displacement and the tangential displacement (the arc length s), namely that $\Delta\theta = \Delta s/r$. This also must be the case for an angular displacement of one revolution: $2\pi = 2\pi r/r$.

Since linear velocity is a vector, we might expect that angular velocity is a vector as well. This is indeed the case. So in which direction does the angular velocity point for the rotating wheel? We use the right-hand rule (RHR) to determine the direction of the angular velocity. Using your right hand, curl your fingers in the direction of the rotation of the wheel; the direction your thumb points is the direction of the wheel's angular velocity. Here, $\boldsymbol{\omega}$ is into the page (computer screen). This may seem weird; after all you might say that the wheel is rotating clockwise. Clockwise is not a good description, as it does not imply a vector-like quantity and the description is not unique. Why is it not unique? If you were on the other side of the page or computer screen, you would say that the wheel is rotating counterclockwise instead!

Can you guess what the relationship between the angular acceleration and the tangential acceleration is? Well, given that acceleration is change in velocity for a given time, $\Delta\mathbf{v}/\Delta t$, you have probably guessed that angular acceleration, called $\boldsymbol{\alpha}$, is the change in angular velocity for a given time, $\Delta\boldsymbol{\omega}/\Delta t$. Given how v and ω are related to each other, it must be that $a = \alpha r$.

Illustration 10.3: Moment of Inertia, Rotational Energy, and Angular Momentum

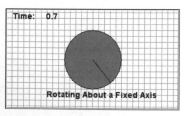

Time: 0.7

Rotating About a Fixed Axis

Many objects rotate (spin) about a fixed axis. Shown is a wheel (a disk) of radius 5 cm and a mass of 200 grams rotating about a fixed axis at a constant rate (position is given in centimeters and time is given in seconds).

In **Illustration 10.2** we discussed how linear speed (velocity) was related to angular speed (velocity), and in the process how angular acceleration is related to the angular velocity ($\boldsymbol{\alpha} = \Delta\boldsymbol{\omega}/\Delta t$). In this Illustration we will discuss kinetic energy of rotation, KE_{rot}, and angular momentum, $\mathbf{L}$.

The easiest way to remember the forms for the kinetic energy of rotation and the angular momentum is by analogy with the kinetic energy of translation and the linear momentum. We recall that $KE = 1/2\, m\, v^2$ and $\mathbf{p} = m\mathbf{v}$. Can you guess what the rotational kinetic energy and angular momentum will look like?

First, what will play the role of v and $\mathbf{v}$ in the rotational expressions? If you said ω and ω you are right. Next we must consider what plays the role of m, and we will be all set. The property of mass describes an object's resistance to linear motion. Therefore, what we are looking for is a property of objects that describes their resistance to rotational motion. This is called the moment of inertia. The moment of inertia depends on the mass of the object, its extent, and its mass distribution. It turns out that for most simple objects the moment of inertia looks like $I = C\, m\, R^2$, where m is the object's mass, R is its extent (usually a radius or length), and C is a dimensionless constant that represents the mass distribution.

Therefore, we have that $KE_{rot} = 1/2\, I\, \omega^2$ and $\mathbf{L} = I\, \boldsymbol{\omega}$. What are this disk's KE_{rot} and $\mathbf{L}$? Well, from **Illustration 10.2** we know that $\omega = 1.256$ radians/s. Since the wheel is a disk, $C = 2$. Therefore, we can calculate the moment of inertia as: 2.5×10^{-4} kg·m². Finally, we have that $KE_{rot} = 1.97 \times 10^{-4}$ J and $L = 3.14 \times 10^{-4}$ J·s (into the page or computer screen). Note that these are small values because I for this disk is small. A 1-m radius and 2-kg mass disk would have a moment of inertia of 1.0 kg·m².

Exploration 10.1: Constant Angular Velocity Equation

By now you have seen the equation: $\theta = \theta_0 + \omega_0 t$. Perhaps you have even derived it for yourself. But what does it really mean for the motion of objects? This Exploration allows you to explore both terms in the equation: the initial angular position by changing θ_0 from 0 radians to 6.28 radians and the angular velocity term by changing ω_0 from −15 rad/s to 15 rad/s.

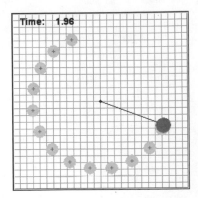

   Answer the following questions (position is given in meters and time is given in seconds).

a. How does changing the initial angular position affect the motion?

b. How does changing the initial angular velocity affect the motion?

Exploration 10.2: Constant Angular Acceleration Equation

By now you have seen the equation: $\theta = \theta_0 + \omega_0 t + 0.5\alpha t^2$. Perhaps you have even derived it for yourself. But what does it really mean for the motion of objects? This Exploration allows you to explore all three terms in the equation: the initial angular position by changing θ_0 from 0 radians to 6.28 radians, the angular velocity term by changing ω_0 from −15 rad/s to 15 rad/s, and the angular acceleration by changing α from −5 rad/s² to 5 rad/s².

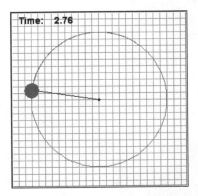

   Answer the following questions (position is given in meters and time is given in seconds).

a. How does changing the initial angular position affect the motion of the object?

b. How does changing the initial angular velocity affect the motion of the object?

c. How does changing the angular acceleration affect the motion of the object?

d. Can you get the object to change direction?

Exploration 10.3: Torque and Moment of Inertia

A mass (between 0.01 kg and 1 kg) is hung by a string from the edge of a massive (between 0 kg and 2 kg) disk-shaped pulley (with a radius between 0.1 and 4 meters) as shown (position is given in meters, time is given in seconds, and angular velocity is given in radians/second).

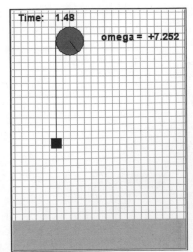

   Set the hanging mass to 0.25 kg, the radius of the pulley to 2 m, and vary the mass of the pulley.

a. How does the magnitude of the angular acceleration of the pulley depend on the mass (and therefore moment of inertia) of the pulley?

b. How does the magnitude of the acceleration of the hanging mass depend on the mass (and therefore moment of inertia) of the pulley?

c. How are your answers to (a) and (b) related?

Set the mass of the pulley to 0.5 kg, the radius of the pulley to 2 m, and vary the hanging mass.

d. How does the magnitude of the angular acceleration of the pulley depend on the hanging mass?

e. How does the magnitude of the acceleration of the hanging mass depend on the hanging mass?

f. How are your answers to (d) and (e) related?

Set the hanging mass to 0.25 kg, the mass of the pulley to 0.5 kg, and vary the radius of the pulley.

g. How does the magnitude of the angular acceleration of the pulley depend on the radius of the pulley?

h. How does the magnitude of the acceleration of the hanging mass depend on the radius of the pulley?

i. How are your answers to (g) and (h) related?

Set the mass of the pulley to 0.5 kg, the hanging mass to 0.25 kg, and the radius of the pulley to 2 m.

j. Determine the acceleration of the hanging mass and the angular acceleration of the pulley.

k. From Newton's second law, determine the tension in the string.

l. How much torque does this tension provide to the pulley?

Exploration 10.4: Torque on Pulley Due to the Tension of Two Strings

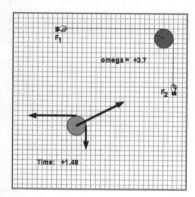

Shown is a top view of a pulley on a table. The massive disk-shaped pulley can rotate about a fixed axle located at the origin. The pulley is subjected to two forces in the plane of the table, the tension in each rope (each between 0 N and 10 N), that can create a net torque and cause it to rotate as shown (position is given in meters, time is given in seconds, and angular velocity is given in radians/second). Also shown is the "extended" free-body diagram for the pulley. In this diagram the forces in the plane of the table are drawn where they act, including the force of the axle.

Set the mass of the pulley to 1 kg, the radius of the pulley to 2 m, vary the forces, and look at the "extended" free-body diagram.

a. How is the force of the axle related to the force applied by the two tensions?

b. How do you know that this must be the case?

Set the mass of the pulley to 1 kg, the radius of the pulley to 2 m, and vary the forces.

c. What is the relationship between F_1 and F_2 that ensures that the pulley will not rotate?

d. For $F_1 > F_2$, does the pulley rotate? In what direction?

e. For $F_1 < F_2$, does the pulley rotate? In what direction?

f. What is the general form for the net torque on the pulley in terms of F_1, F_2, and r_{pulley}?

Set the mass of the pulley to 1 kg, F_1 to 10 N, F_2 to 5 N, and vary the radius of the pulley.

g. How does the angular acceleration of the pulley depend on the radius of the pulley?

Set the radius of the pulley to 2 m, F_1 to 10 N, F_2 to 5 N, and vary the mass of the pulley.

h. How does the angular acceleration of the pulley depend on the mass of the pulley?

i. Given that the pulley is a disk, find the general expression for the angular acceleration in terms of F_1, F_2, m_{pulley}, and r_{pulley}.

Problems

Problem 10.1

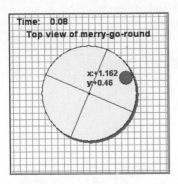

A child sits on a merry-go-round at the position marked by the red circle (position is given in meters and time is given in seconds). What is her angular displacement in radians after 0.44 seconds?

Problem 10.2

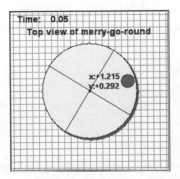

A child sits on a merry-go-round at the position marked by the red circle (position is given in meters and time is given in seconds).

a. What is her average speed and instantaneous velocity?

b. What is her angular speed and angular velocity?

For the instantaneous velocity and angular velocity, you should give a value for the speed and a description of the velocity's direction for any point in time.

Problem 10.3

A quarter and a penny are on a turntable as shown in the animation (position is given in centimeters and time is given in seconds).

a. Which coin has the greater angular speed?

b. What are their angular speeds?

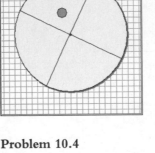

Problem 10.4

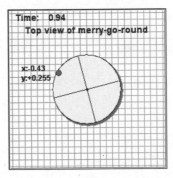

A child sits on a merry-go-round at the position marked by the red circle (position is given in meters and time is given in seconds). What is her angular acceleration?

Problem 10.5

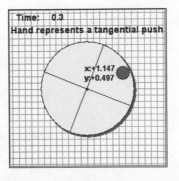

A boy sits on a merry-go-round at the position marked by the red circle. A girl gives the merry-go-round a constant tangential push for 0.2 seconds as shown in the animation (position is given in meters and time is given in seconds). What is the magnitude of the tangential acceleration of the boy while the girl is pushing the merry-go-round?

Problem 10.6

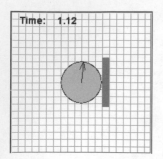

A grinding wheel is rotating at constant speed when an object makes contact with the outer edge as shown in the animation (position is given in meters and time is given in seconds). Friction causes the wheel to stop. What is the angular acceleration of the wheel?

Problem 10.7

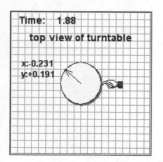

A turntable (a flat disk) of mass 5.0 kg is rotating at a constant speed when your finger makes contact with the outer edge, as shown in the animation (position is given in meters and time is given in seconds). Friction between your finger and the turntable causes the turntable to stop. What is the average torque on the turntable caused by the frictional force?

Problem 10.8

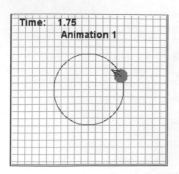

A car starts from rest and accelerates until it is halfway around a circular track. After that time it moves at constant speed (position is given in meters and time is given in seconds). Which animation correctly shows the acceleration vector?

Problem 10.9

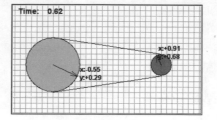

The animation depicts an idealized drive train for a bicycle. A large green disk (i.e., a flat cylinder) is used to rotate a small green disk of the same density and thickness via a massless chain that does not slip (position is given in meters and time is given in seconds). What is the ratio of the kinetic energy of the green disk to the kinetic energy of the red disk (KE_{green}/KE_{red})?

Problem 10.10

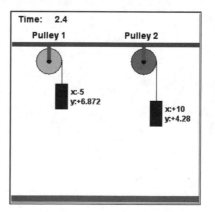

Two identical black masses, m, are hung via massless strings over two pulleys of identical mass M and radius R, but different mass distributions as shown in the animation (position is given in centimeters and time is given in seconds). The bearings in the pulleys are frictionless, and the strings do not slip as they unwind from their pulleys.

a. Which mass has the greater acceleration?

b. Which pulley has the greater moment of inertia?

c. Which pulley has the greater tension acting on it?

d. Which pulley has the greater torque acting on it?

Answer the following in terms of a general formula for either pulley using the following variables: a (the acceleration of the black mass), g, m, M, and R.

e. What is the tension in the string?

f. What is the torque acting on the pulley?

g. What is the moment of inertia of the pulley? Remember that we do not know the pulley's mass distribution.

Problem 10.11

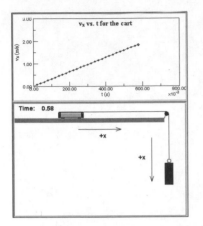

A 1.0-kg cart (not shown to scale) on a low-friction track is connected to a string and a 0.5-kg hanging object as shown in the animation (position is given in meters and time is given in seconds). The pulley has a uniform mass distribution in the shape of a disk and therefore affects the motion of the system.

a. What is the acceleration of the system?

b. What is the tension in the string? (There are two regions of the string to consider.)

c. What is the mass of the pulley?

d. What is the moment of inertia of the pulley?

Note that the coordinates for each object (the positive *x* direction) are already chosen for you.

Problem 10.12

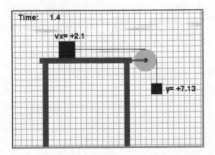

Two masses, one on a table, $M = 2.5$ kg, and one hanging, $m = 1.0$ kg, are attached with a massless string over a pulley as shown in the animation (position is given in meters and time is given in seconds). The table is frictionless, the bearings in the pulley are frictionless, and the string does not slip on the pulley. What is the moment of inertia of the pulley? You may use either torque/force or energy methods.

Problem 10.13

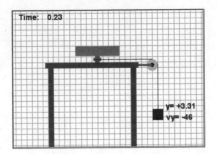

A 2.5-kg rotating red disk is connected by a string over a pulley of mass $m = 1$ kg to a black hanging mass as shown in the animation (position is given in centimeters, time is given in seconds, and velocity is given in centimeters/second). The post the red disk sits on is massless and has frictionless bearings. The string is wrapped around the post and does not slip. The bearings in the pulley are frictionless and the string does not slip. What is the black block's mass? You may use either force/torque or energy methods.

Problem 10.14

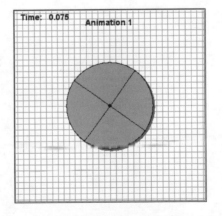

Each animation shows an object rotated about a fixed axis through the center (position is given in meters and time is given in seconds). Every object has the same mass, $m = 2$ kg. Every animation has a black dot in the center indicating the origin of the coordinate system. You are to calculate the angular momentum about this point for each animation. Calculate and rank (from greatest to least) the angular momentum (about the origin) of the objects in each animation.

CHAPTER 11

General Rotations

Topics include cross products, rolling motion, angular momentum of a particle, torque and angular momentum, and conservation of angular momentum.

Illustration 11.1: Cross Product

We talk about the magnitude of the torque as the amount of force perpendicular to the radius arm on which it acts. No radius arm, no torque. Torque is positive (out of the page) if F acts to rotate the object counterclockwise via the right-hand rule (RHR) and negative if F acts to rotate the object clockwise (again, via the RHR).

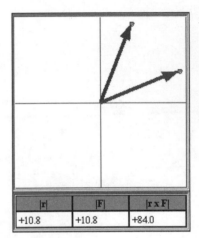

$|\mathbf{r}|$	$|\mathbf{F}|$	$|\mathbf{r} \times \mathbf{F}|$
+10.8	+10.8	+84.0

In order to mathematically describe torque, we must use the mathematical construction of the vector or cross product. Torque is the vector product of the radius vector and the force vector, $\mathbf{r} \times \mathbf{F}$. The magnitude of the torque is $r\,F\sin(\theta)$, and the direction of the torque is determined by the RHR. θ is the angle between the two vectors, and A and B are the magnitudes of the vectors $\mathbf{r}$ and $\mathbf{F}$, respectively. Drag the tip of either arrow (position is given in meters). The **red arrow is r** and the **green arrow is F**. The magnitude of each arrow is calculated as well as the cross product.

The direction of the torque, $\mathbf{r} \times \mathbf{F}$, is determined by the RHR (point your fingers toward $\mathbf{r}$, curl them into the direction of $\mathbf{F}$, and the direction that your thumb points is the direction of the torque. Therefore,

$$\tau = \mathbf{r} \times \mathbf{F} = r\,F\sin(\theta) \text{ with the direction prescribed by the RHR,}$$

where $\mathbf{r}$ is the moment arm on which the force acts and $\mathbf{F}$ is the force.

Illustration 11.2: Rolling Motion

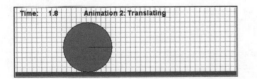

Many everyday objects roll without slipping (position is given in centimeters and time is given in seconds). This motion is a mixture of a pure rotation and a pure translation. The pure rotation is shown in **Animation 1**, while the pure translation is shown in **Animation 2**. So how might we combine the two motions together so that the disk rolls without slipping?

First consider the various points on the surface of the rotating wheel. Since it has a constant angular velocity, every point has the same speed but a different velocity. Consider three special points: the top of the wheel, the wheel's hub, and the bottom of the wheel. The top of the wheel has a velocity $v = \omega R$, and the velocity points to the right. The hub has zero velocity. And the bottom of the wheel has a velocity $v = \omega R$ to the left.

Now consider the pure translation. Every point on the wheel has a velocity v to the right.

So how do we combine the two motions together to get rolling without slipping? If the velocity of the point at the bottom of the wheel—the point that touches the ground—has a velocity of zero with respect to the ground, the wheel will not slip.

Consider the three special points again: the top of the wheel, the wheel's hub, and the bottom of the wheel. We will add the translational velocity to the rotational velocity and see what we get. The top of the wheel has a rotational velocity $v = \omega R$ to the right, which when combined with the translational velocity of v to the right gives us $2v$ to the right. The hub has zero rotational velocity, which when combined with the translational velocity of v to the right gives us v to the right. And finally, the bottom of the wheel has a velocity $v = \omega R$ to the left, which when combined with the translational velocity of v to the right gives us 0!

Therefore, as long as the angular velocity gives us a v that is the same v as the translation, we have rolling without slipping as in **Animation 3**.

Illustration 11.3: Translational and Rotational Kinetic Energy

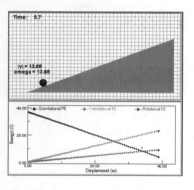

How do we describe rolling without slipping from an energy standpoint? We already know how to represent the kinetic energy of translation: $(1/2)\ mv^2$. We also know how to represent the kinetic energy of rotation: $(1/2)\ I\omega^2$. But what if we have both?

As the ball rolls down the incline, the gravitational potential energy gets transformed into kinetic energy, but how much of each? With rolling without slipping, we found that there is a relationship between the linear velocity and the angular velocity: $v = \omega R$. Given this relationship we know that $KE_{trans} = (1/2)\ mv^2$, while $KE_{rot} = (1/2)\ I\ (v^2/R^2)$. But the moment of inertia always looks like CmR^2, so we find that $KE_{rot} = (1/2)\ C\ mv^2$. Therefore, we find that $KE_{total} = (1+C)\ (1/2)\ mv^2$. The gravitational potential energy gets transformed into the total kinetic energy, and what fraction goes into KE_{trans} or KE_{rot} is determined by the constant C. Specifically,

$$KE_{trans}/KE_{total} = 1/(1 + C) \quad \text{and} \quad KE_{rot}/KE_{total} = C/(1 + C).$$

A ball of radius 1.0 m and a mass of 0.25 kg rolls down an incline, as shown (position is given in meters and time is given in seconds). The incline makes an angle $\theta = 20°$ with the horizontal. Watch the graph of gravitational potential energy and rotational and translational kinetic energy vs. time or position.

Why do you think that the energy vs. time graphs curve, while the energy vs. position graphs are straight lines?

Illustration 11.4: Angular Momentum and Area

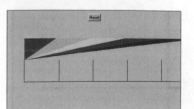

One of the strangest ideas regarding angular momentum is that an object moving in a straight line can have angular momentum. Angular momentum for a particle is given by the cross product $\mathbf{L} = \mathbf{r} \times \mathbf{p}$. Given this, we see that the origin matters for the calculation of angular momentum for a particle.

In the absence of a net external torque acting on a system, a particle's angular momentum remains constant. For this discussion, the particle is free, so angular momentum should be conserved. Is there a different way to state the concept of angular momentum conservation? There may be. Consider the statement, Does a particle sweep out equal areas in equal times (with respect to **any** origin)?

Specifically, in this Illustration does a free particle moving in a straight line sweep out equal areas in equal times?

Press "start" to begin the animation, and let the show begin: A black dot will move freely from left to right. Different colors show the area the particle sweeps out with respect to some fixed point (the origin). Do all the areas have the same size? Click within each area and see what will happen. Certainly from the mathematical equations we know the area of a triangle = width * height/2. All of the areas have the same height and the same width ($= v_x * dt$).

Note: Kepler's second law (see Chapter 12 on gravitation for more details) states, during equal time intervals, the radius vector from the sun to a planet sweeps out equal areas. What does this tell you about the angular momentum of the planets?

Illustration 11.5: Conservation of Angular Momentum

A red mass (1 kg) is incident on an identical black mass (1 kg) that is attached to a massless rigid string so that it can rotate around the origin as shown (position is given in meters and time is given in seconds). At $t = 2.6$ s the red mass undergoes a completely elastic collision with the black mass.

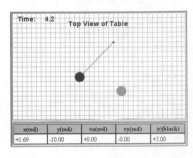

Consider the beginning part of the animation in which a 1-kg red ball is incident on a 1-kg black ball that is constrained to move in a circle. From which point should we measure the red ball's angular momentum? The best place, given the collision with the pendulum, is the point $(0, 0)$, the pivot. This is because we can easily measure the angular momentum of the pendulum about this point. What then is the angular momentum of the red ball before the collision? Certainly it must be changing, since $\mathbf{r}$ changes. No! The angular momentum for a particle is given by the cross product $\mathbf{L} = \mathbf{r} \times \mathbf{p}$, which means we can consider the part of $\mathbf{r}$ that is perpendicular to $\mathbf{p}$ ($rp \sin \theta$, where θ is the angle between $\mathbf{r}$ and $\mathbf{p}$). Since $\mathbf{p}$ is in the negative x direction, the part of $\mathbf{r}$ that is perpendicular to $\mathbf{p}$ is y. Therefore, $|\mathbf{L}| = 50$ kg·m²/s. Note that even though $\mathbf{r}$ changes, y does not. The direction of the angular momentum is found with the RHR and is into the page (which is the negative z direction).

Now what happens to the angular momentum after the first collision? Given that only the black ball moves, we find that $|\mathbf{L}| = mvr = I\omega = 50$ kg·m²/s (again, into the page). The angular momentum is the same value as before the collision. Given that there are no external torques (The pendulum string does not create a torque. Why?), angular momentum is conserved.

What about after the second collision? Well, this is a bit harder. The radius vector $\mathbf{r}$ changes (before the first collision the radius changed, but the part of the radius perpendicular to the momentum was constant). We must use a better definition of the magnitude of $\mathbf{r} \times \mathbf{p}$ than $rp \sin \theta$. In general we get for the z component of the angular momentum: $L_z = (xp_y - yp_x)$. At $t = 16$ seconds, we have $(-5.06)(-1.73) - (-12.51)(-4.69) = -50 = 50$ kg·m²/s (again, into the page).

Note that in general, $\mathbf{A} \times \mathbf{B} = (A_y B_z - A_z B_y)\,\mathbf{i} + (A_z B_x - A_x B_z)\,\mathbf{j} + (A_x B_y - A_y B_x)\,\mathbf{k}$.

Exploration 11.1: Torque

Drag the tip of the force arrow (position is given in meters and force is given in newtons). The **red arrow is the radius**, on which the force acts, and the **dark green arrow is the force**. The light green arrow also represents the force and is there to help illustrate the angle between $\mathbf{r}$ and $\mathbf{F}$.

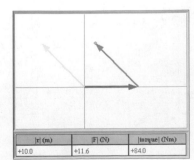

a. When is the cross product zero?

b. What is the angle between $\mathbf{r}$ and $\mathbf{F}$ that goes in $r F \sin(\theta)$?

c. Is there anything missing in this representation of the torque?

d. Does the assignment of $\mathbf{r}$ and $\mathbf{F}$ matter? In other words, if $\mathbf{r}$ was $\mathbf{F}$ and $\mathbf{F}$ was $\mathbf{r}$, would the torque be the same?

Exploration 11.2: Nonuniform Circular Motion

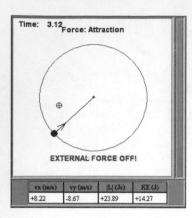

In this Exploration you are looking down at a black ball on a table top. Drag the crosshair cursor (position is given in meters and time is given in seconds) to within 5 m of the 0.2-kg black ball. The cursor will then exert a constant force on the black ball. You may choose either an attractive or a repulsive force. In addition, the black ball is constrained to move in a circle by a very long wire. The blue arrow represents the net force acting on the mass, while the bar graph displays its kinetic energy in joules.

For attraction and repulsion, drag the cursor around to see the net force.

a. At the beginning of the animation (after you press "play" but before you move the cursor), in what direction does the net force point?

b. With this force, does the black ball move? Why or why not?

c. Where must you apply the force in order to make the ball acquire a tangential velocity?

d. Describe the direction of the force that makes the ball acquire the maximum tangential velocity for the force applied.

e. How does the magnitude of the torque relate to the force applied?

f. How does the direction of the torque relate to the force applied?

Exploration 11.3: Rolling Down an Incline

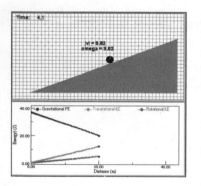

A solid ball of radius 1.0 m rolls down an incline, as shown (position is given in meters and time is given in seconds). The incline makes an angle θ with the horizontal. Adjust the mass (100 g $< m <$ 500 g) and/or the angle ($10° < \theta < 40°$) and watch the graph of gravitational potential energy and rotational and translational kinetic energy vs. time or distance.

Change the angle and the mass of the ball to determine the answers to the following questions.

a. What percent of the initial gravitational potential energy is converted into translational kinetic energy at the bottom of the hill?

b. What percent of the initial gravitational potential energy is converted into rotational kinetic energy at the bottom of the hill?

c. What is the ratio of KE_{rot}/KE_{trans}? What does this number correspond to?

d. How does the ratio of KE_{rot}/KE_{trans} depend on the mass of the ball? On the angle of the incline?

e. How would the animation change if the ball were replaced by a disk of the same radius?

Exploration 11.4: Moment of Inertia and Angular Momentum

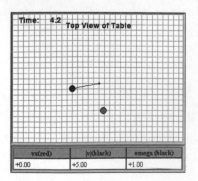

A 1-kg red mass is incident on an identical black mass that is attached to a massless rigid string so that it can rotate around the origin as shown (position is shown in meters and time is shown in seconds). At $t = 2.6$ s the red mass undergoes a completely elastic collision with the black mass.

Watch the animation. You may vary the radius of the pendulum between 2 and 10 m. Answer the first three questions before clicking the "see other variables" check box.

a. As you reduce the length of the pendulum, does the angular speed of the pendulum increase or decrease?

b. From what you know about conservation laws, state whether you think linear momentum, angular momentum, and kinetic energy are conserved during the animation. Why?

c. Set R = 5 m. Calculate the linear momentum, angular momentum (about the origin), and kinetic energy of the system at t = 1, 2, 4, and 5 s.

You may now click the check box.

d. If your answers differ from what you thought, explain why they differ.

Exploration 11.5: Conservation of Angular Momentum

A man is standing beside a 150-kg merry-go-round and suddenly drops a red object onto the merry-go-round (position is given in meters and time is given in seconds). You may change the mass of the object dropped on the merry-go-round and assume that the merry-go-round is a solid, uniform disk.

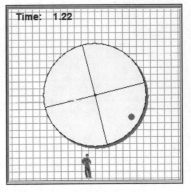

a. What happens to the final angular velocity of the merry-go-round when a heavier object is dropped onto it?

b. Is there a mass that you can add to make the final angular velocity exactly half of the initial angular velocity? If so, what is it?

c. How do your answers to (a) and (b) relate to the conservation of angular momentum?

Problems

Problem 11.1

A wheel rolls as shown in the animation (position is given in meters and time is given in seconds). What is the velocity (with respect to the floor) of a point on the edge of the wheel when it is at the highest point?

Problem 11.2

A wheel rolls without slipping while being pulled by a string wrapped around its circumference, as shown in the animation (position is given in meters and time is given in seconds). Which animation properly depicts the physical situation?

Problem 11.3

A 1-kg object moves down the blue incline and onto the black table as shown in the animation (position is given in meters and time is given in seconds).

a. Determine from the motion of the object whether it rolls without slipping or slides without rolling down the blue incline.

b. If the object rolls without slipping, determine if it is a disk, a hoop, or a sphere.

Problem 11.4

A 1-kg object moves down the blue incline and onto the black table as shown in the animation (position is given in meters and time is given in seconds).

a. Determine from the motion of the object whether it rolls without slipping or slides without rolling down the blue incline.

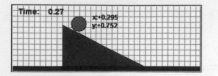

b. If the object rolls without slipping, determine if it is a disk, a hoop, or a sphere.

Problem 11.5

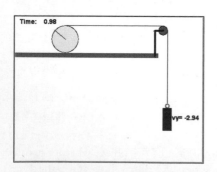

A giant 2.5-kg yellow yo-yo, made of two solid green disks and a massless red hub, is shown (position is given in meters and time is given in seconds). Determine the torque that the string exerts on the yo-yo.

Problem 11.6

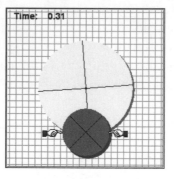

A wheel rolls without slipping while being pulled by a massless string wrapped around its circumference. The string is also attached to a 0.11-kg hanging mass via a massless pulley as shown (position is given in meters and time is given in seconds). If the wheel closely resembles a uniform disk, what is the mass of the wheel? Hint: use energy.

Problem 11.7

Several objects are rotating as shown (position is given in meters and time is given in seconds). Every object, or collection of

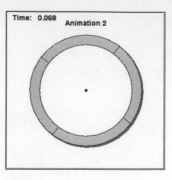

objects, has the same mass, $m = 2$ kg. All mass distributions are uniform and all strings are massless. Every animation has a black dot representing the origin of the coordinate system. You are to calculate the angular momentum about this point for each animation. Calculate the angular momentum (about the origin) of the systems shown.

Problem 11.8

A red disk is dropped onto a rotating yellow disk that has a mass of 20 kg as shown in the animation (position is given in meters and time is given in seconds). What is the mass of the red disk?

Problem 11.9

A puck sliding on an air table collides with another puck of equal mass that is attached to a string as shown in the animation (position is given in meters and time is given in seconds). What quantities are conserved?

Problem 11.10

A 100-gram projectile is incident on a tethered mass on a tabletop (position is given in meters and time is given in seconds). You are viewing the tabletop from above. Assume only conservative forces are acting. Determine the total angular momentum before and after each collision as measured from the pivot point of the pendulum (which is also the origin of coordinates).

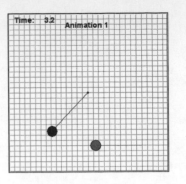

Gravitation

Topics include universal gravitation, Kepler's laws, planetary orbits, and planetary reference frames.

Illustration 12.1: Projectile and Satellite Orbits

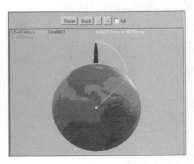

Newton, in his consideration of gravity, realized that any projectile launched from the surface of Earth is, in a sense, an Earth satellite (if only for a short time). For example, in this animation a ball thrown from a tall building sails in a modest orbit that soon intersects Earth not far from its point of launch.

If the ball were fired with a greater initial velocity, it would travel farther. Further increasing the speed would result in ever larger, rounder elliptical paths (they would be elliptical if Earth was not in the way) and more distant impact points. Finally, at one particular launch speed the ball would glide out just above Earth's surface all the way around to the other side without ever striking the ground. At a greater speed the orbit of the ball will be a circle.

At successively greater launch speeds the ball would move in an ever-increasing elliptical orbit until it moved so fast that it would sail off in an open parabolic or (with an even faster launch speed) into a still flatter, hyperbolic orbit, never to come back to its starting point.

This Illustration lets you change the launch speed (but maintain the launch direction) by mouse click. Click "+" to increase launch speed and click "−" to decrease launch speed. Press "Start" to fire the ball. Press "Reset" to change parameters back to default values. The red arrow represents the velocity vector. Click the left mouse button near the tip of the arrow and drag the mouse to change the ball's initial velocity (both direction and speed). Click the right mouse button to stop the animation; press it again to resume. When will the ball start to go around without striking the ground? When will the motion become a circular motion? Try clicking the "full" checkbox and find out what happens.

Illustration 12.2: Orbits and Planetary Mass

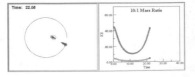

When we consider the elliptical orbits of the planets (Kepler's first law), we assume that the Sun is stationary at one focus of the ellipse. Why does this happen? The mass of the Sun must be much, much greater than the mass of the planets in order for the motion of the Sun to be ignored. But how large does the mass of the Sun need to be in order to achieve this idealized planetary motion? For reference, the Sun is about 1000 times heavier than Jupiter (the most massive planet) and about 100 million (10^8) times more massive than the least massive planet, Pluto.

As you vary the mass ratio in the animation, the mass of the system changes such that the product of the masses, $m_1 m_2$, remains the same. Therefore as you change the mass ratio, the force will remain the same for the same separation between the masses.

The **1000:1 Mass** animation closely resembles the Sun and Jupiter system (the distance is given in astronomical units [A.U.] and the time is given in 10^8 seconds). The green circle is like the Sun, while the red circle is like Jupiter. The force of attraction due to gravity is shown by the blue arrows (not shown to scale), and the relative kinetic energies are shown as a function of time on the graph (note that for this animation the unit for the kinetic energy is not given since we are comparing the relative amount for each object). Also note that the eccentricity of the orbit $e = 0.048$, the perihelion and aphelion distances, and the planet's period closely match those of Jupiter.

In the **100:1 Mass** animation does the "Sun" remain motionless? What about the **10:1 Mass** animation? The **2:1 Mass** animation? The **1:1 Mass** animation? What do you think this means for planetary dynamics in our solar system?

For elliptical orbits, the force due to gravity changes magnitude since the separation changes. But at every instant, the forces of gravitational attraction (the force of the green circle due to the red circle and the force of the red circle due to the green circle) are always the same. This is Newton's third law. It is not too surprising that the law of universal gravitation (described by Newton) contains the third law (also described by Newton).

At the same time, what happens to the kinetic energy of the system as a function of time? It too changes. But why? As the separation between the "Sun" and the "planet" changes, the gravitational potential energy of the system changes too. While the kinetic energy of the system changes, the sum of the kinetic energy and the potential energy of the system must—and does—remain a constant throughout the motion of the objects.

Illustration 12.3: Circular and Noncircular Motion

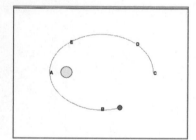

A planet (green) orbits a star (orange) as shown in the two animations. One animation depicts the **Uniform Circular Motion** of a planet and the other one depicts the **Noncircular Motion** of a planet (position is given in 10^3 km and time is given in years). This Illustration will compare the two motions by focusing on the velocity and the acceleration of the planet in each of the animations.

Start the **Uniform Circular Motion** animation of the planet and watch its motion. How would you describe the motion of the planet (consider velocity and acceleration)? The speed of the planet is certainly a constant since the motion of the planet is uniform. But using our usual xy coordinates, the velocity certainly changes with time. Recall that the term velocity refers to both the magnitude and direction. However, if we use the radial and tangential directions to describe the motion of the planet, the velocity can be described as tangential, and the acceleration is directed along the radius (the negative of the radial direction). **Click here** to view the velocity vector (blue) and a black line tangent to the path. **Click here** to view the acceleration vector (red), also. Notice that the acceleration vector points toward the center star.

Start the **Noncircular Motion** animation of the planet and watch its motion. How would you describe the motion? How would you now describe the motion of the planet (consider velocity and acceleration)? The speed of the planet is certainly no longer a constant because the motion of the planet is no longer uniform. Again using our usual xy coordinates, the velocity certainly changes with time; now both the direction and the magnitude change. However, if we use the radial and tangential directions to the path of the planet, the velocity can be described as tangential and the acceleration is directed along the radius. **Click here** to view the velocity vector (blue) and **click here** to view the acceleration vector (red), also. Notice that the velocity and the acceleration are no longer perpendicular for most of the orbit of the planet.

Notice that between points A and C the planet is speeding up, and between points C and A the planet is slowing down. This means that at points A and C the tangential component of acceleration is zero. It turns out that for a planet orbiting a star, if there are no other planets or stars nearby, the acceleration of the planet is directed exactly toward the star whether the motion of the planet is uniform or not.

Illustration 12.4: Angular Momentum and Area

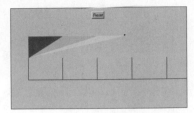

In the absence of a net external torque acting on a system, a particle's angular momentum remains constant. For this discussion, the particle is free, so angular momentum should be conserved. Is there a different way to state the concept of angular momentum conservation? There may be. Consider the statement, *Does a particle sweep out equal areas in equal times* (with respect to **any** origin)? Specifically, in this Illustration, *does* a free particle moving in a straight line sweep out equal areas in equal times?

Press "start" to begin the animation, and let the show begin: A black dot will move freely from left to right. Different colors show the area the particle sweeps out with respect to some fixed point (the origin). Do all the areas have the same size? Click within each area and see what will happen. Certainly from mathematical equations we know the area of a triangle = width $*$ height/2. All of the areas have the same height and the same width ($= v_x * dt$).

Kepler's second law states, during equal time intervals, the radius vector from the sun to a planet sweeps out equal areas. What does this tell you about the angular momentum of the planets? What does this tell you about the motion of the planets?

Illustration 12.5: Kepler's Second Law

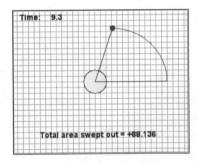

A planet orbits a star under the influence of gravity (distance is given in astronomical units [A.U.] and time is given in years; the total area swept out by the planet's orbit is given in A.U.^{2}). The animation begins from the point of aphelion, the point where the planet is farthest from the star. The planet's orbit is elliptical, and its trail is shown as it orbits the star. Kepler's second law states that the planets sweep out equal areas in their orbits in equal times. What does this mean for the planet's orbit? If the planet had a circular orbit, the planet would undergo uniform circular motion and Kepler's second law is just a statement of equal speed; it confirms the statement of uniform circular motion. For elliptical orbits, therefore, the planet's motion must not be uniform.

Starting at $t = 0$, run the animation for 3 years (not real time, animation time!). How much area has been swept out by the planet in this time interval? There is 28.43 A.U.^{2} swept out. What about from 3 to 6 years? Again 28.431 A.U.^{2} is swept out. Does it matter where you are in the orbit? No. Try it for yourself. While the planet is closer to the star, its speed increases; when the planet is farther away from the star, its speed decreases.

So what does Kepler's second law really tell us? The sweeping out of equal areas is equivalent to telling us that angular momentum is conserved! We know that if angular momentum is conserved (see Chapter 11), there is no net torque. Here the only force on the planet is gravity, and gravity cannot create a torque because the radius arm and the force are always in the same direction.

Illustration 12.6: Heliocentric vs. Geocentric

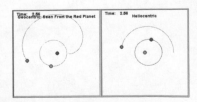

Does Earth orbit around the Sun or does the Sun orbit Earth? For a very long time people thought that Earth was stationary (as the argument goes, otherwise the birds would be ripped from their perches!) and the Sun orbited Earth. This belief is where we get the terms sunrise and sunset. But the Sun does not orbit Earth; it is the other way around. In addition, the motion of the planets, as seen from the reference frame of the **Sun** (the heliocentric reference frame) is rather simple. But from the perspective of each individual planet (the geocentric reference frames of the **Inner Planet** and the **Outer Planet**) the motions of the other planets are rather complicated. The geocentric view is exactly what we see on Earth when we look at the Sun and the other planets of the Solar System.

In this Illustration two planets (the red circle is the inner planet and green circle is the outer planet) orbit a central star (the black circle) as shown in the animation. Along with the animation from the star's reference frame, the heliocentric point of view, two other animations show the motion as seen from each of the planets' reference frames, the geocentric points of view.

As you view the animation, keep in mind that in the **Inner Planet** animation, if Earth is the red planet, the green planet behaves like Mars as seen from Earth, while in the **Outer Planet** animation, if Earth is the green planet, the red planet behaves like Venus as seen from Earth.

Exploration 12.1: Different x_o or v_o for Planetary Orbits

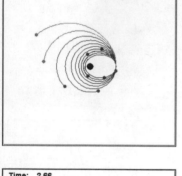

This Exploration shows 10 identical planets orbiting a star. The initial position of the planets can be set at $t = 0$ time units when the planets are on the x axis. The difference in orbital trajectory, therefore, is due to the planets' initial velocities (in this animation $GM = 1000$).

a. As you vary the initial positions of the planets, how do the orbital trajectories change?

b. Find a planet with circular motion. What is the period for this motion?

c. What happens to the orbit when x gets really small?

d. What happens to the orbit when x gets really large?

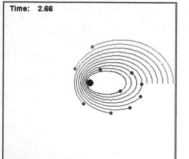

This part of the Exploration shows 10 identical planets orbiting a star. The initial velocity of the planets can be set at $t = 0$ time units when the planets are on the x axis.

e. As you vary the initial velocities of the planets, how do the orbital trajectories change?

f. Find a planet with circular motion. What is the period for this motion?

g. What happens to the orbit when v gets really small?

h. What happens to the orbit when v gets really large?

Exploration 12.2: Set Both x_o and v_o for Planetary Orbits

This Exploration shows a planet orbiting a star. The initial position in the x direction and the initial velocity in the y direction of the planet can be set at $t = 0$ time units when the planet is on the x axis. The difference in orbital trajectory, therefore, is due to the planet's initial position and velocity (in this animation $GM = 1000$).

a. As you vary the initial velocity of the planets, how do the orbital trajectories change?

b. What happens to the orbit when x_0 gets really small (keep $v_{0y} = 10$)?

c. What happens to the orbit when x_0 gets really large (keep $v_{0y} = 10$)?

d. What happens to the orbit when v_{0y} gets really small (keep $x_0 = 5$)?

e. What happens to the orbit when v_{0y} gets really large (keep $x_0 = 5$)?

f. Find the condition for circular motion.

g. For circular motion, what is the period?

h. During each of your investigations, what was happening to the angular momentum as time passed? Why?

i. Make $x_0 = 10$. Then for small v_0, what type of orbit occurs?

j. For $x_0 = 10$, what v_0 makes the orbit circular?

k. As you increase v_0 ($x_0 = 10$), the orbit changes shape. What shape does it have just beyond the speed required for circular orbit?

l. As you increase v_0 ($x_0 = 10$) even further, you eventually reach a condition of "escape." Use energy considerations to predict what this escape velocity should be.

m. For any circular orbit, predict (and then check on the graphs) how the magnitude of potential energy compares to kinetic. Likewise for escape velocity.

n. For a given orbit, you should note that the angular momentum remains constant. How does this relate to the other quantities in the table (under the simulation window), and discuss what is meant by the angle "theta."

When you get a good-looking graph, right-click on it to clone the graph and resize it for a better view.

Exploration 12.3: Properties of Elliptical Orbits

A planet (green) orbits a star (yellow) as shown in the animation.

On a piece of paper sketch vectors for the velocity, radial component of acceleration, and tangential component of acceleration for the planet. The length of the vectors should be indicative of the magnitude of the vectors.

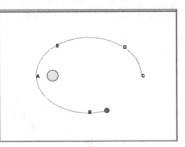

a. Rank the points, A—E, according to the speed of the planet at that point.

b. Rank the points, A—E, according to the gravitational potential energy of the planet.

c. Rank the points, A—E, according to the kinetic energy of the planet.

d. Rank the points, A—E, according to the total energy of the planet.

e. At which of the points, A—E, is the planet's acceleration in the same direction as the velocity?

f. What can you say about the direction of the planet's acceleration at any point on its path? Would you call this acceleration a tangential acceleration or a radial acceleration?

Click here to view the velocity vector (blue) and acceleration vector (red). Compare what you see to your answers (a)–(f).

Exploration 12.4: Angular Momentum and Energy

A planet (with a mass equal to that of Earth) orbits a star as shown in the animation (position is given in A.U. and time is given in years). Along with the animation is a graphical depiction of the energy of the planet. Three curves are shown: in black, the total effective potential energy; in blue, the gravitational potential energy; and in red, the effective rotational potential energy represented by the term: $L^2/2mR^2$. The teal line represents the total energy of the planet as a function of distance to the central star, R.

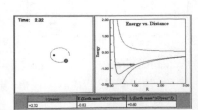

a. What happens to the red curve as the initial speed of the planet is changed?

b. What happens to the blue curve (the gravitational potential energy) as the initial speed of the planet is changed?

c. What happens to the teal curve (the total energy) as the initial speed of the planet is changed?

Now consider the total energy and angular momentum calculated in the table. Look at the circular, bound, and unbound cases.

d. How do the values for total energy and angular momentum change when the type of orbit is changed?

e. Can you find a general rule for whether an orbit is bound?

f. Feel free to explore different values of the initial velocity.

When you get a good-looking graph, right-click on it to clone the graph and resize it for a better view.

Problems

Problem 12.1

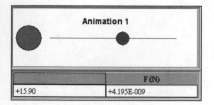

+15.90 | F (N) +4.195E-009

A 100-kg mass can be moved around (by click–dragging it) near an unknown mass as shown in the four animations (distance is given in meters and force is given in newtons). The table shows how the force on the unknown mass changes due to the position of the 100-kg mass.

a. Which animation is physical? Why?

b. For that animation, what is the mass of the unknown mass?

Problem 12.2

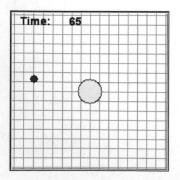

A planet orbits a star (yellow) as shown in the animation (position is given in $R_{EarthSun}$ and time is given in Earth days). Determine the mass of the star.

Problem 12.3

Two planets orbit a star (red), whose mass is $M = 7 \times 10^{25}$ kg as shown in the animation (position is given in 10^6 km and time is given in Earth years).

a. What is the net force on the central star? Write an expression, not a numerical answer.

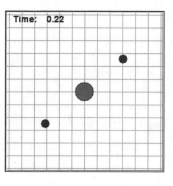

b. What is the net force on each planet? Write an expression, not a numerical answer.

c. Determine the mass of the orbiting planets.

Problem 12.4

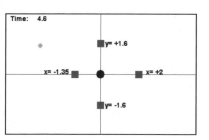

A planet has an initial velocity in the y direction that gives it a slightly elliptical orbit around a star as shown in the animation (position is given in 10^6 km and time is given in years).

a. Determine the mass of the star.

b. What is the correct orbital radius to give circular motion? Drag the planet at $t = 0$ to determine this radius. You may also drag the gray blocks around to use them as temporary or permanent markers.

Problem 12.5

The animation purports to model a solar system. However, the planet shown is not in a circular or an elliptical orbit (position is given in 10^4 km and time is given in years).

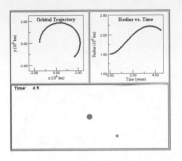

a. Could this represent a physical situation?

b. Why or why not?

When you get a good-looking graph, right-click on it to clone the graph and resize it for a better view.

Problem 12.6

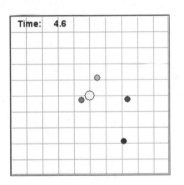

The animation purports to model a solar system. However, one of the planets does not obey all of Kepler's laws for this solar system (position is given in 10^6 km and time is given in Earth years). Identify that planet. (The sun is yellow.)

Problem 12.7

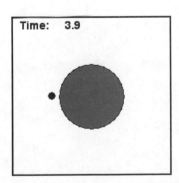

A satellite orbits a planet as shown in the animation (position is given in 10^3 km and time is given in Earth days). If the satellite's orbit above the planet were doubled, what would its speed have to be in order for it to stay in a circular orbit?

Problem 12.8

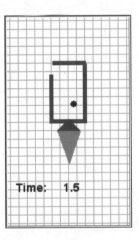

The animation shows a comet orbiting a star (position is given in 10^6 km and time is given in Earth years). Determine the mass of the star.

Problem 12.9

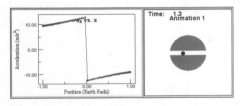

A rocket accelerates upward while a ball is fired into the opening in the rocket as shown in the animation (position is given in meters and time is given in seconds). The rocket and the ball are far from any massive object. For an observer in the rocket, what acceleration does the ball appear to have?

Problem 12.10

A very wealthy individual proposes to dig a hole through the center of Earth and run a train (the small black circle) from one side of Earth to the other as shown in the animation (position is given in Earth radii and time is given in seconds). Which of the animations correctly depicts the motion of the train? Ignore frictional effects and treat Earth as a uniform mass distribution.

Statics

Topics include equilibrium (stacking bricks, diving board, etc.) center of mass, and center of gravity.

Illustration 13.1: Equilibrium on a Ramp

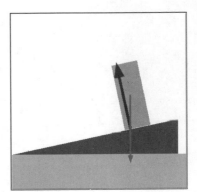

A uniform block of wood sits in equilibrium on a ramp as shown in the animation (position is given in meters). The slider allows you to control the height of the ramp and therefore the angle of the ramp. The red vector represents the weight of the block, and the blue vector represents the normal force of the ramp on the block (*Note: The normal force is shown with a fatter arrow so you can see it better*). The force of static friction is an important feature of this problem, but its vector is not shown so we can focus on the weight and the normal force.

Consider the ramp when it is flat. All parts of the ramp are touching the bottom surface of the block. Therefore, the normal force does not act at just one point; actually, it is distributed across the bottom surface of the block. This is known as a distributed load. Despite this fact, we draw one vector, the normal force, to represent the resultant perpendicular component of the force of the ramp on the bottom surface of the block. But where should we draw the normal force of the ramp on the block? We draw the normal force at a location such that the torque due to this one force vector is equal to the net torque due to the distributed load of the ramp on the block. Therefore, when the ramp is flat, we draw the normal force of the ramp on the block as if it acts at the middle of the base of the block.

Before increasing the height of the ramp, predict what will happen to the location of the normal force of the ramp on the block. Will it stay in the same place or will it shift? If it shifts, will it shift toward the front edge or toward the back edge of the block? Now, increase the height of the block to 0.35 m. Notice the position where the normal force is drawn. Is this what you predicted?

There is a certain angle of the ramp that is too steep for the block to remain in equilibrium. If static friction is great enough to keep the block from sliding, then at this angle the block will tip; there will be an unbalanced torque to cause the block to "rotate." Where will the normal force on the block act when the ramp is at this angle? Increase the height of the ramp to its maximum value. Is this what you predicted?

At this angle the block would tip over. However, in **Animation 2** you can increase the height of the ramp past the point where the block should tip. The animation will show you the normal force needed to keep the block from tipping. You can see how ridiculous this looks. Physicists would say that this animation is *unphysical*.

What do you notice about the point where the line of action of the normal force intersects the line of action of the weight? Try proving that the point where the line of action of the weight intersects the base of the block will be the same point where the normal force acts. Be sure to consider the conditions for static equilibrium.

Illustration 13.2: Center of Mass and Gravity

The first animation shows two blocks of equal mass (position is given in meters). The center of mass of the system is shown by a red dot and its position is calculated for

you. Drag the right-hand green block to the right or to the left. What do you notice about the location of the center of mass as you change the block's position?

Now suppose the two blocks have unequal masses as shown in **Animation 2**. Is the center of mass at the center of the system? By viewing the location of the center of mass, you know which block is more massive. So which one is more massive?

How can we calculate the ratio of the mass of the blue block to the mass of the red block? If you only have two blocks, then the ratio of the distances of each block from the center of mass is related to the ratio of their masses. Therefore, by measuring the distance from each block to the center of mass, you can calculate the ratio of their masses.

In terms of the location of the two objects, the center of mass is located at

$$X_{cm} = (x_1 m_1 + x_2 m_2)/(m_1 + m_2)$$

for a one-dimensional, two-object system.

A concept similar to that of the center of mass is that of the center of gravity. In fact the two are often used interchangeably. The center of mass is defined above; the center of gravity is defined as the point on a system where gravity can be considered to act. The center of gravity takes into account the fact that the force of gravity—and therefore the acceleration due to gravity—is different for different heights above the surface of Earth. For this Illustration the center of mass is equivalent to the center of gravity. Only if the system is really large might the acceleration due to gravity be different at different parts of the system, thereby causing the center of gravity to differ from the center of mass.

Illustration 13.3: The Force and Torque for Equilibrium

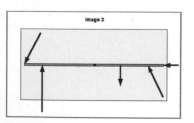

A rigid rod of uniform mass is shown on top of a frictionless table as shown in **Animation 1**. The black circle is the location of the center of mass (position is given in meters and torque is given in newton meters).

Click **Animation 2** to see the forces acting on the rod (the weight and normal force cancel each other out and are not shown; they are into and out of the page, respectively). Assume that the lengths of the force vectors are indicative of the magnitudes of the forces in newtons. If these are the only forces acting on the rod, is the rod in equilibrium?

If these are the only forces acting on the rod, then it is not in equilibrium because adding up the force vectors shows that the net force on the rod is not zero. Since the net force is not zero, the center of mass of the rod will have an acceleration, and therefore its velocity will change. In addition, adding up the torques on the rod shows that the net torque on the rod about the center of mass is not zero. The rod will have a changing angular velocity out of the page.

Suppose that we want the rod to be in equilibrium. What additional force must we apply to the rod?

Consider the conditions of static equilibrium. The net force on the rod must equal zero. If you add up all of the forces presently acting on the rod (as shown in **Animation 2**), the sum is not zero. Therefore, we must apply another force to the rod that is the negative of the sum of the other forces on the rod.

At what location should this additional force be applied?

To be in static equilibrium, the net torque on the rod must equal zero. Therefore, the torque due to this additional force on the rod must equal the negative of the sum of the torques of the other forces presently acting on the rod. Knowing the torque and the force needed for the rod to be in equilibrium, you can calculate the location where the force should be applied to the rod.

Now, in **Animation 3** you get to add a force to get the rod into equilibrium. Adjust the magnitude and direction of the blue force vector and place it at the correct location on the beam. Then **check your answer**. You will see a red vector for the net force and a green calculation of the net torque (this is in the z direction, which is positive out of the page). If the net force is zero, then its vector will be zero and will not be seen. If the rod is in equilibrium, the net force vector will vanish and the torque calculation will be zero, and you calculated your answers correctly. If not, recheck your calculations, adjust the magnitude, direction, and location of the blue force vector and check your answer again.

Illustration 13.4: The Diving Board Problem

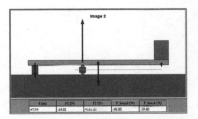

A 2-kg box sits 0.3 m from the right end of a board of negligible weight in **Animation 1** and a board of weight 10 kg in **Animation 2**. Two supports (support 1 and support 2) exert forces on the left and right ends of the board, as indicated by the two force vectors (position is given in meters). The arrows represent the relative sizes of the force vectors, but their length does not represent their actual magnitudes (the actual value of the forces, as well as the separation between the supports, is shown in the table). The board is 6 m long and support 1 is 0.3 m in from the left edge.

Consider the situation in **Animation 1** in which the board has a negligible weight. How does the force of each support on the board ($F1$ corresponds to the right support and $F2$ corresponds to the left support) depend on where the box is located? You can drag the second support from left to right to view the forces of the supports on the board. When the board is of a negligible mass, there are three forces that act on the board: the weight of the box, and the forces of the supports. As you move support 2 around, what do you notice? When the movable support is as far right as it can go, the force of the first support is zero and the force of the second support cancels the force of the box. This seems logical, but why this force? An $F1 = 10$ N and an $F2 = 9.6$ N, would work too, right? Well, yes and no. It would certainly make the sum of the forces on the board equal to zero, but what about the sum of the torques? This sum would not be equal to zero no matter where you measured the torques from. Only an equal and opposite force acting at the same position as the weight of the box will keep the board in equilibrium. As you move the second support to the left, note that both forces exerted by the supports get bigger, and that the force that the first support exerts is negative. We can understand this effect by measuring the torque on the board from the first support. Since the force is in the y direction and the radius arm is in the x direction, the magnitude of the torque is rF. The radius arm for the weight of the box is always 5.4 m. The radius arm for the second support changes as you drag it to the left. Therefore, as you drag it to the left the force the support exerts must increase in order to keep rF the same magnitude as (and in the opposite direction of) that of the box. Once the right force is determined from the torque, the force due to the first support can be determined by requiring the sum of the forces on the board to be zero.

Now, consider the situation in **Animation 2** in which the board has a weight of 10 kg. How does the force of each support on the board depend on where the box is located? You can drag the second support from left to right to view the forces of the supports on the board. When the board has a mass, there are four forces that act on the board: the weight of the box, the weight of the board, and the forces of the supports. Much of the same analysis described above still is valid here; you just need to factor in one more force and therefore one more torque. Again drag the support around and pay close attention to the forces. What happens to the forces when the separation between the supports is 3.15 m? How about 2.7 m?

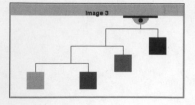

Exploration 13.1: Balance a Mobile

The center of gravity is at the same location as the center of mass for systems in which the acceleration due to gravity is virtually equal for all points in the system. Therefore, in this Exploration we will talk about and calculate the center of mass.

   The goal when making a mobile is for the center of mass to be below the string attached to the ceiling. Otherwise, there will be a net torque on the mobile about its center of mass until this condition is met. Consider a mobile made of two blocks as shown in the animation. The mass of the blue block is 0.050 kg. Assume that the rods and string connecting the blocks are light enough to be neglected (position is given in meters).

a. What is the mass of the green block in order for the mobile to be balanced in **Animation 1**?

To answer this question, consider the conditions of equilibrium. The net torque on the horizontal rod about the point where it is attached to the string connected to the ceiling must equal zero. Therefore, the magnitude of the torque on the rod due to the tension in the right string must be equal to the magnitude of the torque on the rod due to the tension in the left string.

b. Suppose you want to replace the green block with another two-block system just like the first one, but with less massive blocks as shown in **Animation 2**. What are the masses of the red and orange blocks?

This is very similar to the previous question, except that you are applying the conditions of equilibrium to the rod connecting the red and orange blocks. However, when you solve the equation, *net torque* = 0, you have a problem. The masses of BOTH the red and orange blocks are unknown. Before, for the green and blue blocks, you knew the mass of the blue block and could solve for the mass of the green.

c. You need another relationship between the red and orange blocks to help you out. Since we replaced the green block with the red and orange ones, how are the masses of these three blocks related?

d. Ok, this is getting fun. Suppose that you now replace the orange block with another identical two-block system as shown in **Animation 3**. What are the masses of the yellow and purple blocks?

e. Where is the center of mass of the system of four blocks?

Since you know the masses of the blocks, measure the x, y coordinates of each block and calculate the coordinates of the center of mass.

f. Now click on the animation to locate the point that you just calculated. Is it directly beneath the string that connects the mobile to the ceiling? It should be!

Note that adding a new system of blocks to the left end each time did not change the x coordinate of the center of mass; however, it did cause the y coordinate to decrease since each system of blocks hung lower and lower. However, shifting the y coordinate of the center of mass did not change the equilibrium status of the mobile.

Exploration 13.2: Static Friction on a Horizontal Beam

You hold a piece of wood by pushing it horizontally against a wall as shown in the animation (position is given in meters).

a. What forces act on the wood? Draw a free-body diagram for the wood showing the forces at their proper locations. Compare your diagram to the one shown in **Animation 2**.

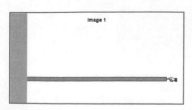

b. What is the force in the $+y$ direction that counteracts the weight of the wood in this example? Note that this force is parallel to the surface of the wall and wood where they are in contact.

c. In this case, do you know if the static frictional force of the wall on the wood is equal to its maximum value?

In **Animation 3** you can adjust the magnitude of the push by clicking and dragging on the white circle at the tip of the vector representing the force of your hand on the wood. The maximum frictional force (shown as a **red vector**) adjusts accordingly. At the instant where the actual frictional force (**black vector**) equals the maximum frictional force (**red vector**), the beam will still be in equilibrium. In this case this is the least force that you can push the wood and have it remain in equilibrium. If you push it with less force, the meter stick will fall.

In **Animation 4** you can adjust the magnitude of the push so that it is less than the minimum push required for the wood to remain in equilibrium. If the maximum possible static frictional force is less than the actual frictional force needed for equilibrium, the piece of wood will fall. In the animation, if you adjust $f_{s\,max} < f_s$, the resulting situation depicted in the animation is *unphysical* since the piece of wood will actually fall.

Exploration 13.3: Distributed Load

A box sits on a board of negligible weight. Two supports exert forces on the left and right ends of the board (position is given in meters). The arrows represent the relative sizes of the two force vectors, but their length does not represent their actual magnitudes.

a. How does the force of each support on the board depend on where the box is located? You can drag the box from left to right to view the forces of the supports on the board.

Suppose the box is exactly halfway between the center and the right support.

b. What is the ratio of the magnitude of the force of the right support on the board to the magnitude of the force of the left support on the board?

Consider a situation in which the board has significant weight.

c. What then might the force vectors look like when the box is sitting above one of the supports, say for example the left support? Check using **Animation 2**.

d. What is the ratio of the weight of the board to the weight of the box in this case?

Exploration 13.4: The Stacking of Bricks

How can you stack four uniform bricks (or blocks or meter sticks) one on top of another so that they extend as far over a table as possible and yet remain stable? **Rules:** You can drag and move bricks horizontally with your mouse (position is given in tenths of inches; hence each brick is a foot long). The stability of each brick is color coded:

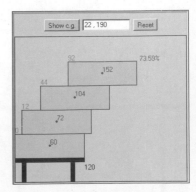

- **green:** the brick is in stable equilibrium
- **yellow:** the center of gravity of the brick, or a group of bricks, is right above the edge of the supporting brick
- **red:** the brick is unstable; it will fall in real life.

The center of gravity for each brick is shown as a small blue dot. The current mouse position (relative to the top left edge of the table) is shown in the upper part of the animation in the *Text Field*. If you press the "show c.g." button, the center of gravity for the brick subsystems (top brick, top two bricks, top three bricks and all four bricks, respectively) will be shown as a small circle with an arrow. The length of the arrow is proportional to the gravitational force for each balanced subsystem. In addition,

> The left edge of each brick is in red
>
> The position of the center of gravity of each brick is in blue
>
> **The position of the center of gravity is in black when "show c.g." is selected**

a. What is the stability condition for an object overhanging a table?

b. How should each brick be positioned relative to the brick underneath it? Be explicit and refer to the center of gravity. Hint: Start at the top and work down.

c. Can the top brick have its entire length beyond the edge of the table?

d. Challenge: Come up with a mathematical description for the overhang of each individual brick and the total overhang.

Problems

Problem 13.1

A box of uniformly distributed mass sits in equilibrium on a ramp as shown in the animation (position is given in meters).

a. Suppose you wish to increase the angle of the ramp. What is the maximum angle of the ramp so that the box does not tip?

b. What is the minimum coefficient of static friction between the box and ramp in order for the box to not slide when the ramp is at the angle measured in part (a)?

Problem 13.2

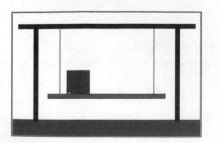

A steel block sits on a board that is held by two strings. The strings are attached to a crossbar that is held by two ringstands on a laboratory table. Assume that all objects have uniformly distributed mass, which means that the center of mass of each object is at its geometric center. The masses are as follows: The block has a mass of 2.0 kg, the board has a mass of 0.50 kg, and the crossbar has a mass of 1.0 kg (position is given in meters).

a. What is the tension in each string?

b. What is the force of each ringstand on the crossbar?

Problem 13.3

The side view of a truck on a level road is shown in the animation. The mass of the truck is 1230 kg. Suppose the resultant force of the road on the front set of tires is 4000 N, in the upward direction of course (position is given in meters).

a. What is the resultant force of the road on the rear set of tires?

b. What is the horizontal distance from the front axle to the center of mass of the truck?

Problem 13.4

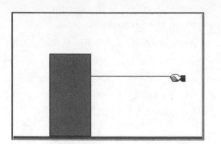

A 10-kg box sits on a table as shown in the animation. A rope is attached to the box, and you pull the rope to the right with a certain force. Assume that the coefficient of static friction is great enough so that the box doesn't slip first (position is given in meters).

a. What minimum force will make the box tip?

b. What is the minimum value of the coefficient of static friction for the box NOT to slide before tipping?

c. If you would like to apply a greater force to the box, yet minimize the risk of tipping, what should you do?

Problem 13.5

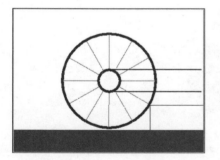

The rear wheel of a bicycle is in equilibrium as it rests against the corner of a curb as shown in the animation. For purposes of this problem, neglect the force of the bicycle frame on the wheel. The wheel is not touching the ground, and its mass is 0.40 kg. Neglect the mass of the bicycle chain (position is given in meters).

a. Assuming that the top and bottom chain tensions are equal, what is the tension in the chain?

b. What is the magnitude and direction of the force of the curb on the wheel?

Problem 13.6

A string is attached between the end of a narrow uniform beam and the wall. Friction on the beam keeps the left side from falling. The magnitude of the weight of the beam is w (position is given in meters).

a. What is the ratio of the tension in the string to the weight, T/w?

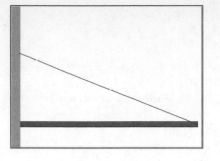

b. What is the magnitude and direction of the force of the wall on the beam?

c. What is the minimum coefficient of static friction required for the left end of the beam not to slip?

Problem 13.7

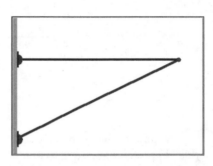

Two uniform rods are hinged to a wall and screwed to each other as shown in the animation. The mass of the blue rod is 2.0 kg and the mass of the black rod is 1.5 kg (position is given in meters).

a. What are the magnitude and direction of the force of the top hinge on the black rod?

b. What are the magnitude and direction of the force of the bottom hinge on the blue rod?

c. What are the magnitude and direction of the force on each rod due to the screw attaching the rods? Your answer should include two forces, the force of the screw on the blue rod and the force of the screw on the black rod.

Problem 13.8

A seesaw, usually in equilibrium, is shown in the animation. A little girl named Melody sits on the seesaw and thereby applies a force of 200 N to the seesaw as indicated by the red vector

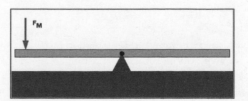

labeled F_M. The seesaw weighs 500 N (position is given in meters).

a. Where should her dad sit on the seesaw in order to keep the system in equilibrium if he applies a bulky 900 N to the seesaw when sitting on it?

b. What is the force of the axle on the seesaw, assuming that the only other force applied to the seesaw besides Melody and her dad is the force of the axle?

c. When Melody's dad wants to practice a circus act by applying a 900 N force to the far right edge of the seesaw, Melody's mom immediately rushes to her aid by applying a downward force on the seesaw to keep the system in equilibrium. If the force of Mom on the seesaw has a magnitude of 750 N, at what location on the seesaw is it applied?

Problem 13.9

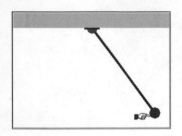

A long pendulum is made of a pendulum bob of mass 10.0 kg attached to a lightweight cable. By pushing horizontally on the pendulum bob, you keep it in equilibrium. Note: The hand is not drawn to scale; it appears larger than its actual dimensions (position is given in meters).

a. What is the force of your hand on the pendulum bob?

b. What is the force of the cable on the pendulum bob (i.e., tension)?

c. Suppose you wish to make it easier on yourself by applying the minimum force necessary to keep the pendulum bob in equilibrium, while maintaining its position as shown in the animation. In this case, what are the magnitude and direction of the minimum force of your hand on the pendulum bob, and what would then be the tension in the cable?

Problem 13.10

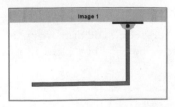

An emergency lever (the red, sideways, L-shaped object) is designed to rotate clockwise about an axle in a hinge (the gray half circle) as shown in the animation. Assume the lever is rigid, is made of uniform material, and has a mass of 0.20 kg.

The axle (the black circle) exerts a frictional force on the lever (position is given in meters).

a. What is the magnitude and direction of the torque on the lever due to friction between the lever and axle?

b. What is the magnitude and direction of the net force of the axle on the lever?

c. In trying to turn the lever, you apply a force of magnitude 5.0 N in the direction shown in the following animation. **Show Animation with Applied Force Vector**. But the lever remains in equilibrium. What is the torque on the lever due to friction between the lever and axle, and what are the magnitude and direction of the net force of the axle on the lever?

d. Suppose you keep the magnitude of the applied force the same (5.0 N) but wish to apply a greater torque to the lever. What could you do differently?

e. Suppose the maximum torque on the lever due to friction at the axle is 10.0 N·m. You decide to apply a force straight downward at the left end of the lever. What is the minimum value of the magnitude of the force in order to just barely turn the lever?

Problem 13.11

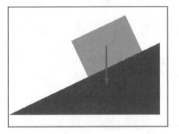

A wood block of uniformly distributed mass sits in equilibrium on a ramp as shown in the animation. Its mass is 0.20 kg (position is given in meters).

a. If you replace the "load" of the ramp on the bottom surface of the block with a single normal force acting at a certain distance from the front edge of the block, what are the magnitude and direction of this force and at what location does the force act?

b. What is the force of static friction on the block?

Problem 13.12

Four spheres are shown in the animation. A blue sphere is half as massive as a red one, and a purple sphere is twice as massive as a red one. Where should the purple one be placed in order for the center of gravity to be at the location of the black dot (position is given in meters)?

Problem 13.13

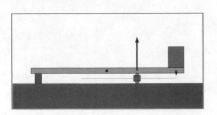

A 2-kg box sits 0.3 m from the right end of a board of unknown weight. Two supports exert forces on the board (position is given in meters). You can drag the second support from the left to the right to view how that force changes with its position. The arrows represent the relative sizes of the force vectors for the movable support and the box, but their length does not represent their actual magnitudes (the actual value of the forces, as well as the **separation** between the supports, is shown in the table). The board is 6 m long and support 1 is 0.3 m in from the left edge. Determine the mass of the board.

Static Fluids

Topics include pressure, density, pressure at a depth, Pascal's principle, Archimedes' principle, and buoyancy.

Illustration 14.1: Pressure in a Liquid

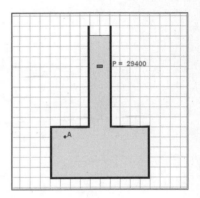

With fluids, instead of discussing forces, we usually talk about pressure, which is defined as the force per unit area or $P = F/A$. This is because the direction of the force a liquid exerts on its container depends on the shape of the container (force is normal to the surface of the container) and the size of the container. Pressure is not a vector (no direction) and does not depend on the size of the container (position is given in meters and pressure is given in pascals).

Move the pressure indicator in the tube and note the pressure readings (the pressure is only measuring the effect of the liquid as described below). Let's discuss why pressure increases as a function of depth. Assume the blue liquid is water (density 1000 kg/m³). Pick a point to measure the pressure somewhere in the upper tube. If the dimension of the container into the screen (the dimension you cannot see) is 1 m, what is the volume of water above the point you picked? What is the mass and thus the weight of the water at that point? For example, consider a depth of 3 m. The pressure is 29,400 N/m². The volume of water above this point is a cylinder of volume 9.4 m³. The mass of the water is the volume times the water's density, or 9,400 kg, and therefore the weight of the water is 92,120 N.

What is the force downward at that point? Well, take the weight and divide by the cross-sectional area of the column of water at that point, which is 3.14 m². This pressure should be equal to the pressure reading. The units of pressure are N/m² = pascals (abbreviated Pa).

Strictly speaking, this is the gauge pressure, not the absolute pressure, because we assumed $P = 0$ at the top of the water column when the pressure (due to the atmosphere) is actually around 1×10^5 Pa. The absolute pressure then would be the pressure at the top due to the atmosphere added to the pressure due to the weight of the water. All of this comes together in the equation:

$$P = P_0 + \rho g y,$$

where P_0 is the pressure at the top, ρ is the density of the liquid, g is acceleration due to gravity and y is the depth of the liquid.

What will be the pressure at point A? Add **a second pressure indicator** to check.

Illustration 14.2: Pascal's Principle

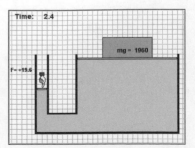

The animation shows a model of a hydraulic lift (position is given in centimeters, force is given in newtons, and time is given in seconds). A gas or liquid, usually oil (yellow in this animation), fills a container with pistons.

Notice that because the pressure is the same at the same height in the fluid, the force required for the hand to support the mass is much less than the weight of the mass (10 times less). This is because the areas of the gray circular "lids" on the top of the oil are different by a factor of 10. Now calculate the pressure exerted by the green mass and the pressure exerted by the force arrow. They should be the same. The ratio of the areas is the ratio of the forces required for equilibrium. Enter a new value for the mass (between 100 and 300 kg) and try it.

Now, play the animation. As the arrow moves down, the mass moves up (but the motion is too small to accurately measure how far it moves up). How far up should the mass move and why? The volume of fluid pushed out of the left tube must be the same amount that goes into the right cylinder.

What is the work done by the force arrow? What is the work done on the mass? The work done by the force arrow is exactly the work done on the mass. So what is the advantage of this lift? The advantage is that it allows a small force to raise something much more massive (e.g., lifting up a car in a repair garage). One can think of this as a small force over a large displacement doing the same amount of work as a large force over a small displacement. The force arrow could not directly lift the green mass, but using the hydraulic lift, it can.

In this setup we neglected the change in pressure as a function of depth in the liquid. We assumed the pressure was still the same at the two gray lids on top of the oil at the end of the animation as it was at the beginning of the animation. At the end of the animation, however, the force arrow should actually be larger than that listed because it is some distance below the liquid level of the right side at the end of the animation.

Illustration 14.3: Buoyant Force

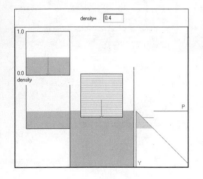

The buoyant force on an object (whether or not it floats) is due to the pressure difference between the bottom of the object and the top of the object. If the object is going to be buoyed up, the bottom of the object is subjected to a greater pressure than the top of the object. Remember that pressure is force/area and so, if the pressure at the bottom of an object (submerged under a liquid) is greater than the pressure on top (still in air), there is a net upward force.

This animation shows a block lowered into a liquid and then floating. The density of the block can be changed by click-dragging in the block on the upper left of the animation (click-drag at the white-gray interface of the box). The graph in the lower right of the animation shows the pressure difference (P) as a function of depth (Y). When the net upward force, the buoyant force, is equal to the weight of an object (the white and gray striped block), it floats in the gray liquid. In addition, the amount of fluid displaced when the block is in the fluid is shown in the block to the left.

Assume that the gray liquid is water (density 1000 kg/m^3). The density of the block is a fraction of the density of water. So, to start with, the block has a density of 0.4 * 1000 kg/m^3 = 400 kg/m^3. Note that approximately 40% of the block is submerged when it is floating. Assume the dimensions of the block are 1 m (and it is a cube). First, let's find the net force on the block when it is floating by finding the pressure on the block.

Since pressure as a function of depth in a liquid is $\rho_{\text{liquid}} g y$ (where ρ is the density of liquid, g is the acceleration due to gravity, and y is the depth in the liquid), what is the pressure at the bottom of the block? It is $P_{\text{atm}} + \rho_{\text{liquid}} g y$ where $y = 0.4$ m. What is the pressure at the top of the block? It is just P_{atm}. Therefore, the liquid exerts a total force of $\Delta P A = \rho_{\text{liquid}} g y A = 400$ N, which must also be the block's weight if it is in equilibrium. Why do we neglect the force on the sides of the block due to water pressure?

The animation also shows the spillover of water into a second container (to the left). What is the volume of water in this second container? Using the density of water, what is the weight of the water? Note that this is equal to the buoyant force because, if the block were removed and the water in the spillover tray (to the left) were put back into the main water container, that water would be supported, so the pressure difference supports that weight of water. This means that although the buoyant force is due to the pressure difference at different depths, it is also equal to the weight of the water displaced by the object, as expressed in the equation below:

$$F_B = \rho_{\text{liquid}}\, g V_{\text{liquid displaced}} = \rho_{\text{liquid}}\, g V_{\text{submerged part of object}}$$

If the buoyant force is equal to the weight of the object, the object floats. Change the density of the object by click-dragging the mouse on the base of the red arrow in the top left-hand box and then let the animation run. Repeat the calculations to show that the buoyant force is equal to the weight of the object.

What happens if the object is pushed into the water below where it would naturally float (away from equilibrium)? Try using the mouse to drag the floating block farther down in the water. What type of motion do you observe, and why? Think about the forces acting on the block.

Illustration 14.4: Pumping Water up from a Well

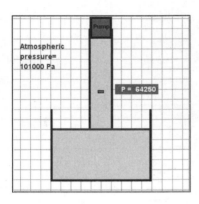

Why can't you pump water up out of a well deeper than 10.3 m? You may not have even known of this limitation! We can answer this question by considering the change in pressure in a liquid as a function of depth (position is given in meters).

First, what is the pressure at the top of the water in the well (the dark blue line)? It is just atmospheric pressure. Measure the pressure of the water in the tube at the pump by dragging the pressure gauge there. Change the well depth by dragging down the dark blue line and see what happens. Varying this level increases the height to which the pump must pump the water. When the well is deeper than 10.3 m (i.e., when the height of the water column up to the pump is greater than 10.3 m), what is the problem? Can the pump pressure ever be less than zero? No. The best a pump could do is to create a vacuum, which would be $P = 0$ Pa. In reality, the pump could never reach $P = 0$ Pa since there would be vapor pressure in the top of the tube.

Note that this animation is exactly how you use a straw to get liquid out of a glass. You reduce the pressure in your mouth (from atmospheric pressure) and the liquid goes up the straw and into your mouth.

What would happen if you tried to pump out a less dense material (like oil)? (Remember that $P = P_0 + \rho g y$). Could you have a deeper well or would it require a shallower well? **Try it**.

Exploration 14.1: Floating and Density

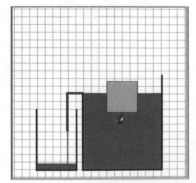

How can a boat made out of a material more dense than water float? The block has a mass of 0.185 kg (position is given in centimeters). If this block is a *cube*, what is the density of the block? Note that since it is greater than water (1000 kg/m^3) the block sinks as shown in the animation.

We **reshape the block** so that it has the same depth into the screen, but is wider and taller with walls that are 0.21 cm thick.

a. When the animation runs, what is the volume of water displaced (the dimension of the water container into the screen that you cannot see is 10 cm)?

b. Using the density of water (1000 kg/m^3), find the mass of the water displaced. Show that it is equal to the mass of the reshaped block. Thus, the block floats.

c. Another way to think about this is that in its new shape the block has an effective density (total mass/total volume) less than that of the water. Divide the mass (0.185 kg) by the new volume to find the new effective density of the block.

d. How does the effective density compare to the density of water?

The weight (mass * 9.8 m/s^2) of the water displaced (even if the displaced water leaves the container) is equal to the buoyant force on the block. In the case of a floating object, the buoyant force is equal to the weight of the floating object.

Exploration 14.2: Buoyant Force

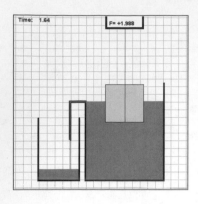

When an object is put into a liquid, it experiences a buoyant force that is equal to the weight of the liquid the object displaces. The force on the wire is given as the block is slowly lowered into the liquid (position is given in centimeters and force is given in newtons). You can change the mass of the block between 0.125 kg and 0.375 kg and the density of the liquid between 500 kg/m³ and 1000 kg/m³. The object is in static equilibrium when the clock stops.

a. What is the weight of the block and the tension in the string when the block is in the liquid? Therefore, what is the value of the buoyant force? The buoyant force and the tension in the string (the force on the support wire) act upward and the weight acts down.

b. What is the volume of the block in the liquid—either the submerged part of the block if the block is partially submerged when you paused it or the entire block if it is completely submerged (the dimension of the block that is into the screen is 5 cm)?

c. What is the volume of the water that is displaced by the block (the dimension of both water containers into the screen is 10 cm)? Verify that this is equal to the answer in (b).

d. What is the mass of the liquid displaced? What is the weight of the liquid displaced? Check that this is equal to the buoyant force.

e. Pick two different masses and densities and verify that the buoyant force is equal to the weight of the water displaced.

Exploration 14.3: Buoyancy and Oil on Water

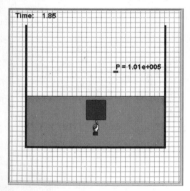

This Exploration will address the buoyant force in more depth (pun intended). Specifically, what happens if we put an object in two "layers" of fluids? Assume the brown block is a *cube* (position is given in meters and pressure is given in pascals).
Note: The format of the pressure is written in shorthand. For example atmospheric pressure, 1.01 × 10⁵ Pa, is written as 1.01e + 005.

Move the pressure indicator and measure the pressure at the bottom of the wooden block and at the top of the block.

a. If the block is a cube, what is the force on the block due to the water (buoyant force)?

b. What, then, is the weight of the block? What is the density of the block?

c. Another method: How much (what percentage) of the block is submerged? Check that the density of the block is that same percentage of the density of water (1000 kg/m³).

Now consider what would happen if we put the block in an oil with a different density.

d. Predict what you expect will happen if we put the block in an oil with a density of 700 kg/m³.

e. **Try it**. Was your prediction correct? Explain.

f. What is the pressure at the bottom of the block and at the top of the block? What is the buoyant force on the block in the oil?

Now, suppose the wood block is put in a mixture of water on the bottom with oil on the top (the oil floats on the water and doesn't mix with the water).

g. What do you expect will happen? Why?

h. **Try it**. Is more or less of the block submerged in water in this case compared with the block simply floating in water (without oil)? Why?

i. One way to look at what happened is to measure the pressures. Find the pressure at the bottom of the block and at the top of the block.

j. What is the pressure difference and thus the net buoyant force on the block?

k. In order for a block to float only in water (with air on top), to get the same pressure difference to support the block, why does the block need to be lower in the water? (Think about the density of air compared with the density of oil and, therefore, the change in pressure with depth in air and in the oil.)

Another way to look at this is to compare the buoyant forces.

l. In comparison with the block floating in water only, has the buoyant force increased, decreased, or stayed the same?

m. What is the volume of water that the block displaces?

n. What is the weight of that water?

o. What is the volume of oil that the block displaces?

p. What is the weight of the displaced oil?

q. How do those two compare with the weight of the block?

Problems

Problem 14.1

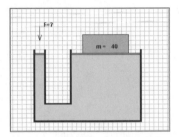

The animation shows a model of a hydraulic lift. The gray areas are circular lids on top of the yellow fluid inside the lift (position is given in centimeters).

a. What force is required on the left side to support the 40-kg mass?

b. If the mass is lifted up 1 cm, how far down does the fluid on the left need to be pushed?

Problem 14.2

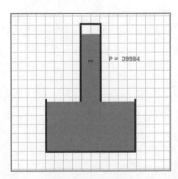

A tube contains a column of mercury while the bottom container of mercury is open to the atmosphere to form a mercury barometer (position is given in tenths of meters and pressure given in pascals). What is the atmospheric pressure?

Problem 14.3

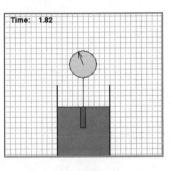

Find the density of the object being immersed in the water bucket. The initial reading on the spring scale is 19 N. One full revolution of the spring scale represents a change of 10 N.

Problem 14.4

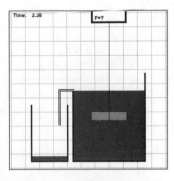

The blue liquid is an oil with $\rho = 850 \ \text{kg/m}^3$ (position is given in centimeters and the dimension of the oil containers into the screen is 20 cm).

a. If the mass is 150 g, what is the tension on the wire at the following times: 0.4 s, 1.5 s, and 4 s?

b. What is the gauge pressure at the top of the mass at $t = 1.5$ s and $t = 4$ s?

c. Some students find the answers to questions (a) and (b) (the tension in the wire and the gauge pressure) inconsistent. Explain why they are consistent with each other.

Note that gauge pressure is the difference in pressure due to the surface of the water (the absolute pressure would be the pressure due to the atmosphere at the surface plus the pressure due to the water).

Problem 14.5

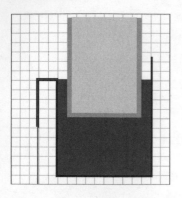

How much more mass can this "boat" sitting in water hold and still float? The dimension of the "boat" into the screen is 8 cm (position is given in centimeters and time is given in seconds).

Problem 14.6

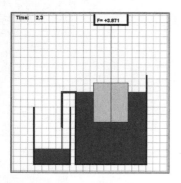

A block is lowered into a liquid as shown (position is given in centimeters, time is given in seconds, and force is given in newtons). The dimension of both containers into the screen is 20 cm. The dimension of the block into the screen is 10 cm. What is the density of the liquid?

Problem 14.7

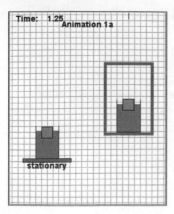

Shown is a block of wood floating in a bucket of water. The bucket is placed in an elevator as shown in the animation (position is in meters and time is in seconds).

a. Elevator in free fall: If the picture on the left represents the orientation of the wood when the elevator is stationary, which animation correctly depicts the new orientation of the wood while the elevator is in free fall as shown (assume a broken cable)?

b. Elevator rising: If the picture on the left represents the orientation of the wood when the elevator is stationary, which animation correctly depicts the new orientation of the wood while the elevator is moving as shown?

Problem 14.8

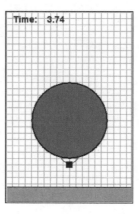

As the air inside a hot-air balloon is heated, the density of the air inside the balloon decreases and the balloon expands (see **Chapter 20** and the Kinetic Theory and Ideal Gas Law Illustrations for a detailed explanation). The animation shows a hot-air balloon ascending with constant acceleration (position is given in meters and time is given in seconds). If the balloon fabric and basket have a combined mass of 300 kg, what is the density of the air inside the balloon? (Neglect the volume of the basket). The density of the air outside the balloon is 1.3 kg/m^3.

Problem 14.9

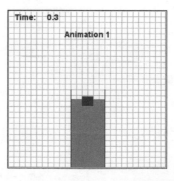

An ice cube melts in a glass of water as shown in the animation (position is given in centimeters and time is given in minutes). Which animation correctly shows what the final water level will be? Explain.

Problem 14.10

The animation is color coded as follows: Blue is water, red is oil, and brown is a wood block initially floating at the interface. A pump, which starts at $t = 1$ s, removes the oil. Which animation is physical? (In other words, which animation obeys the laws of physics?) Explain.

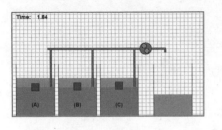

CHAPTER 15

Fluids in Motion

Topics include fluids in motion, continuity equation, and Bernoulli's equation.

Illustration 15.1: The Continuity Equation

A continuity equation is simply a way to express that "what goes in must come out." This simply means that as a fluid goes from a pipe of one diameter to another, the fluid flow changes. Assume an ideal fluid (position is given in meters and pressure is given in pascals). The dark blue in the animation is a section of water as it flows through the pipes from left to right (assume they are cylindrical, that is, the vertical distances in the animation correspond to the diameter of a circular cross section). Notice that as the water enters the narrower pipe, it goes faster. How long does it take the dark blue region to cross a line in the wide region and a line in the narrow region? The volume of the dark blue region divided by this time is the volume flow rate for each region. It should be the same for both regions because whatever goes in (per second) must come out (per second). We express this with the continuity equation Av = constant, where A is the cross-sectional area and v is the speed of the fluid flow (What are the units of Av? They should be volume/time). When we couple this with Bernoulli's equation (conservation of energy),

$$P + 1/2\rho v^2 + \rho g y = \text{constant},$$

where P is the pressure, ρ is the density of the fluid, v is the speed of the fluid flow, and y is the height of the fluid (you can, of course, pick any point to be $y = 0$ m), we find a change in pressure as well. In this case, because the pipe is horizontal, y is the same, so we simply use $P + 1/2\rho v^2$ = constant, so as the speed increases, the pressure decreases. Note the pressure readings by sliding the pressure indicator along the center of the pipe.

Note: The format of the pressure is written in shorthand. For example, atmospheric pressure, 1.01×10^5 Pa, is written as $1.01e + 005$.

Illustration 15.2: Bernoulli's Principle at Work

A reservoir of water has a hole in it at a height that you can adjust (position is given in meters). Notice what happens to the water flow out of the opening. Assume an ideal fluid. How do you describe what happens to the initial speed of the water depending on the height of the hole? The velocity of the water out of the opening can be determined according to Bernoulli's equation, which compares two points in the fluid:

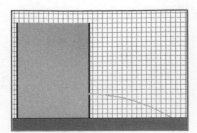

$$P_1 + 1/2\rho v_1^2 + \rho g y_1 = P_2 + 1/2\rho v_2^2 + \rho g y_2 = \text{constant},$$

where P is the pressure, ρ is the density of the fluid, v is the speed of the fluid flow, and y is the vertical position from $y = 0$ m (you can, of course, pick any point to be $y = 0$ m—this is equivalent to picking any spot to be the zero of potential energy, but once you have picked the $y = 0$ m spot you must be consistent).

Consider point 1 to be the top of the fluid and point 2 to be the point where the fluid leaves the hole. Given this assignment, we can easily see that $P_1 = P_2 = P_{atm}$. At the top of the reservoir the water is essentially stationary, so $v_1 = 0$ m/s there. This means that Bernoulli's equation simplifies to

$$\rho g y_1 = 1/2 \rho v_2^2 + \rho g y_2 \quad \text{or that} \quad v_2^2 = 2 g(y_1 - y_2) = 2 g \Delta h,$$

where Δh is the height of the water above the hole (not the height of the opening, although the two are related).

Illustration 15.3: Ideal and Viscous Fluid Flow

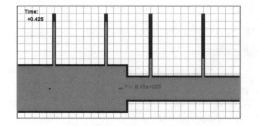

Bernoulli's equation describes the conservation of energy in an ideal fluid system such as the one shown in this animation (position is given in tenths of meters and pressure is given in pascals). The vertical tubes are open to the atmosphere. Notice that the water level is lower to the right, indicating a lower pressure. Why is the pressure lower in the narrower tube? Notice also that the pressure only changes when going from a wider to a narrower tube. The pressure indicator measures the gauge pressure, not the absolute pressure (gauge pressure is pressure above atmospheric pressure). When there is viscosity (that is the fluid sticks together a bit so there is some friction), but still a smooth (laminar) flow, the pressure drops along the length of a pipe. **Try it**. For viscous flow, notice that, to get the same volume per unit time (Av = volume/time, where A is the cross-sectional area and v is the speed of the fluid flow), the pressure drops more in the narrower tube than in the wider tube. The equation governing the flow is Poiseulle's equation, $Av = \pi R^4 \Delta P / 8 \eta L$, where R is the radius of the tube, L is the length of the tube, ΔP is the pressure difference and η is the viscosity of the fluid.

Note: The format of the pressure is written in shorthand. For example, atmospheric pressure, 1.01×10^5 Pa, is written as 1.01e + 005.

Illustration 15.4: Airplane Lift

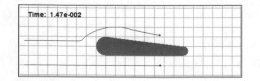

The animation shows the cross section (side view) of a model airplane wing with air moving past the wing (position is given in centimeters and time is given in seconds). Where is the speed of the air greatest? Where will the pressure be higher? How does this explain the lift on an airplane? Using Bernoulli's principle we can find the pressure difference between the top and the bottom wing and it is

$$\Delta P = \rho(v_{above}^2 - v_{below}^2)/2$$

First, find the speed of the air above (once above the wing, the air speed is constant) and below the wing. We can find the average speed easily as the displacement over the time interval, and we find that v_{below} = 950 cm/s = 9.5 m/s and v_{above} = 990 cm/s = 9.9 m/s.

Now we can calculate the pressure difference using our results for the air speed and the density of air, ρ = 1.3 kg/m^3. We find that in this case ΔP = 5 Pa. If the surface area of the wing is 0.1 m^2, what is the net force (lift) on this wing? Since $P = F/A$, we find that the net force will be the pressure difference times the area or 0.5 N.

The reason an air flow pattern develops that yields different speeds on the top and the bottom is that the air flowing around the wing moves into nonideal fluid flow. Initially, since the air on top has farther to travel, the air on the bottom of the wing gets to the back of the wing and moves up to "fill" this space, but this instability causes a turbulent wake that eventually allows a new, more stable, air-flow pattern such as the one shown, where air-particles that travel across the top (the longer distance) go faster. For a greater difference in pressure, the wing is tilted up (the angle of tilt is called the angle of attack), and this increases the lift.

Note: The format of the time is written in shorthand. For example, a time of 6.00 × 10^{-3} s, is written as 6.00e-003.

Exploration 15.1: Blood Flow and the Continuity Equation

Blood flows from left to right in an artery with a partial blockage. A blood platelet is shown moving through the artery. How does the size of the constriction (variable from 1 mm to 8 mm from each wall) affect the speed of the blood flow? Assume an ideal fluid (position is given in millimeters and pressure is given in torr = mm of Hg). We can use the continuity equation and Bernoulli's equation to understand the motion:

$$\text{Continuity: } Av = \text{constant} \quad \text{Bernoulli: } P + 1/2\rho v^2 + \rho gy = \text{constant}$$

With a 2.0-mm constriction,

a. What is the platelet's speed before and after it passes through the constriction?

b. What is the platelet's speed while it passes through the constriction?

Set the constriction to 8.0 mm.

c. Does the speed of the platelet before it reaches the constriction increase, decrease, or not change?

d. With the 8-mm constriction, is the speed of the platelet in the constriction faster, slower, or the same as with the 2-mm constriction?

e. Assume that the blood vessel and the blockage are cylindrical (circular cross-sectional area for both). Measure the radius of the artery and the radius of the flow area where the blockage is. Verify the equation of continuity to compare the 2-mm and 8-mm cases.

Now compare the 2-mm and 8-mm cases.

f. What is the pressure inside of and outside the constriction (use the white box to measure pressure)?

g. Does the pressure decrease or increase in the region where the blockage is?

h. This result, (g), is surprising to many students, so let's figure out why: At the instant the platelet travels from the wide region to the narrower constricted region, what is the direction of acceleration?

i. What, then, is the direction of the force that the platelet feels?

j. What region should have a larger pressure?

k. Do the same analysis for the platelet as it leaves the constricted region and goes back to the unblocked artery (sketch a diagram to show the direction of acceleration and force).

l. Verify that Bernoulli's equation holds inside and outside the constricted region for the 2-mm and 8-mm cases (760 Torr = 760 mm of Hg = 1.013×10^5 Pa). The density of blood is 1050 kg/m^3.

Exploration 15.2: Bernoulli's Equation

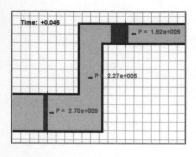

Bernoulli's equation describes the conservation of energy in an ideal fluid system. Assume an ideal fluid (position is given in meters and pressure is given in pascals). The dark blue in the animation is a section of water as it flows into the region marked by the horizontal line and the corresponding water that must move out of the region in the top right. We will explore the connection between Bernoulli's equation and conservation of energy.

Note: The format of the pressure is written in shorthand. For example, atmospheric pressure, 1.01×10^5 Pa, is written as $1.01e + 005$.

The relationship between the speed and dimensions of the water going in compared with the water leaving is governed by the continuity equation (what flows in must flow out unless there is a leak in the pipes!): $Av =$ constant, where A is the cross-sectional area and v is the speed of the liquid. Assume the pipes are cylindrical.

a. What is the volume of both blue regions (should be the same)?

b. What is the speed of the water in the left pipe?

c. What is the cross-sectional area of the left pipe?

d. What is the speed of the water leaving the region (in the right pipe)?

e. What is the cross-sectional area of the right pipe?

f. Does the continuity equation hold?

As the water travels through this pipe system, work must be done on the fluid to raise it up and to increase its speed. The work done must be equal to the change in kinetic plus potential energy.

g. Given the pressure (you can move the red pressure indicators), find the force (from the water behind it) on the lower left dark blue region.

h. What is the work done by that force for the duration of the animation?

i. Similarly, find the force on the upper right dark blue region that opposes the motion.

j. What is the work done by that force for the duration of the animation (note that the displacement and force are in opposite directions, so this is negative work)?

k. What is the net work done, then, during the duration of the animation, on the water in the middle region?

l. Calculate the difference in kinetic energy of the dark blue regions. Note that since the volume is the same, the mass is the same. (The density of water is 1000 kg/m^3.)

m. Calculate the difference in potential energy of the center of mass of the dark blue regions. Does the net work equal the difference in kinetic energy plus the difference in potential energy?

This is all described by Bernoulli's equation.

n. Show that the net work is equal to $(P_{left} - P_{right}) \, Avt$.

o. Show that the net change in kinetic energy is $(1/2) \, \rho \, Avt(v_{right}^2 - v_{left}^2)$.

p. Show that the net change in potential energy is $\rho \, gAvt(y_{right} - y_{left})$.

P is the pressure, ρ is the density of the fluid, v is the speed of the fluid flow, A is the cross-sectional area, t is the time, and y is the height of the fluid. Combining these three terms, we get

$$(P_{left} - P_{right}) = (1/2) \, \rho \, (v_{right}^2 - v_{left}^2) + \rho g(y_{right} - y_{left}),$$

or Bernoulli's equation, $P + 1/2\rho v^2 + \rho g y = $ constant, so that Bernoulli's equation is simply another way to restate conservation of energy.

Exploration 15.3: Application of Bernoulli's Equation

Adjust the height of the water in the reservoir and notice what happens to the water flow out of the opening. Assume an ideal fluid (position is given in meters). We can use Bernoulli's equation (i.e., conservation of energy for fluids) to understand what happens, $P + 1/2\rho v^2 + \rho g y = $ constant, where P is the pressure, ρ is the density of the fluid, v is the speed of the fluid flow, and y is the height of the fluid (you can, of course, pick any point to be $y = 0$ m).

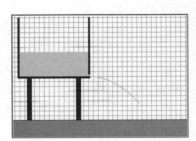

The amount of water leaking out is small during the animation. So the height effectively stays constant during the time this animation is running (to a good approximation).

a. Use Bernoulli's equation to find the pressure at the bottom of the reservoir. Pick a height of water in the reservoir. The pressure above the water is atmospheric pressure (1.01×10^5 Pa). What is the pressure of the water at the bottom of the reservoir? (Note that for both of these cases, $v = 0$ m/s).

b. Use Bernoulli's equation at the bottom of the reservoir to find the speed of water flow out of the reservoir. Equate Bernoulli's equation somewhere in the middle of the bottom of the reservoir (where $v = 0$ m/s) to the water flowing out of the opening (where P is atmospheric pressure) and note that the value of y is the same for both cases.

c. Using this value for the initial x velocity of the water, calculate where the water will land and verify that it does land at the proper spot. Repeat this procedure for another value of the reservoir height.

Problems

Problem 15.1

Blood flows in an artery with a partial blockage as shown in the animation (position is given in centimeters and time is given in seconds). Assume the blood can be treated as an ideal fluid. A blood platelet is shown moving through the artery. Which of the animations properly represents the motion of the platelet as it moves through and past the blockage? Explain.

Problem 15.2

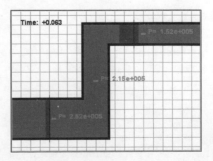

Assume an ideal fluid (position is given in meters and pressure is given in pascals). The dark brown in the animation represents a section of liquid as it flows into a region marked by the horizontal line and the corresponding water that must move out of the region in the top right. What is the density of the liquid? *Note: The format of the pressure is written in shorthand. For example, atmospheric pressure, 1.01×10^5 Pa, is written as 1.01e + 005.*

Problem 15.3

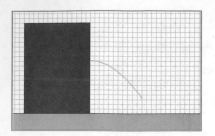

A wooden tank of water whose top is open to the atmosphere is shown (position is given in meters). Assume an ideal fluid. What is the water level in the tank?

Problem 15.4

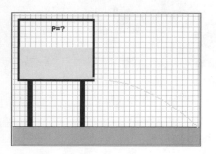

A tank of water is under pressure. What is the pressure at the top of the tank? Assume an ideal fluid (position is given in meters).

Problem 15.5

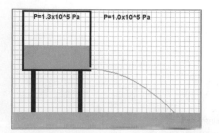

What is the density of the fluid in this reservoir? Assume an ideal fluid (position is given in meters and pressure is given in pascals).

Problem 15.6

Assume an ideal fluid (position is given in meters and pressure is given in pascals). The dark blue in the animation is a section of water (density 1000 kg/m³) as it flows through the pipes (assume they are cylindrical; that is, the vertical distances in the animation correspond to the diameter of the circular cross section). The pressure indicator can slide along the center of the pipes.

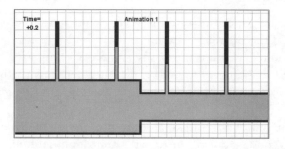

a. What will the length of the dark blue region be in the narrowest tube?

b. How fast will it go in the narrowest tube?

c. What is the pressure in that tube?

Note: The format of the pressure is written in shorthand. For example, atmospheric pressure, 1.01×10^5 Pa, is written as 1.01e + 005.

Problem 15.7

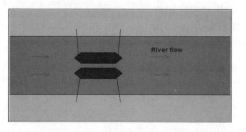

Which of the animations, if any, depicts a possible physical situation for ideal fluid flow? Explain what is wrong with the animations that are not physically possible. Assume an ideal fluid in each case (position is given in tenths of meters). The vertical tubes are open to the atmosphere. Assume all tubes are cylindrical.

Problem 15.8

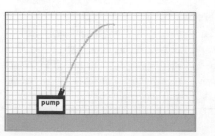

The animation shows an overhead view of two boats loosely moored to the banks of a river (position is given in meters).

a. Explain why the boats move together as seen in the animation.

b. If instead of a left-to-right flow of the river, the river water flowed from right to left, how would the animation change?

Problem 15.9

What is the gauge pressure of the pump of this water fountain in order for it to pump the water as shown? Treat the water as an ideal fluid. Assume the exit of the pump is where the water leaves the fountain (position is given in centimeters). The density of water is 1000 kg/m^3.

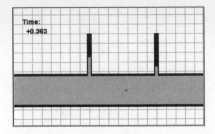

Problem 15.10

What is the viscosity of the fluid? The density of the fluid is 900 kg/m^3 (position is given in tenths of meters). The vertical tubes are open to the atmosphere. Assume the pipes are cylindrical.

Periodic Motion

Topics include simple harmonic motion, springs, pendula, Fourier series, damped and driven motion, and resonance.

Illustration 16.1: Representations of Simple Harmonic Motion

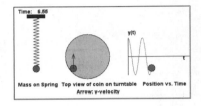

In 1610 Galileo discovered four moons of Jupiter. Each moon seemed to move back and forth in what we would call simple harmonic motion. What was Galileo really seeing? He was seeing the essentially uniform circular motion of each moon, but he was looking at the motion edge-on. We can use what Galileo was experiencing to hand-wave some properties of simple harmonic motion by using an analogy with uniform circular motion. Consider the above animation (position is given in meters and time is given in seconds).

First, let us look at position as a function of time. The point on the circle marked by the red ball is always at the same radius, R. If we look at the y position as a function of time we see that $y = R \cos(\omega t)$, and the x position is $x = R \sin(\omega t)$. How do we know this? We can decompose the radius vector into components.

What about the velocity? Well, we know it is tangent to the path of the ball and, since the motion is uniform, the magnitude of the velocity is constant and equal to ωR. We can break up the velocity vector into components. We find that $v_y = -\omega R \sin(\omega t)$ and $v_x = \omega R \cos(\omega t)$ and that both are functions of time. Watch the animation to convince yourself that this decomposition is correct as a function of time. If we know a bit of calculus, we can take the derivative of the position with respect to time. We again find that $v_y = -\omega R \sin(\omega t)$ and $v_x = \omega R \cos(\omega t)$.

We also know that the acceleration is a constant, v^2/R, and points toward the center of the circle. We can again decompose this acceleration as $a_y = -\omega^2 R \cos(\omega t)$ and $a_x = -\omega^2 R \sin(\omega t)$. Again, if we know a bit of calculus, we can take the derivative of the velocity with respect to time. We again find that $a_y = -\omega^2 R \cos(\omega t)$ and $a_x = -\omega^2 R \sin(\omega t)$. Note that since this is simple harmonic motion, there must be a relationship between the position and the force. Since force must be a linear restoring force and, since force is also mass times acceleration, we must have that $ma = -k\,x$ or that $a(t) = -(k/m)$ and $x(t) = -\omega^2 x(t)$, which is the case if we compare our functions for $y(t)$ and $x(t)$ to $a_y(t)$ and $a_x(t)$.

For simple harmonic motion we change two things, $R \rightarrow A$ where A is called the amplitude, and we only consider one direction, in this example the y direction. This yields $y = A \cos(\omega t)$, $v = -\omega A \sin(\omega t)$, and $a = -\omega^2 A \cos(\omega t)$. Simple harmonic motion requires a linear restoring force, an equilibrium position, and a displacement from equilibrium.

Illustration 16.2: The Simple Pendulum and Spring Motion

When we think about simple harmonic motion we think about a mass on a spring. This is the prototype motion and is the easiest to deal with since k, the spring constant, is the proportionality factor between F and $-x$. However, there is another standard example of simple harmonic motion that is all around us, that of pendulum motion. A pendulum is nothing more than a heavy object (the pendulum bob) hanging from a very light string (if the string's mass is large enough, we have a compound pendulum and the string must be considered). Consider **Animation 1**. Here the length of the string is 15 m and the mass of the pendulum bob is 1 kg (position is given in meters, angle is given in radians, and time is given in seconds). When we analyze the forces acting on the pendulum bob (drag the pendulum bob from its equilibrium position and press "play"), we find that the **force of gravity** and the **force of tension** act. Since the string does not stretch, the part of the gravitational force opposite to the tension (the components of the weight are shown in light green) must cancel the force of the tension in the string. This leaves a net force perpendicular to the tension and parallel to the path of the pendulum. Show **Animation 1 with pendulum bob path**. When we do the calculation, we find that the net force on the pendulum bob goes like

$$F_{net} = -mg\sin(\theta),$$

which at first glance does not look at all like simple harmonic motion. But what happens when the angle θ is small? Well, $\sin(\theta) \approx \theta$ for small enough θ; therefore, $F_{net\ small\ angles} = -mg\,\theta$.

Drag the pendulum bob to a large angle and see how the two net forces (any angle vs. small angle) deviate at large angles. The motion of the pendulum is shown according to the actual force, $F_{net} = -mg\sin(\theta)$, and not the small angle approximation, $F_{net} = -mg\,\theta$, although both are shown on the graph. Therefore the period of the pendulum is the actual period. *When you get a good-looking graph, right-click on it to clone the graph and resize it for a better view.*

Since we are using radians, $x = \theta L$, and the net force for small angles can be written as $F_{net\ small\ angles} = -(mg/L)\,x$, where the proportionality factor between F and $-x$ is now mg/L. For small enough angles (when $\sin(\theta) \approx \theta$) we have simple harmonic motion.

Now consider both the motion of a pendulum and the motion of a mass attached to a spring by looking at **Animation 2**. In this animation the pendulum is the same as Animation 1 (the net force on the bob is shown as a green arrow), the spring has a spring constant of 1.30666 N/m, and the mass of the red ball attached to the spring is 2 kg (the net force on the red ball is represented by the blue arrow). It may seem strange that we have chosen such an oddly precise value for the spring constant. Drag the pendulum to about 0.15 radians and drag the mass on the spring to some initial amplitude (it does not matter what this values is, but for simplicity choose 2.3 m) and play the animation. What do you notice about the graph? Do you see why the spring constant was carefully chosen? These values were chosen to tune the motion of the two systems to be the same:

$$\omega_{mass-spring} = (k/m)^{0.5} = \omega_{pendulam}(k_{effective}/m)^{0.5} = (g/L)^{0.5}.$$

Now reset this animation and drag the pendulum bob to 0.75 radians and the mass on the spring to 10.3 m and play the animation. What happens now? By looking at **Animation 1**, can you say why this is? Notice as time goes on that the two motions now deviate from each other. Large-amplitude pendulum motion is no longer simple harmonic motion.

Illustration 16.3: Energy and Simple Harmonic Motion

In this Illustration we shall look at energy and simple harmonic motion of both a pendulum and a mass on a spring. We shall consider small-amplitude motion for the pendulum since this will yield simple harmonic motion (see **Illustration 16.2** for details). In addition, like **Illustration 16.2**, we have chosen the mass of the pendulum bob to be 1 kg and the length of the pendulum to be 15 m, while choosing the mass

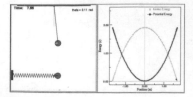

of the ball on the spring to be 2 kg and the spring constant to be 1.30666 N/m (position is given in meters, angle is given in radians, and time is given in seconds). These values tune the motion of the two systems to be the same:

$$\omega_{\text{mass-spring}} = (k/m)^{0.5} = \omega_{\text{pendulum}} = (k_{\text{effective}}/m)^{0.5} = (g/L)^{0.5}.$$

In the following animations we will show graphs of the kinetic and potential energy of the mass-spring system but will not show the kinetic and potential energy of the pendulum. However, the kinetic and potential energy of the pendulum will look the same with *exactly half* the kinetic and potential energy (and therefore half the total energy) of the mass-spring system. Why half? For the mass-spring system, kinetic energy is $(1/2\ mv^2)$ and the potential energy is $(1/2\ kx^2)$, and for the pendulum the kinetic energy of the bob is $(1/2\ mv^2)$, and the potential energy is $(1/2\ k_{\text{effective}}x^2)$. In this Illustration, since the mass on the spring is twice the mass of the pendulum bob, the mass-spring system will always have twice as much kinetic energy as the pendulum bob. Since the spring constant for the mass-spring system is twice the effective spring constant for the pendulum ($k_{\text{effective}} = m_{\text{pendulum}}g/L = 0.6533$ N/m), the mass-spring system will always have twice as much potential energy as the pendulum bob.

When you get a good-looking graph, right-click on it to clone the graph and resize it for a better view.

Consider **Animation 1**, which shows the graph of kinetic and potential energy vs. position. What can you say about the total energy of the system? It is a constant and about 1.89 J. The energy starts out all potential and at the equilibrium position the energy is all kinetic. At maximum compression the energy is all potential again. Given that the total energy is kinetic plus potential, we have that

$$E = 0.5\ mv^2 + 0.5\ kx^2 = 0.5\ kx_{\text{max}}^2 = 0.5\ mv_{\text{max}}^2 .$$

Now consider **Animation 2**, which shows the graph of kinetic and potential energy vs. time. Notice how the two graphs are different in their functional form.

The graphs in **Animation 1** have the form of $0.5\ kx^2$ (the potential energy) and the form of $A - 0.5\ kx^2$ (the kinetic energy), where A is a constant, the total energy. In this animation the total energy is 1.89 J. The form of the kinetic energy can be understood from the energy function shown above. The potential energy is $0.5\ kx^2$, which is proportional to x^2. The kinetic energy can be written in terms of the total energy and the potential energy as $E - 0.5\ kx^2$.

The graphs in **Animation 2** have the form of $\cos^2$ (the potential energy) and the form of $\sin^2$ (the kinetic energy) since both trigonometric functions are a function of time. Why? We know from simple harmonic motion that if the object is initially displaced from equilibrium with no initial velocity that

$$x = x_0 \cos(\omega t) \quad \text{and} \quad v = -\omega x_0 \sin(\omega t).$$

Given the form of the kinetic energy and the potential energy, we have that

$$KE(t) = 0.5\ kx_0^2\ \sin^2(\omega t) \quad \text{and} \quad PE(t) = 0.5\ kx_0^2\ \cos^2(\omega t),$$

where we used $\omega^2 = k/m$ to simplify the kinetic energy. Therefore, the total energy will always add up to $0.5\ kx_0^2 = 1.89$ J.

Illustration 16.4: Forced and Damped Motion

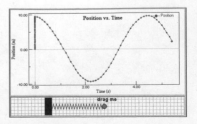

A 1-kg mass on a spring is shown (position is given in meters and time is given in seconds) initially at its equilibrium position. Various parameters related to the spring and the initial conditions of the motion are also given. Once a variable or the velocity box is changed, you must reinitialize your choice by clicking the "set values, then drag the ball" button. Once you have clicked this button, drag the ball to the initial position you desire (by default it starts at its equilibrium position) and then press "play."

When you get a good-looking graph, right-click on it to clone the graph and resize it for a better view.

We have thus far considered ideal motion of a mass on a spring: a perfect spring obeying Hooke's law and no additional varying force or damping. This Illustration will discuss what happens to a mass on a spring subject to a varying additional force and/or a damping force. Specifically, the damping force is $-bv$ and the driving force is $F_0 \cos(\omega t)$.

First, what is the natural frequency of oscillation of the mass? Look at the animation with no additional forces or damping. Drag the ball to 3 m and let go. Pause the animation and measure the period (about 4.45 seconds from peak to peak). The frequency is one over this, or 0.225 Hz. The angular frequency is $2\pi f$ or 1.41 rad/sec. Since the angular frequency squared (here 2) is equal to the ratio of k/m, we know that $k = 2$ N/m.

What happens to the motion of the mass when a driving force is turned on? Try it and find out. Vary the angular frequency of the driving force. What happens when the angular frequency of the oscillation is close to or far from that of the driving force? How sensitive is the motion to this parameter? When the natural and driving frequencies are the same, it is called *resonance*.

There are three types of damped motion you should also investigate:

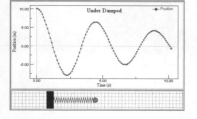

- **Under Damped**: the damping is so small there are many oscillations before motion is stopped.
- **Over Damped**: the damping is rather large; the motion takes a long time to get back to equilibrium.
- **Critically Damped**: a special case in which the time to get back to equilibrium is minimized.

Illustration 16.5: The Fourier Series, Qualitative Features

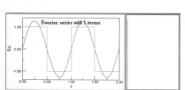

We have thus far only looked at simple periodic motion that can be described by a single sine or cosine. This may seem like a horrible mistake. Most periodic functions are tremendously complicated. Have we been doing something wrong by focusing on only sines and cosines? Well, actually not. ANY periodic function can be represented as a sum of sines or cosines! We find that sometimes we may need an infinite number of them, but nonetheless we are able to describe any periodic phenomena, no matter how complicated, in this way.

Consider a sawtooth (position is given in meters) function that is periodic with $L = 1$ (it is shown over two periods since it is easier to see the function this way). In the animation the amplitude is a function of x, but it could have been a function of time. Select "play the Fourier series of the sawtooth." The gray function is the actual sawtooth, while the red function is the total approximate sawtooth from a Fourier series (if you did not change n, the animation shows the $n = 1$ term only). Change n, the number of sine functions that will be added together to approximate the sawtooth, and see how the red function changes. The green sine function is the current term that is added to get the total red function. On the right is a representation of the relative amount of each sine function as it is added to the total. You may add as many as 35 terms. Also note that at the point where the sawtooth kinks, there is always overshoot (this is called the Gibbs phenomenon).

Now look at the square wave. It turns out that the $n = 2, 4, 6, \ldots$ terms do not contribute to the sum. Verify this by watching the animation for $n = 35$.

When you get a good-looking graph, right-click on it to clone the graph and resize it for a better view.

Illustration 16.6: The Fourier Series, Quantitative Features

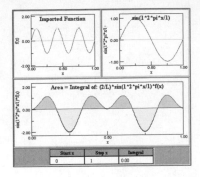

Fourier's theorem states that any periodic function (whether periodic in position or time) can be represented as a sum of sine waves. We find that sometimes we may need an infinite number of sines, but nonetheless we may describe any periodic phenomena this way. In this animation we investigate odd periodic functions in position with Fourier's theorem.

ANY odd periodic function of x (a period of L between 0 and L as opposed to between $-L/2$ to $L/2$) can be described in terms of a Fourier series as:

$$f(x) = \Sigma \, A_n \sin(n2\pi x/L),$$

where in this animation $L = 1$. A_n is the result of an integral that represents the overlap between the original function and a particular Fourier component (one term in the Fourier series represented by the integer n). In order to get this to exactly work out, there must be a $2/L$ (in our case just a factor of 2 since $L = 1$ here) included in the integral. Verify that this is necessary by predicting A_3 for the function $\sin(3 * 2 * \mathrm{pi} * x)$ and then use the animation as a check.

Remember to use the proper syntax, such as $-10 + 0.5 * t$, $-10 + 0.5 * t * t$, *and* $-10 + 0.5 * t\char`^2$. *Revisit* **Exploration 1.3** *to refresh your memory.*

Try various odd functions to see the result of the integral, A_n. Consider the following functions (you may copy and paste them in directly):

the Sawtooth Wave in Illustration 16.5	the Square Wave in Illustration 16.5
$(1 - 2 * x) * \mathrm{step}(1 - x) + (1 - 2 * (x - 1)) * \mathrm{step}(x - 1)$	$\mathrm{step}(0.5 - x) - \mathrm{step}(x - 0.5) * \mathrm{step}(1 - x) + \mathrm{step}(x - 1)$

Exploration 16.1: Spring and Pendulum Motion

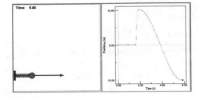

The animations depict the motion of a mass on a spring and a pendulum, respectively.

a. For Animation 1, plot the period of the motion vs. amplitude. Drag the ball from equilibrium, varying the amplitude from 1 m to 10 m in 1-m steps.

b. For Animation 2, plot the period of the motion vs. amplitude. Drag the ball from equilibrium, varying the amplitude from 0.1 radian to 1.0 radian in 0.1-radian steps.

c. What can you say about the period's dependence on amplitude for each animation?

Exploration 16.2: Pendulum Motion and Energy

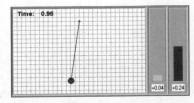

A 4-kg mass is a pendulum bob and undergoes periodic motion (position is given in meters and time is given in seconds). Also shown are bar graphs representing the **kinetic** and **gravitational potential** energies in joules.

Use the animation and the bar graphs to guide your answers to the following questions.

a. What is the period of the motion?

b. What is the amplitude of the motion (in radians)?

c. From the motion and the bar chart, how do you know energy is conserved?

d. Is this simple harmonic motion?

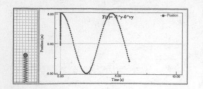

Exploration 16.3: Simple Harmonic Motion With and Without Damping

Enter a value for the damping coefficient, the spring constant of the restoring force, or check the "show velocity" box, then press the "set parameters, then drag the ball" button. When you have done this, drag the ball into position and press "play" to run the animation (position is given in meters and time is given in seconds). *When you get a good-looking graph, right-click on it to clone the graph and resize it for a better view.*

a. Find the mass of the ball by using your knowledge of simple harmonic motion.

b. Enable the velocity graph. Does the velocity lead or lag the position graph during simple harmonic motion?

c. How do the frequencies compare if the restoring force is $-2 * y$, $-4 * y$, and $-8 * y$ N/m? You may right-click on the graph to create a copy at any time.

Now focus on the damping coefficient and how it affects the motion.

d. Set the restoring force to $-2 * y$ and the initial displacement from equilibrium to 5 m. Vary b from 0 to 2 N·s/m in steps of 0.25 N·s/m. What can you say about the frequency of motion as a function of b?

Exploration 16.4: Pendulum Motion, Forces, and Phase Space

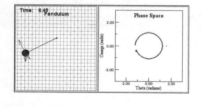

A 1-kg pendulum bob is attached to a 9.8-m massless string to form a pendulum (position is given in meters and time is given in seconds). A graph of angular velocity **(rad/s)** vs. angle **(rad)** is shown. This graph is sometimes called a "phase space" representation of the motion. In addition,

- the red arrow represents the total force
- the blue arrow represents the force of gravity
- the green arrow represents the velocity

The phase-space representation of motion is just another way to describe an object's motion (like a position vs. time graph). For example, when would the phase-space representation of the motion be circular? Well, x and v would have to have the same frequency, be out of phase with each other by $\pi/2$ radians (or 90°), and x_{max} and v_{max} would have to have the same magnitude. This occurs with simple harmonic motion when $\omega = 1$ rad/s.

You must first select the "drag pendulum" button, drag the pendulum bob into place, and then press "play" to begin the animation for a different initial angle.

a. Given the information above and the information depicted in the animation, when does the pendulum approximate simple harmonic motion?

b. Determine the maximum angle for approximate simple harmonic motion from the animation.

c. We have considered a special case of simple harmonic motion, $\omega = 1$ rad/s. What would the phase-space diagram look like for simple harmonic motion with a general ω?

Exploration 16.5: Driven Motion and Resonance

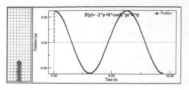

Enter a value for the magnitude of the driving force and its frequency, the spring constant of the restoring force, or check the "show velocity box," then press the "set parameters, then drag the ball" button. When you have done this, drag the ball into position and press "play" to run the animation (position is given in meters and time is given in seconds). *When you get a good looking-graph, right-click on it to clone the graph and resize it for a better view.*

a. Find the mass of the ball by using your knowledge of simple harmonic motion.

b. Enable the velocity graph. Does the velocity lead or lag the position graph during simple harmonic motion?

Set the restoring force to $-2 * y$ and the initial displacement from equilibrium to 0 m. Also set the magnitude of the driving force to -1 N. Vary the driving frequency between 0.10 Hz and 0.20 Hz in 0.01-Hz steps. You may want each animation to run a while before determining the maximum amplitude.

c. Draw a graph of the maximum amplitude of motion as a function of the frequency.

d. What frequency gives the maximum amplitude?

Note that the mass is not allowed to oscillate past about 22 m.

Exploration 16.6: Damped and Forced Motion

A mass can be driven by an external force in addition to an internal restoring force and friction. Specifically, $F_{net} = F_{restore} + F_{friction} + F_{driving}$, where the default values are

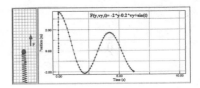

$$F_{restore} = -2 * y, \quad F_{friction} = -0.2 * vy, \quad \text{and} \quad F_{driving} = \sin(t).$$

You can change these default values as you see fit. *Remember to use the proper syntax, such as* $-10 + 0.5 * t, -10 + 0.5 * t * t,$ *and* $-10 + 0.5 * t\verb|^|2.$ *Revisit* **Exploration 1.3** *to refresh your memory.*

a. Find the mass. Hint: consider a linear restoring force.

b. Change the restoring force to $-y - 0.1 * y * y$. Is the motion periodic? Is it harmonic? What about $-y - 2.0$?

c. Design your own force that produces periodic, but not necessarily harmonic, motion.

d. Drive the mass at resonance and explain the behavior of the position graph. How does the behavior change with and without friction?

Drive the system (use a linear restoring force of $-1 * y$ and initially no friction) with a function that switches a constant force on and off. This can be achieved with the step function: $step(\sin(t/4))$. The step function is zero if the argument is negative and one if the argument is positive. The given function, $step(\sin(t/4))$, will therefore produce a square wave with amplitude of one and an angular frequency of one quarter. Note that the total force you should use is $-1 * y + step(\sin(t/4))$. Start the mass in its original position; do not drag it.

e. Draw a graph of the force vs. time superimposed on the position vs. time graph.

f. Why does the system oscillate, stop, and oscillate again?

g. Does this behavior occur at any other frequencies? For example, notice that the function $step(\sin(t/4.5))$ produces qualitatively different behavior. Why is this?

Note that the mass is not allowed to oscillate past about 22 m.

Exploration 16.7: A Chain of Oscillators

Twenty-nine damped harmonic oscillators are driven by an external force, $\sin(t)$. Each oscillator can be thought of as a mass connected to the floor with a spring. The masses are not connected to each other in any way. One spring has been shown for demonstration purposes.

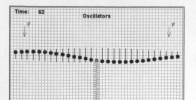

The center oscillator, shown in red, is in resonance with the external force. It has a natural frequency of oscillation of $\omega = 1$ rad/s. Oscillators to the left have a spring with a lower spring constant, while those on the right have a larger spring constant. This animation shows how this collection of oscillators responds to the driving force.

The animation starts with all oscillators at rest. The oscillators then begin to move up and down in phase with the driving force during the first few cycles. This motion is, however, transient, and differing amplitudes and phases soon manifest themselves. Since oscillators to the right of the center have a higher resonance frequency, they begin to lead the driving force, while those to the left of center begin to lag. Although the above oscillators are not connected, this phase shift gives the appearance of a traveling wave. After a few hundred oscillations the transient behavior has dissipated, and a resonance curve appears since the amplitude and phase of each oscillator approach their steady state behavior.

a. Find an example of a resonance curve (amplitude vs. frequency) in your textbook. How does the motion of the masses relate to the resonance curve you found in your book? Hint: Look at both amplitude and phase.

b. What effect does the damping coefficient have on the motion of the masses?

c. Assume that the mass of each ball is 1 kg and that the spring constant for the center spring is 1 N/m. By how much does the spring constant change between neighboring springs?

Problems

Problem 16.1

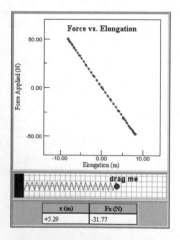

The spring can be stretched by click-dragging the blue ball as shown in the animation (position is given in meters and time is given in seconds). Once you have dragged the blue ball into position, click the "play" button to show the motion of the blue ball.

a. Over what range of compression and stretching is Hooke's law valid?

b. Find the elastic limit of the spring.

c. Determine the spring constant of the spring.

d. Determine the mass of the blue ball.

e. Over what range of compression and stretching is the motion of the spring simple harmonic?

Problem 16.2

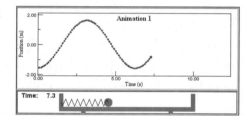

A ball on an air track is attached to a compressed spring as shown in the animation (position is given in meters and time is given in seconds).

a. Determine which graph properly shows the position of the ball as a function of time.

b. Determine the frequency and period of the motion.

c. Write down the equation for $x(t)$.

d. If the mass of the ball is 2 kg, what is the spring constant?

Problem 16.3

A 1-kg ball on an air track is attached to a compressed spring as shown in the animation (position is given in meters and time is given in seconds).

a. Determine which graph properly shows the velocity of the ball in the x direction as a function of time.

b. Write down the equation for $v_x(t)$.

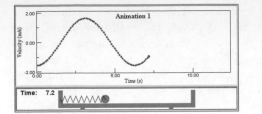

c. What is the total mechanical energy of the system?

Problem 16.4

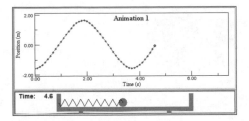

A ball on an air track is attached to a compressed spring as shown in the animations (position is given in meters and time is given in seconds). Each of the five graphs **CORRECTLY** shows a different property of the motion of the ball. Determine whether the red ball undergoes simple harmonic motion, and state which graph(s) tell you this.

Problem 16.5

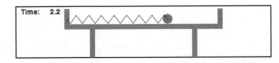

A 500-gram red ball on an air track is attached to a compressed spring (at $x = 0$ m the spring is unstretched) as shown in the animation (position is given in meters and time is given in seconds). Determine the spring constant of the spring (assume $v = 0$ m/s at the beginning and end of the animation).

Problem 16.6

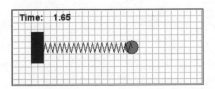

A 200-gram mass is vibrating at the end of a spring as shown (position is given in centimeters and time is given in seconds).

a. What is the spring constant?

b. What is the total mechanical energy of the system?

c. What is the maximum velocity of the ball?

Problem 16.7

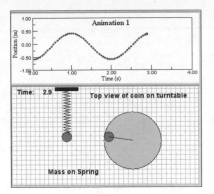

The animation shows the analogy between circular motion (coin on a turntable) and simple harmonic motion (hanging mass on a spring). Given the animation (position is given in meters and time is given in seconds), which graph properly denotes position vs. time for a horizontal spring synchronized with the turntable?

Problem 16.8

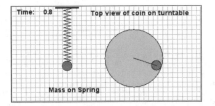

The animation shows the analogy between circular motion (coin on turntable) and simple harmonic motion (hanging mass on a spring). Given the above animation (position is given in meters and time is given in seconds), what is the maximum speed of the hanging mass?

Problem 16.9

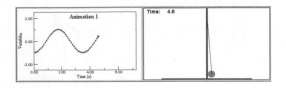

A ball on a string oscillates as shown in the animation (position is given in meters and time is given in seconds).

a. Determine which graph properly shows the position of the ball in the x direction as a function of time.

b. Determine which graph properly shows the velocity of the ball in the x direction as a function of time.

c. Determine which graph properly shows the acceleration of the ball in the x direction as a function of time.

Take data from the graph and answer the following:

d. Write down the equation for $x(t)$.

e. Write down the equation for $v(t)$.

f. Write down the equation for $a(t)$.

g. Write down the equation for $v(x)$.

Problem 16.10

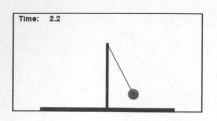

A pendulum is allowed to oscillate in an accelerating elevator as shown in the animation (position is given in meters and time is given in seconds). Determine the effective acceleration due to gravity by analyzing the motion.

Problem 16.11

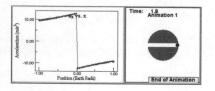

A very wealthy individual proposes to dig a hole through the center of Earth and run a train (the small black circle) from one side of Earth to the other, as shown in the animation (position is given in Earth radii and time is given in seconds). Which of the animations correctly depicts the motion of the train? Ignore frictional effects and treat Earth as a uniform mass distribution.

Problem 16.12

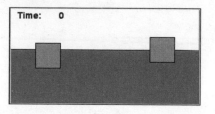

Two identical cubes $(l = 10 \text{ cm})$ are floating in water $(\rho = 1000 \text{ kg/m}^3)$. The one on the left is in equilibrium. The one on the right is initially displaced from equilibrium.

a. What is the condition for equilibrium?

b. For the oscillating cube, what is the net force acting on the cube?

c. For the oscillating cube, what is the period of oscillation?

d. Determine the mass of the cubes.

Waves

Topics include transverse and longitudinal waves, waves on a string, superposition, resonance, and group and phase velocity.

Illustration 17.1: Wave Types

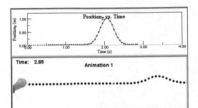

The four animations represent a particle description of three waves on a string and a wave on a spring (position is given in meters and time is given in seconds). For the waves on a string the motion of the red circle is shown as a function of time.

Animation 1 and **Animation 2** depict transverse waves (**Animation 1** shows a wave pulse and **Animation 2** shows the creation of a sinusoidal traveling wave). The waving is in the y direction, while the wave propagation (the direction of the wave velocity) is in the x direction. If you have ever done "the wave" at a football or a basketball game you have been a part of a transverse wave! ("The wave" is a special example of a traveling wave called a pulse, in that every part of the medium that supports the wave does not always wave.) Note that the individual particles that make up the string go up and down, yet do not move in the x direction (just as during the wave you just stand up and then sit down). Watch the red particle in each animation and also view the graph showing the red particle's motion in the y direction.

Animation 3 represents a longitudinal wave. An example of a longitudinal wave is sound. In a longitudinal wave, the waving of the medium (here the string particles) is in the direction of the propagation of the wave. Watch the red particle in this animation and also view the graph showing the red particle's motion in the x direction. Notice that it oscillates back and forth instead of up and down. **Animation 4** represents a wave on a spring. Is it a transverse or longitudinal wave? It is both! Can you tell why this is so?

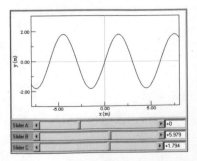

In **Animation 5** water waves are depicted by showing the individual motion of the water molecules (position is given in meters and time is given in seconds). What type of wave is depicted by the animation?

Illustration 17.2: Wave Functions

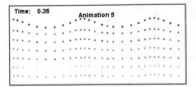

A traveling wave is shown in black at time $t = 0$ seconds (position is given in meters). Three sliders are given that change certain properties of the wave. In general, we would write the wave function for a right-moving wave as

$$y(x, t) = A \cos(k x - \omega t + \varphi) = A \cos((2\pi/\lambda) x - (2\pi/f) t + \varphi).$$

However, we are looking at the wave at $t = 0$ and we cannot determine the wave speed or frequency (where $v = \lambda f = \omega/k$), so we just have that

$$y(x, t) = A \cos(k x + \varphi) = A \cos((2\pi/\lambda) x + \varphi).$$

Which slider changes which quality of the wave? Well, there are three sliders and three parameters in the wave function. Try each slider and see what happens. Slider A controls the phase shift, φ, since it shifts the function to the left or right. Slider B controls the wavelength of the wave and therefore the wave number k, since $k = 2\pi/\lambda$. Clearly Slider C controls the amplitude, A, of the wave function.

If what was discussed above has made sense, you should be able to identify the wave parameters (find the value of the phase shift, wavelength and amplitude) using the sliders for **this wave function** (shown in red).

Illustration 17.3: Superposition of Pulses

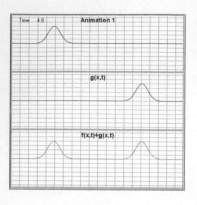

One of the most interesting phenomena we can explore is that of a superposition of waves. In this Illustration we consider a superposition of two traveling pulses, while in **Illustration 17.4** and **Illustration 17.7** we consider the superposition of two traveling waves. (In **Illustration 16.5** and **Illustration 16.6** we considered the addition of multiple periodic functions in a Fourier series).

A superposition of two waves is nothing more than the arithmetic sum of the amplitudes of the two underlying waves. We can represent the amplitude of a transverse wave by a wave function, $y(x, t)$. Notice that the amplitude, the value y, is a function of position on the x axis and the time. If we have two waves moving in the same medium, we call them $y_1(x, t)$ and $y_2(x, t)$, or in the case of this animation, $f(x, t)$ and $g(x, t)$. Their superposition, arithmetic sum, is written as $f(x, t) + g(x, t)$.

This may seem like a complicated process, so we often focus on the amplitude at one point on the x axis, say $x = 0$ m (position is given in meters and time is given in seconds). So now let's consider **Animation 1**, which represents waves traveling on a string. The top panel represents the right-moving Gaussian pulse $f(x, t)$, the middle panel represents $g(x, t)$, the left-moving Gaussian pulse, and the bottom panel represents what you would actually see: the superposition of $f(x, t)$ and $g(x, t)$. As you play the animation, focus on $x = 0$ m. Until the tail of each wave arrives at $x = 0$ m, the amplitude there is zero. Watch what happens during the time that the two waves overlap. They add together in the way you would expect. As time goes on, the waves "separate" and move along the string as if they had not "run into" each other.

What does the superposition in **Animation 2** look like at $t = 10$ s? The two waves add together and exactly cancel there. As time goes on, the waves "reappear" (they were always there) and move along the string as if they had not "run into" each other.

Illustration 17.4: Superposition of Traveling Waves

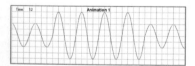

In **Illustration 17.3** we considered the superposition of two traveling pulses. In this Illustration we consider a superposition of two traveling sinusoidal waves. (In **Illustration 16.5** and **Illustration 16.6** we consider the addition of multiple periodic functions in a Fourier series).

Let's begin by considering **Animation 1**, which represents two waves traveling on a string (position is given in meters and time is given in seconds). As you play the animation, focus on $x = 0$ m. Until each wave arrives at $x = 0$ m, the amplitude there is zero. Watch what happens during the time that the two waves overlap, $t \geq 8$ s. They add together in the way you would expect. Given that the two waves always have the opposite amplitude at $x = 0$ m, the superposition of the two traveling waves at $x = 0$ m will always be zero. This point that never moves is called a node. The resulting wave is called a standing wave. It is created when we have two identical waves traveling in opposite directions in a particular medium (here the medium is a string, but we can set up standing waves in air as well).

What does the superposition in **Animation 2** look like at $t \geq 8$ s? The two waves add together and exactly cancel at $x = 0$ m. As time goes on, the waves "reappear" (they were always there) and move along the string as if they had not "run into" each other. Given that the two waves always have the same amplitude at $x = 0$ m, the superposition of the two traveling waves at $x = 0$ m will always be changing. This point is called an anti-node. The resulting wave is still a standing wave. Note that it is shifted in comparison to **Animation 1**.

In **Animation 3** we have a traveling wave that is incident on a wall located at $x = 15$ m. The wave travels and is then reflected by the wall. By reflected we mean that the direction of propagation of the wave changes (the right-moving wave is now a left-moving wave), and its amplitude is now the negative of what it was before. So we now have a right-moving wave and a left-moving wave that resemble the superposition shown in **Animation 1**. In that animation the node was at $x = 0$ m; here in **Animation 3** the node is at $x = 15$ m.

Illustration 17.5: Resonant Behavior on a String

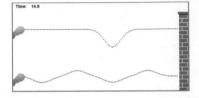

Thus far we have considered either a traveling wave or a traveling pulse. The wave traveled off to infinity unencumbered by any barrier. Here we consider a pulse on two strings, but the strings are pulsed multiple times. To complicate matters the two pulse frequencies are different. Run the animation and consider the results (position is given in meters and time is given in seconds).

How can we understand what is happening? First we notice that the pulse is reflected at the wall. Second we notice the effect of good and bad timing. Which animation has the good timing and which one the bad timing?

In the bottom animation the timing is awful! The waves add up in a way that does not yield a maximum wave amplitude. All we get is a jumbled-up mess.

The top animation shows the effect of good timing. All of the pulses add constructively to the returning reflected wave to give the largest wave amplitude possible. Whenever we get successive contributions to the wave adding in this way, we call it a resonance. It is like pushing a swing. If you push a swing at just the right frequency (good timing), large amplitude motion will result. If you apply the same force, but at a different frequency (bad timing), not a lot usually happens. In order to get a large amplitude you must push at the same frequency as the natural frequency of the swing.

Illustration 17.6: Plucking a String

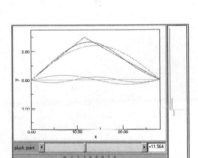

A green string of length $L = 28$ cm (position is given in centimeters) is shown plucked to $x = 6$ cm and $y = 3$ cm. The unstretched position of the string is shown in gray. Changing the slider changes this plucking point along the length of the string in the x direction (the y point of the pluck remains the same). You may also look at the Fourier components that make up the green stretched string by clicking on an n value. The relative size of these sine waves is depicted by the graph on the right.

We have thus far looked at using a Fourier series to describe an arbitrary periodic wave (see Illustration 16.5 and Illustration 16.6). For the plucked string, we must consider a different way to add up waves to get the Fourier series. Here we must consider any wave that is zero at the ends of the string (since the plucked string, like a standing wave, has ends that are tied down). Therefore, we find that our plucked string can be described in terms of a Fourier series as

$$f(x) = \Sigma A_n \sin(n\pi x/L),$$

where in the animation $L = 28$ cm (see **Illustration 16.5** and **Illustration 16.6** for more details on the periodic case).

When you get a good-looking graph, right-click on it to clone the graph and resize it for a better view.

Illustration 17.7: Group and Phase Velocity

So what do we mean by the velocity of a wave? This may seem like a simple question. When we talk about a wave on a string (or a sound wave) we can talk about the velocity as $v = \lambda f$. We can rewrite this expression in terms of the wave's wave number,

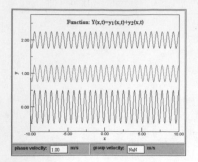

k, and angular frequency, ω, given that $\lambda = 2\pi/k$ and that $f = 2\pi/\omega$. We therefore find that $v = \omega/k$. We note here that the velocity of the wave is also fundamentally related to the medium in which the wave propagates.

But what happens when we want to add several traveling waves together? In this case we are interested in several waves traveling in the same direction. We can change the wave number and angular frequency for each wave, but we must ensure that the wave speeds are identical. In this animation we add the red wave to the green wave to form the resulting blue wave (position is given in meters and time is given in seconds).

Consider what happens when we change k_1 to 8 rad/m and ω_1 to 8 rad/s. Note the interesting pattern that develops in the superposition. Notice that there is an overall wave pattern that modulates a finer-detailed wave pattern. The overall wave pattern is defined by the propagation of a wave envelope with what is called the group velocity. The wave envelope has a wave inside it that has a much shorter wavelength that propagates at what is called the phase velocity. For these values (of k and ω) the phase and group velocities are the same.

Now consider $k_1 = 8$ rad/m and $\omega_1 = 8.4$ rad/s. What happens to the wave envelope now? It does not move! This is reflected in the calculation of the group velocity. The finer-detailed wave has a phase velocity of 1.02 m/s. Now consider $k_1 = 8$ rad/m and $\omega_1 = 8.2$ rad/s. The group velocity is now about half that of the phase velocity (certain water waves have this property). Now consider $k_1 = 8$ rad/m and $\omega_1 = 7.6$ rad/s. The group velocity is now about twice that of the phase velocity.

For a superposition of two waves the group velocity is defined as $v_{\text{group}} = \Delta\omega/\Delta k$ and the phase velocity as $v_{\text{phase}} = \omega_{\text{avg}}/k_{\text{avg}}$. In general, the group velocity is defined as $v_{\text{group}} = \partial\omega/\partial k$ and the phase velocity as $v_{\text{phase}} = \omega/k$.

So what velocity do we want? The physical velocity is that of the wave envelope, the group velocity. For waves on strings we got lucky; the phase and group velocities are the same (these are harmonic waves).

Exploration 17.1: Superposition of Two Pulses

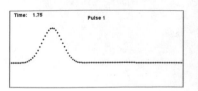

One of the most interesting phenomena we can explore is that of a superposition of waves. Each panel shows an individual wave that is traveling on a string.

If these two waves are traveling on the same string, draw the superposition of the two pulses between $t = 0$ and $t = 20$ s in 2-s intervals for each animation (position is given in meters and time is given in seconds).

When you have completed the exercise, check your answers.

Exploration 17.2: Measure the Properties of a Wave

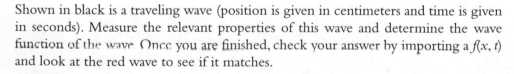

Shown in black is a traveling wave (position is given in centimeters and time is given in seconds). Measure the relevant properties of this wave and determine the wave function of the wave. Once you are finished, check your answer by importing a $f(x, t)$ and look at the red wave to see if it matches.

Exploration 17.3: Traveling Pulses and Barriers

A string can be approximated by many connected particles as shown in the animations (position is given in meters and time is given in seconds). This Exploration considers a pulse on a string and looks at the motion of the individual particles that make up such a string. **Pulse 1** shows a Gaussian pulse incident from the left, while **Pulse 2** shows a Gaussian pulse incident from the right. Notice how the particles never really move in the x direction, yet the information in the pulse does travel across the screen.

In the other two animations the pulse is incident from the left and hits either a **Hard** or a **Soft** barrier. The hard-barrier example is depicted by the hand that represents a string whose end is tied down; the soft-barrier example represents a string with one end free.

a. During the hard-barrier example, what is the direction of the force that is exerted on the hand?

b. During the hard-barrier example, what is the direction of the force that is exerted on the string?

c. Describe the differences between the waves reflected at the two barriers (**Hard** or **Soft**). Explain those differences.

Exploration 17.4: Superposition of Two Waves

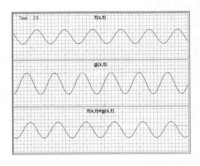

The top two windows display waves that are traveling simultaneously in the same nondispersive medium: string, spring, air column, etc. (position is given in meters and time is given in seconds). The wave in the bottom window is the superposition (algebraic sum) of the two component waves in the upper windows. The superposition is what you would actually see. You wouldn't see the component waves. You can adjust the amplitude, wavelength, and wave speed for $g(x, t)$ (the middle window). For the waves described (traveling in the same medium), the two waves could have different amplitudes and wavelengths, but they must have the same speed (you will need to adjust the wave speed of $g(x, t)$ appropriately).

a. Why must the two waves have the same speed? (Think in terms of what influences wave speed in the medium.)

b. For each $f(x, t)$, determine the amplitude, wavelength, frequency, and wave speed of the wave. Check your answer by making $g(x, t)$ identical to $f(x, t)$.

c. Determine the amplitude, wavelength, and wave speed of the wave, $g(x, t)$, that will make $f + g$ a standing wave.

Exploration 17.5: Superposition of Two Waves

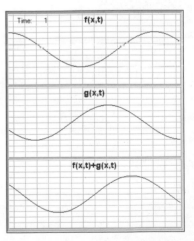

The top two windows display waves that are traveling simultaneously in the same nondispersive medium: string, spring, air column, etc. (position is given in meters and time is given in seconds). Note that the two waves are traveling at the same speed in opposite directions and that they have the same amplitude and wavelength. It is, of course, possible that the two waves could have different amplitudes and wavelengths. However, the waves that we are studying must have the same speed.

The wave in the bottom window is the *superposition* (algebraic sum) of the two *component* waves in the upper windows. The superposition is what you would actually see. You wouldn't see the component waves.

a. Why must the two waves have the same speed? (Think in terms of what influences wave speed in the medium.)

b. Stop the top wave and measure its wavelength in units of divisions along the horizontal axis. Sketch the wave, showing the two points between which you measured the wavelength.

c. Now measure the period of the top wave in time units. Describe your method for doing this.

d. Calculate the speed of the top wave. Show your work.

e. Assume that the bottom wave shown represents the displacement of a string. What is the longitudinal speed of a point on the string?

f. Assume that the bottom wave shown represents the displacement of a string. Is there a time when the transverse speed of the string is zero?

g. What relationship, if any, do the speeds in (d), (e), and (f) have to one another?

Exploration 17.6: Make a Standing Wave

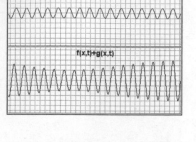

Find a wave function, $g(x, t)$, that will produce a standing wave with a node at $x = 0$ m, i.e., at the center (position is given in meters and time is given in seconds). You may want to pause the animation before you click-drag the mouse to read position coordinates.

Problems

Problem 17.1

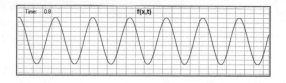

Find the frequency of the wave shown in the animation (position is given in centimeters and time is given in seconds).

Problem 17.2

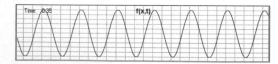

Find the velocity of the wave shown in the animation (position is given in centimeters and time is given in seconds).

Problem 17.3

The animation shows disturbances on two identical strings (position is given in centimeters and time is given in seconds). What is the tension in the second string if the tension in the first string is 500 N?

Problem 17.4

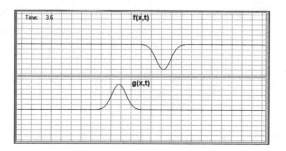

The animation shows disturbances on two identical strings (position is given in centimeters and time is given in seconds). At $t = 2.5$ s, which of the following, if any, statement(s) is (are) true regarding the superposition of the two waves?

a. Their sum adds up to zero.

b. Their sum adds up to twice that of the original waves.

c. Their sum is as if only one of the original waves is there.

d. Their sum has a large peak, a depression, and then another large peak.

Problem 17.5

The animation shows disturbances on two identical strings (position is given in centimeters and time is given in seconds). At $t = 2.0$ s, which of the following, if any, statement(s) is (are) true regarding the superposition of the two waves?

a. Their sum adds up to zero.

b. Their sum adds up to twice that of the original waves.

c. Their sum is as if only one of the original waves is there.

d. Their sum has a large peak, a depression, and then another large peak.

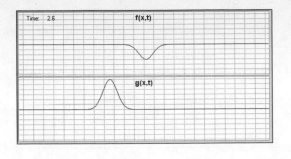

Problem 17.6

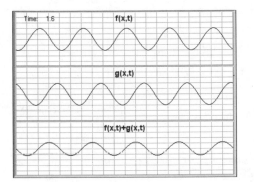

The animation shows how two waves can add together to produce a standing wave on a string (position is given in centimeters and time is given in seconds). The third panel represents the string. The waves in the first two panels have been superimposed to produce the wave in the third panel.

Which of the following, if any, statement(s) is(are) true?

a. Waves never pass through the point $x = 0$ cm on the string since this point never moves.

b. The string is perfectly straight when the maxima in the first two panels overlap.

c. There is an instant in time when the string does not move.

d. The string is moving fastest when the maxima in the first two panels overlap.

Problem 17.7

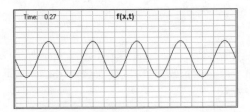

The animation shows a standing wave on a string (position is given in centimeters and time is given in seconds). With what speed do waves propagate on this string?

Problem 17.8

The animation shows a standing wave on a string (position is given in meters and time is given in seconds). If the tension in the string is 4 N, determine the mass of the string.

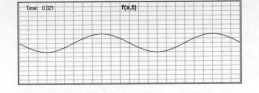

Problem 17.9

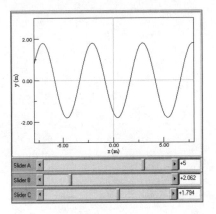

Shown in black is a wave (position is given in meters). Three sliders are given that change certain properties of the wave.

a. Which slider changes which property of the wave?

b. Use the sliders to identify the wave parameters for **this wave function** (shown in red).

Problem 17.10

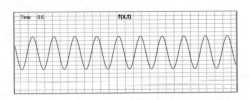

a. Measure the frequency, wavelength, and period of the wave (position is given in centimeters and time is given in seconds).

b. Verify that the speed of the wave crest is wavelength times frequency.

c. Write down a formula for the wave as a function of both distance and time. That is, write a formula for y(x, t).

Problem 17.11

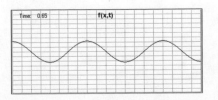

The animation shows a standing wave on a string (position is given in meters and time is given in seconds).

a. What is the speed of waves traveling on this string?

b. Assume you are standing at the point $x = 2$ m. Sketch the height of the wave at this point as a function of time.

c. Write an equation for the height of the wave as a function of time at the points $x = 0$ m and $x = 2$ m.

d. Write down a formula for the wave as a function of both distance and time. That is, write a formula for $y(x, t)$.

Problem 17.12

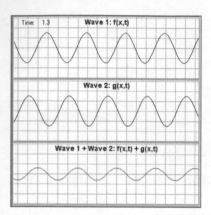

Two traveling waves (the top two panels) are depicted on a string (position is given in meters and time is given in seconds). They are traveling in opposite directions and add to a standing wave as depicted in the bottom panel.

a. What are the wavelength, frequency, and velocity of the initial two waves?

b. What are the wavelength, frequency, and velocity of the resulting standing wave?

Problem 17.13

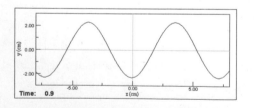

Shown in black is a traveling wave (position is given in centimeters and time is given in seconds). Measure the relevant properties of this wave and determine the wave function of the wave.

Problem 17.14

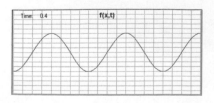

The animation shows a portion of a standing wave on a taut string (position is given in centimeters and time is given in seconds).

a. What is the speed of a wave traveling to the right on this string?

b. Write an equation for the height of the string as a function of both position and time. That is, write a formula for $y(x, t)$.

Problem 17.15

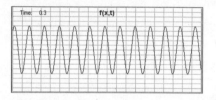

The animation shows a portion of a standing wave on a taut string (position is given in centimeters and time is given in seconds).

a. Measure the frequency, wavelength, and period of the wave in the animation.

b. Write a formula for the position as a function of time for a small section of string located at $x = 0$ cm and a formula for $x = 2$ cm.

c. Write a formula for the velocity as a function of time for a small section of string located at $x = 0$ cm and a formula for $x = 2$ cm.

d. Sketch the velocity of the string as a function of x at $t = 0$ cm. That is, show how each small section of string is moving at $t = 0$ s.

Problem 17.16

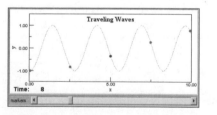

The animation marks sections of a taut string with small circles. You can change the number of small circles (the wave markers) by dragging the slider. Consider a traveling sinusoidal wave on this string.

a. Describe the motion of a small section of this string. Does a section of string ever move to the right or left?

b. Write an equation, $f(t)$, that describes the motion of a small section of the string shown.

c. Compare the motion of two different small sections of string. What is the same and what is different?

d. If the wave function is not shown, how many markers are needed to clearly discern the sinusoidal nature of the wave function? What mistake are you likely to make if you use too few markers?

e. Write an equation, $f(x, t)$, that describes this wave.

Problem 17.17

Sketch the displacement of each wave at the point $x = 0$ cm as a function of time (position is given in centimeters and time is given in seconds).

Time: 3.2   Animation 4

How do your sketches change if you measure the waves at $x = 2$ cm?

Sound

Topics include two-dimensional and three-dimensional waves, sound, beats, Doppler effect, and resonance in open and closed pipes.

Illustration 18.1: Representations of Two-Dimensional Waves

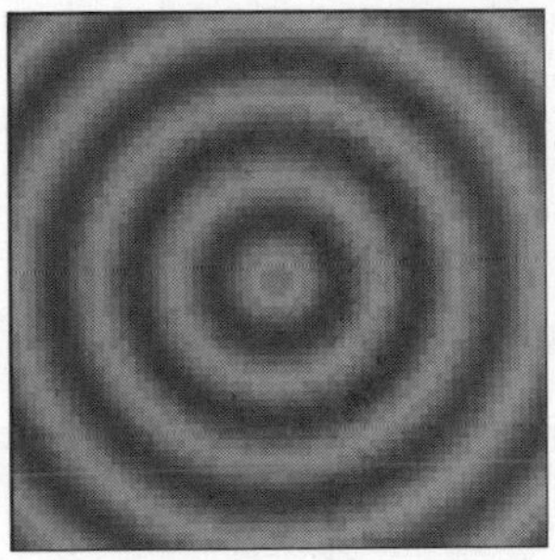

When we have an oscillating source on the surface of a body of water, a wave is generated that travels out in circular wave fronts in two dimensions. The amplitude of the wave (the actual direction of the waving) is in a direction that is perpendicular to the surface of the water. So how do we represent such a wave?

One way to represent such a wave is in two dimensions where the amplitude of the wave is represented by grayscale. When the wave has a positive amplitude, the color is white, when the amplitude is zero the color is light gray, and when the amplitude is negative the color is black. This is shown in the animation (position is given in centimeters and time is given in seconds).

Another way to represent a traveling wave in two dimensions is in three dimensions. After all, there are three dimensions to consider: the propagation (which accounts for two dimensions) and the direction of waving. Click the "check to see three-d mode" check box, then click the "set values and play" button to see the three-dimensional representation of the wave.

Which representation do you like? In which one is it easier to "see" the wave's motion? While the three-dimensional representation is the more realistic representation, the pure two-dimensional representation that uses grayscale is certainly easier to view and determine the properties of the wave phenomena.

Illustration 18.2: Molecular View of a Sound Wave

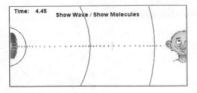

A sound wave is a longitudinal wave. In a longitudinal wave the waving of the medium (here the air molecules) is in the direction of the propagation of the wave. In **show wave/hide molecules** mode, we see a speaker, which is the source of the sound wave, and wave fronts propagating toward the detector (the man's ear). This is the way we usually think of sound waves: originating from a source and propagating toward a detector. But what is really going on in the medium as the sound wave goes by?

In **show wave/show molecules** mode, we see the individual air molecules, which are the medium in which the wave travels, and the waving of the medium. Consider the motion of the red air molecule. The molecule oscillates back and forth about its equilibrium position. If we were describing the sound wave in terms of the individual molecules, we would call the wave a displacement wave. It turns out that the amplitude of the displacement wave is only on the order of 10^{-6} m! The other way to describe the sound wave is in terms of the pressure wave that travels to the right. The pressure wave fluctuates ever so slightly about atmospheric pressure.

Illustration 18.3: Interference in Time and Beats

This animation depicts the superposition of two sound waves. A **red wave** with $\lambda = 3.43$ m and a frequency of $f = 100$ Hz is added to a **green wave**, and the resulting wave is shown in **blue** (position is given in meters and time is given in seconds).

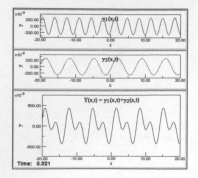

Now change the frequency of the green wave to 120 Hz. What is the new wavelength of the green wave? You should right-click on the graph to clone it and resize the graph to make it easier to make measurements. The result is 2.86 m. Did you really need to make a measurement to get the wavelength? Since the speed of sound is a constant and $v = \lambda f$, we also know that $\lambda = 343/f$. Notice the animation always does the right thing: As you change the frequency of the green wave, the wavelength also changes to maintain the wave speed of 343 m/s.

Now consider what happens when the two waves add together to create the blue wave. Enter several frequencies for the **green wave** and see the resulting **blue waves**. What do you see? When the two frequencies are the same, the resulting blue wave looks like the original waves but with twice the amplitude. But the resulting wave is more interesting when the frequencies (and therefore the wavelengths) do not quite match. Consider the resulting wave when the green wave's frequency is 120 Hz. If you were at $x = 20$ m, you would hear the sound wave getting louder and softer, louder and softer with time. When you hear this pattern, you are hearing beats. The time in between the loud sounds (or conversely the soft sounds) can be measured and is 0.05 seconds. This corresponds to a beat frequency of 20 Hz. This is precisely the difference in the frequencies! What happens when the green wave's frequency is now 80 Hz? We get the same period and therefore the same beat frequency of 20 Hz. Therefore we find that the beat frequency is $f_{\text{beat}} = |f_1 - f_2|$.

So what is going on? Look at the underlying waves. They go in and out of phase with each other as a function of time. At one instant they exactly add together (constructively interfere); at another time they exactly cancel (destructively interfere). We therefore can say that the phenomenon of beats is due to an interference in time.

Try this out for yourself. You may vary the green wave's frequency between 50 and 150 Hz.

Illustration 18.4: The Doppler Effect

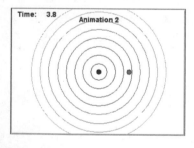

In this Illustration we consider what happens when the source of sound is moving either toward or away from a detector at rest (position is given in meters and time is given in milliseconds). At the same time we can consider what happens when the detector is moving toward a sound source at rest.

What we notice from everyday experience is that if the source of the sound is moving toward us, the frequency we hear increases. If it is moving away, the frequency we hear drops. If we are moving toward a source, the frequency we hear increases and if we are moving away, the frequency we hear drops. The reason for the difference between when the observer is moving and the source is moving is due to how the detected sound waves change in each case.

__Animation 1__ depicts what happens when the source of the sound wave and the detector of the sound wave are both stationary. Notice that the wavelength of the sound wave is 1.7 m and its period is 0.5 ms, which corresponds to a frequency of 200 Hz.

When the observer is moving, as in __Animation 2__, the sound waves emitted from the source are undisturbed. The wavelength does not change as observed by the moving observer. He or she just comes across more/less wave fronts per time ($[vt \pm v_D t]/\lambda t$) when moving toward/away from the source and consequently sees a change in frequency.

For the case in which the source is moving, shown in __Animation 3__, the frequency (time in between wave fronts) and wavelength change. The wave fronts are emitted much closer together/farther apart ($\lambda' = vT \mp v_S T = [v \mp v_S]/f$) as the source is moving toward/away from us. __Animation 4__ represents the sound wave of a source moving according to a linear restoring force (simple harmonic motion).

We may write both these cases together, with v_S as the velocity of the source and v_D as the velocity of the observer or detector, as

$$f' = f[v \pm v_D]/[v \mp v_S].$$

Hence when the source is stationary and the observer/detector is moving $f' = f[v \pm v_D]/v$, and when the detector/observer is stationary and the source is moving $f' = fv/[v \mp v_S]$. Here the upper signs indicate a velocity toward and the lower signs represent a velocity away.

When the source is moving at the speed of sound, the emitted wave travels forward at the same speed as the source. The sound waves build up, causing a sonic boom as shown in **Animation 5**. For things like supersonic airplanes, we get a double boom— a boom from the front of the plane and a boom from the back. The resulting waves pile up, all on top of each other, and create at first a huge increase in pressure and then a huge decrease in pressure before the return to normal atmospheric pressure.

Illustration 18.5: The Location of a Supersonic Airplane

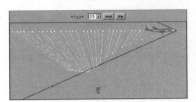

If an airplane is flying faster than the speed of sound, it will produce a shock wave called a sonic boom. In the animation, consider an airplane that is flying from point A to point B. A listener, the ear, is located at point C. We consider how the airplane's speed and the position of the listener affect when the sound from the airplane's engines is heard by the listener.

In the animation, you can change the airplane's speed, which we call v. The speed of sound is fixed at 343 m/s and we will represent it as v_s (shown as v_s in the animation). Press "play" to begin the animation. In addition,

- You can drag the ear across the screen to change its location.
- The program draws sound wave paths to the listener.
- The animation pauses when the sonic boom arrives at the listener; the animation can be resumed by clicking the right mouse button.
- The color of the paths of the sound waves changes to blue when those sound waves reach the listener. The order in which the sound from different paths arrives at the listener is shown as numbers located at the point that the sound was produced.
- Press "reset" for default values.

Consider a sound generated by the airplane when it is at some point A traveling toward some point B along the straight path AB. The listener hears the sound as the airplane flies toward point B (AB > AC). DC is the path of a subsequent sound generated at some point D and traveling to the listener at point C.

Consider a few time intervals ($\Delta t = |\Delta \mathbf{x}|/v$). The time it takes sound to travel from A to C is AC/v_s, the time it takes the plane to move from A to D is AD/v, and the time it takes sound to travel from D to C is DC/v_s.

Now, how does the time interval AC/v_s compare to the time interval $AD/v + DC/v_s$? In other words, which event happens first: the sound emitted at A reaches C or the sound emitted at D reaches C?

First consider an airplane traveling at less than or equal to the speed of sound. $AD/v + DC/v_s > AC/v_s$ because the path ADC is longer than the path AC. The best you can do is when the time interval for AD is the smallest it can be, which is when $v = v_s$. In this case comparing the two time intervals is equivalent to comparing the two paths. Clearly, ADC > AC. When $v \ll v_s$, the situation is worse and the time interval for the path ADC is even longer. Therefore, you would hear the sound from the airplane when it was at A before you heard it from when it was at point D.

Now consider what you will hear if a supersonic airplane flies over you ($v > v_s$). Again, what you hear is dependent on whether $AD/v + DC/v_s$ is greater than, less than, or the same as AC/v_s. If v is large enough, the extra path difference, AD, accounts for a smaller and smaller time interval, and since DC < AC we may hear the sound emitted at D before hearing the sound emitted at A. Try it in the applet above. Set v

and move the ear around. Notice when you "hear" the sounds from the airplane by looking at the numbers that show the ordering of the events.

Exploration 18.1: Creating Sounds by Adding Harmonics

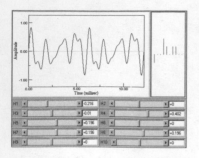

Begin by choosing the first harmonic (represented by the H#, with H1 being the fundamental or first harmonic) and drag the slider to add that harmonic to the total wave. As you do this, note that the frequency remains the same, but the amplitude slowly decreases. Continue to decrease the value of H1 so that it is negative. Notice that the negative sign simply inverts the shape of the sound wave. Therefore, the slider controls the amplitude and phase (0 or π only) of the harmonic of the sound wave. In addition to the overall wave form, the relative size of the components of the wave is shown in the graph on the right.

a. Measure the fundamental's period.

b. What is the fundamental frequency?

H	Case A	Case B	Case C	Case D
1	1.000	1.000		1.000
2			0.500	0.500
3	−0.111	0.333		0.333
4			0.250	0.250
5	0.040	0.20		0.20
6			0.166	0.166
7	−0.020	0.142		0.142
8			0.125	0.125
9	0.0123	0.111		0.111
10			0.100	0.100

Consider the following values for the harmonics:

c. What wave patterns develop from these values?

d. Can you write down a mathematical formula describing each case? (Hint: it is a sum.)

Exploration 18.2: Creating Sounds by Adding Harmonics

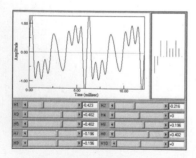

Begin by setting the slider for H5 (Harmonic 5) to the value of 1. You should hear a pure tone.

a. What is the frequency of the fifth harmonic?

Now slowly decrease the value of H5 to zero. As you do this, note that the frequency remains the same, but the amplitude slowly decreases. Continue to decrease the value of H5 so that it is negative. Notice that the negative sign simply inverts the shape of the sound wave. Therefore, each slider controls the amplitude and phase (0 or π only) of the harmonic of the sound wave.

b. What happens to the sound you hear as the slider value goes from +1 to zero?

c. What happens to the sound as the slider goes from 0 to −1?

d. Can you hear a difference in the sound at +1 and −1?

At this point you should have determined that loudness is related to amplitude. Now let's figure out what determines the pitch (musical note). Set H5 to zero. (You can do this using the slider or by typing zero into the text box beside the slider.) Use the slider to turn on some other harmonics. The sound you will hear is the sum of all harmonics that are on, so if you want to hear an individual harmonic, you will need to set all other values to zero.

e. Based on your experimenting, what determines the pitch of a tone? Specifically, is this quantity larger or smaller for high vs. low notes?

Now comes the really cool part. *How does an electronic keyboard mimic the sounds of individual instruments?* To understand this, we first need to understand why (mathematically) a trumpet sounds different from a clarinet. When you had only one harmonic turned on, you heard a pure tone. When you play a note on a trumpet, you do not create a pure tone. You set the trumpet vibrating, and this sets up resonant standing waves (pure tones). More than one resonant standing wave can exist at any one time. All of the resonant standing waves (harmonics) will add together, and what you hear is the sum of all these individual pure tones. The relative magnitudes of the individual harmonics are different for the clarinet and the trumpet, which is why the two instruments sound different even when they play the same note.

H	Clarinet	Trumpet
1	0.91	0.53
2	0.51	1
3	0.71	0.94
4	0.86	0.95
5	1	0.66
6	0.71	0.58
7	0.54	
8	0.2	
9	0.18	
10		

Try the following values for the harmonics. Do the resulting tones sound like the clarinet and trumpet? Well, sort of. You should be able to hear some similarities to the instruments they are supposed to represent, and you should be able to tell that the sounds are different, but it does not exactly sound like the real thing.

f. Can you think of some reasons why the sound you produced is not exactly like the real thing?

Exploration 18.3: A Microphone between Two Loudspeakers

A microphone is placed between two loudspeakers (position is given in centimeters and time is given in seconds). The speakers are connected to two sources of sound that have variable frequencies, f_1 and f_2. The graph shows the sound waves arriving at the microphone, as a function of time, from each speaker and also shows the sum of the two waves. Change the frequency of either sound source ($25\ Hz < f_1, f_2 < 30\ Hz$), and watch the changing interference between the two sound waves. Study the phenomenon of beats and verify that the beat frequency is correct.

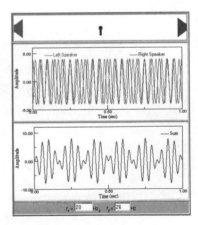

a. What happens as the two frequencies get closer together?
b. What happens as the two frequencies get farther apart?
c. Does it matter which speaker has the higher frequency?
d. What happens if the two frequencies are identical?

Remember that it is the **difference** between the two sound frequencies that determines the beat frequency.

Exploration 18.4: Doppler Effect and the Velocity of the Source

This example shows the Doppler effect. The black dot represents the source of the sound wave and travels with a speed set by the slider. That speed is given in terms of the speed of sound; hence $v = 1$ corresponds to the speed of sound.

Vary the speed of the source from zero to the speed of sound and then to the maximum value allowed by the slider. Watch the animation and answer the questions below.

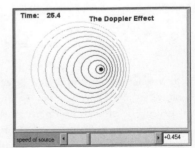

a. How does the pattern of the wave fronts change according to v_{source}?
b. For $v_{source} > v_{sound}$ (slider values > 1) how does the V-shaped shock wave vary according to v_{source}?

Exploration 18.5: An Ambulance Drives by with Its Siren On

When the ambulance is moving (position is given in meters and time is given in seconds), use the animation to guide your answers to the following questions.

a. How does the wavelength of the sound wave change in relation to the woman at the right?

b. How does the frequency of the sound wave change in relation to the woman at the right?

c. How does the wavelength of the sound wave change in relation to the man at the left?

d. How does the frequency of the sound wave change in relation to the man at the left?

e. How does the wavelength of the sound wave change in relation to the patient in the ambulance?

f. How does the frequency of the sound wave change in relation to the patient in the ambulance?

Problems

Problem 18.1

The animation is a slow-motion representation of a cross section of a sound wave propagating in Lucite. A detector, the orange square, is placed in the pipe and properly measures the pressure (position is given in meters and time is given in seconds). What is the speed of the sound wave? Note: The animation runs for 0.1 s. Press "reset" to reload the animation.

Problem 18.2

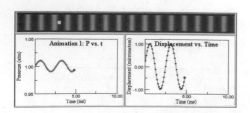

The animation represents a cross section of a sound wave propagating in a very long pipe. A detector, the orange square, is placed in the pipe and properly measures the pressure (position is given in meters and time is given in milliseconds). Which of the graphs properly represents the displacement of the air molecules in the pipe?

Problem 18.3

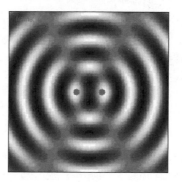

Why are there no dead spots in the sound distribution (position is given in centimeters and time is given in seconds) when either the **left** or the **right** source is transmitting, but there are multiple dead spots when **both** sources are transmitting?

Problem 18.4

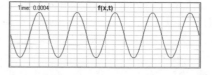

The animation shows a standing wave on a string (position is given in centimeters and time is given in seconds). If this string is on a musical instrument, what wavelength sound is produced by the standing wave?

Problem 18.5

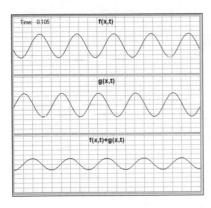

The animation shows a standing wave on a stringed musical instrument (position is given in centimeters and time is given in seconds). If the tension in this string is doubled and the string stays in its fundamental mode, what frequency sound is produced by the new standing wave?

Problem 18.6

The animation shows a superposition of two waves on identical strings (position is given in meters and time is given in seconds). What is the difference in frequency between the two waves?

Problem 18.7

A man and a woman are in front of the White House as an ambulance drives by with its sirens on (position is given in meters and time is given in seconds).

The three animations play the possible siren sound heard by three individuals: the man, the woman and the ambulance driver.

a. The sound in Animation 1 is heard by whom?

b. The sound in Animation 2 is heard by whom?

c. The sound in Animation 3 is heard by whom?

Problem 18.8

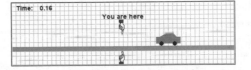

The animation represents a cross section of a three-dimensional sound wave propagating away from a moving source (time is given in seconds).

a. In which of the animation(s) does the source travel slower than the speed of sound?

b. In which of the animation(s) does the resulting sound wave travel the fastest?

Problem 18.9

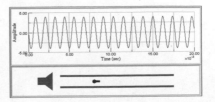

You are standing beside a highway as a police car with its siren on drives by, as shown in the animation (position is given in meters and time is given in seconds). Which of the animations represents what you would hear?

Problem 18.10

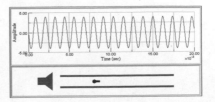

You are standing beside a highway as a police car with its siren on drives by, as shown in the animation (position is given in meters and time is given in seconds). If the frequency of the siren is 800 Hz, determine the change in frequency you will hear as the police car goes by.

Problem 18.11

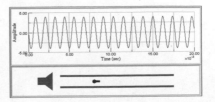

Using a speaker, a standing sound wave has been set up inside a tube. A movable microphone lies inside the tube (position is given in centimeters and time is given in seconds). The graph shows the sound recorded by the microphone as a function of time. Move the microphone back and forth to study the changing amplitude of the sound it receives.

a. For what microphone position(s) does the amplitude of the sound go to zero? What is such a location called?

b. For what microphone position(s) is the amplitude of the sound a maximum? What is such a location called?

c. From the locations of the nodes, determine the wavelength of the sound waves.

d. From the graph, determine the frequency of the sound waves.

e. Using the wavelength and the frequency, find the velocity of the sound waves in the tube.

Problem 18.12

This animation shows a standing wave in an open pipe (position is given in centimeters and time is given in seconds).

a. In which harmonic, n, is the air in the pipe oscillating?

b. Determine the frequency of the musical tone produced by the pipe in this situation.

c. Determine the fundamental frequency of the pipe (lowest frequency of resonance).

Problem 18.13

This animation shows a standing wave in an open pipe (position is given in centimeters and time is given in seconds).

a. In which harmonic, n, is the air in the pipe oscillating?

b. Determine the frequency f_n of the tone produced by the pipe in this situation.

c. Determine the frequency of the eighth harmonic f_8.

d. Determine the speed of sound v.

Problem 18.14

The animation shows a standing wave in a half-open pipe (position is given in centimeters and time is given in seconds).

a. In which harmonic, n, does the air in the pipe oscillate?

b. Determine the frequency f_n of the tone produced by the pipe in this situation.

c. Determine the fundamental frequency f_1 (the lowest frequency of resonance in the pipe).

d. Determine the speed of sound v.

Problem 18.15

This animation shows a standing wave in a half-open pipe (position is given in centimeters and time is given in seconds).

a. Determine the frequency f_n of the tone produced by the pipe in this situation.

b. Now the pipe is cut into two pieces of equal length. Determine the fundamental frequency of each of the two pieces.

Problem 18.16

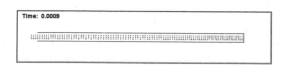

This animation shows a standing wave in a half-open pipe (position is given in centimeters and time is given in seconds).

a. In which harmonic, n, is the air in the pipe oscillating?

b. Determine the wavelength for this oscillation.

Heat and Temperature

Topics include temperature, thermal expansion, specific heat, calorimetry, conduction, convection, and radiation.

Illustration 19.1: Specific Heat

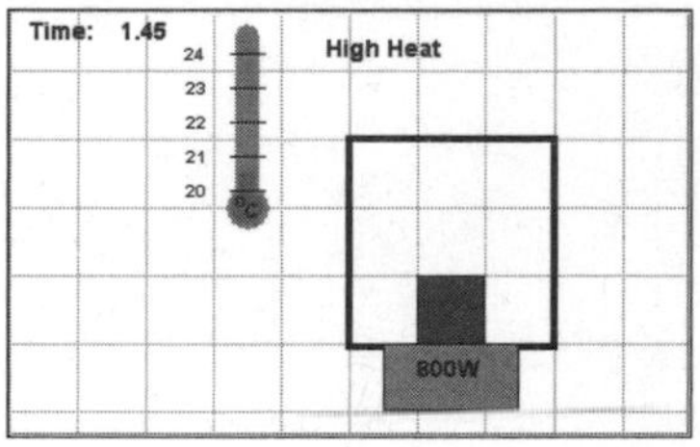

Specific heat (which is sometimes also called the specific heat capacity) describes how much heat is required to increase the temperature of a given quantity of material. In this Illustration a blue mass sits in an insulated oven (time is given in minutes and temperature is given in degrees Celsius). Assume that the block absorbs all the heat from the heater. Not surprisingly, a higher-powered heater (the amount of heat delivered/second) will result in a higher temperature of the blue mass during the same time interval. Notice that as you change the mass of the object, the temperature change is different (for a given power of the oven). The quantitative description of this is given by the equation

$$Q = mc\,(T_f - T_i),$$

where Q is the heat, m is the mass, c is the specific heat, and T is the temperature (with subscripts indicating final and initial temperatures). Note that if you double the mass, for the same total heat delivered, the *temperature change* will be cut in half. Different materials have different values of specific heat (or specific heat capacity). Water has a much higher specific heat than copper, for example. This is why it does not take long for a copper kettle on the stove to increase in temperature in comparison with the water inside. Furthermore, with a full kettle of water, the water is more massive, so it also takes longer to reach an acceptable final temperature (usually around 100°C to boil).

Note that the specific heat usually has units of joules/(kg·C°), where C° represents a *change in temperature* (your text book may or may not follow this notation).

Illustration 19.2: Heat Transfer, Conduction

Heat transfers via three mechanisms: convection, radiation, and conduction. This Illustration briefly describes these mechanisms, but the animation focuses on conduction (temperature is given in degrees Celsius).

Convection is the transfer of heat energy through the motion of a gas (or liquid): Heated air expands and rises, displacing cooler air, which moves downward and is then heated and rises again, setting up "convection currents."

Radiative heat transfer occurs when an object absorbs/emits electromagnetic radiation and gains/loses energy (see **Illustration 19.3**).

Conduction, as shown in the animation, is the transfer of heat within a material due to a temperature difference across the object (think of a spoon in hot coffee). We describe materials that transfer heat easily (more heat/time) as having a high conductivity (e.g., metal spoon) and those that do not (e.g., cloth) as having low conductivity. The reason for having insulation in a house, for example, is to reduce the conductivity of the walls so that it requires less power to keep the inside of a house at a given temperature, even though the outside is much colder or warmer. Use the

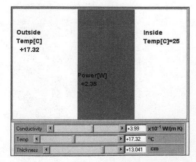

animation to change the conductivity, the temperature on the outside of the "wall," and/or the thickness of the "wall" to see the power loss. The power loss is the power required to heat or cool the inside of the house.

Illustration 19.3: Heat Transfer, Radiation

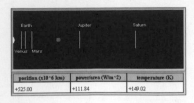

Heat from the Sun is transferred via radiation to planets. A planet, in turn, radiates energy back out into space. A planet reaches its equilibrium temperature when the power delivered to it from the Sun is equal to the power it radiates. The power a planet radiates is given by the following equation:

$$P = \sigma \varepsilon A T^4,$$

where σ is the Stefan–Boltzmann constant $(5.67 \times 10^{-8} \text{ W/m}^2 \cdot \text{K}^4)$, ε is the emissivity (1 for a "blackbody" absorber/emitter; 0 for a perfect reflector), A is the surface area $(4\pi R^2)$, and T is the temperature. The power/area delivered to a planet varies as the inverse square of the distance from the Sun. Note that the effective area of a planet that the Sun's radiation hits is πR^2, where R is the radius of the planet. However, the total area over which the planet radiates is equal to its surface area, which is the surface area of a sphere $(4\pi R^2)$ of radius R.

If we neglect the planet's atmosphere (which reflects some of the light from the Sun and traps some of the radiation from the planet's surface), we can predict the temperature of the planet. Drag the red planet in the animation to different distances from the sun and see the various surface temperatures that result. Notice that when the red planet is at Earth's position, its temperature is below Earth's true average temperature of 287 K. Once the effect of the atmosphere is taken into account, the power delivered to Earth's surface is reduced further (since the atmosphere reflects some light). What keeps Earth from being a frozen planet? The greenhouse effect, in which gases in the atmosphere do not allow some of the radiation (in the infrared) that Earth radiates to escape from the atmosphere, does that. This radiation is trapped in Earth's atmosphere, thus warming Earth up to its current average temperature. As "greenhouse" gases increase in the atmosphere, Earth's average temperature will increase.

Exploration 19.1: Mechanical Equivalent of Heat

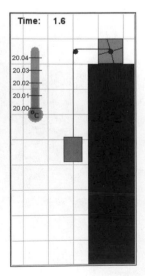

As the 100-kg red mass drops, the paddle turns in the liquid and the liquid heats up. Joule used a version of this device to determine the equivalence between heat and work. You will run the animation to do the same (position is given in meters, time is given in seconds, and temperature is given in degrees Celsius). The temperature of the liquid is given by the thermometer shown.

The dimension of the container that holds the blue liquid that you cannot see (into the screen) is 0.1 m. The density of the liquid is $13,600 \text{ kg/m}^3$.

a. What is the volume of the liquid?

b. What is the mass of the liquid?

c. During the animation, what is the change in temperature of the liquid?

d. If it takes 33 calories to raise 1 kg of the liquid 1°C, how much heat goes into the liquid?

e. What is the change in kinetic energy of the falling red mass?

f. What is the work done by gravity on the mass (in joules)?

g. The work in (f) goes into frictional heating of the liquid (as the paddles turn through the liquid).

Therefore, how many calories are equal to 1 joule?

Exploration 19.2: Expansion of Materials

A rod is fixed at one end. In the animation you see both the rod and a magnified view of the right end (position is given in meters, time is given in minutes, and temperature is given in kelvin). As you increase the temperature, notice that the rod increases in length. This Exploration will help you develop a quantitative relationship for the increase in the length of the rod, as a function of the initial length and the temperature change, that holds for all materials.

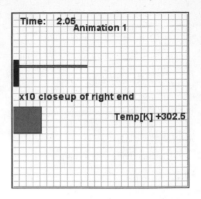

Note that the ×10 closeup means that a reading (in meters) really is in tenths of meters.

a. For Animation 1, if you double the length, what happens to the change in length?

b. Repeat (a) for the material in Animation 2. How do the two results compare?

c. How does changing the final temperature change the expansion? (If you double the *change in temperature*, what happens to the change in length?)

d. What general expression can you now write for the change in length as a function of the temperature change and initial length?

The difference between the two materials is described by a different coefficient of linear expansion, α. For the material in Animation 1, α is 30×10^{-6}/K, while for the material in Animation, 2α is 20×10^{-6}/K.

When heated, a solid (even a thin rod as above) expands in all three dimensions. The equation for the volume expansion is similar to the linear expansion case with the coefficient of expansion approximately equal to 3α.

e. Why didn't you see the expansion of the rod in the other dimensions?

Exploration 19.3: Calorimetry

When two objects at different temperatures are in thermal contact with each other, they will eventually reach the same temperature. Heat will flow from the warmer object to the cooler one until they are at the same temperature (temperature is given in kelvin and heat is given in joules). Using the equation for heat absorbed or released with a temperature change, $Q = mc\,(T_f - T_i)$, where Q is the heat, m is the mass, c is the specific heat, and T is the temperature (with subscripts indicating final and initial temperatures), we can determine the specific heat of an object.

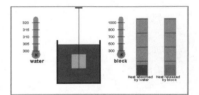

In the animation a piece of metal at a high temperature is placed in water at a lower temperature and, if the water is well insulated so we can assume essentially no heat loss to the environment, the final temperature of the water/metal combination depends on the mass of the water, the mass of the metal, and the specific heat of each. Try changing the initial temperature of the metal and the mass of the metal. By equating the heat lost by the object to the heat gained by the water, we can calculate the specific heat of an unknown object or the final temperature of the system. There are 10 kg of water, and the specific heat of water is 4.186 kJ/kg·K. The specific heat of the block is 0.39 kJ/kg·K.

a. For a 1-kg block and an initial block temperature of 800 K, use the equation above to calculate the heat absorbed by the water and the heat released by the block when they reach the final temperature.

b. What is the scale for the heat on the bar graphs? In other words, what unit of heat does each mark correspond to (10 kJ, 100 kJ, 200 kJ, etc.)?

c. If $m = 3$ kg and the initial temperature of the block is 1000 K, equate the heat released to the heat absorbed to predict the final temperature. Run the animation to check your prediction of both the final temperature and the heat released and absorbed.

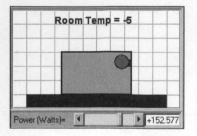

Exploration 19.4: Heat Balance

The animation depicts an incubator for chicks. Inside the box is a heat lamp with varying power. The incubator is made of material with a thermal conductivity of 0.15 W/m·K and of 2-cm thickness. The dimension of the box you cannot see (the depth of the box going into the screen) is 0.3 m (position is given in tenths of meters and temperature is given in degrees Celsius). As you change the power of the heat lamp, notice how the inside equilibrium temperature changes, as does the energy/time (power) lost by conduction through the walls to the outside.

The box must also be coated in a shiny reflective material (foil) so that there are no significant contributions through radiative heat-exchange processes. The only significant energy-exchange process is conductive.

a. The animation shows an instantaneous change in temperature as you change the heater power. Explain why this is nonphysical (the temperature does not change instantly as shown). What determines how long it will actually take for the system to reach equilibrium?

b. When the heater runs at 50 W, calculate the energy lost through conduction using $P = (kA/x)\,\Delta T$, where k is the thermal conductivity (0.15 W/m·K in this case), A is the surface area of the box, x is the thickness of the box material, and ΔT is the temperature difference between the inside of the incubator and the outside environment. This should be equal to the power delivered by the heater (50 W).

c. By equating the energy/time into the box (from the heat lamp) to the energy lost from the box (by conduction out through the walls), predict the power required from the heat lamp to keep the box at 27°C.

d. Check your prediction by varying the heat lamp power. Note that if there are chicks in the incubator, they will radiate heat as well, and so the heat required from the heat lamp will be reduced.

Problems

Problem 19.1

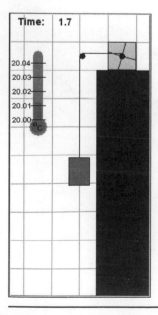

As the 100-kg red mass drops, a paddle turns in a liquid and the liquid heats up (position is given in meters and time is given in seconds). The dimension of the container that holds the liquid that you cannot see (into the screen) is 0.1 m. The density of the liquid is 920 kg/m³. What is the heat capacity of the liquid in the animation?

Joule used a version of this device to determine the equivalence between heat and work.

Problem 19.2

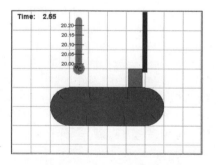

The red block gets stuck on a conveyor belt (position is given in meters and time is given in seconds). Assume (rather unrealistically) that only the block heats up (and not the conveyor belt). If the 4.3-kg red block is aluminum (specific heat of 236 J/kg·K) and the thermometer shows the temperature of the block, what is the coefficient of friction between the block and the conveyor belt?

Problem 19.3

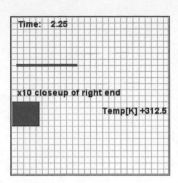

A rod is resting on a surface (you see the top view) and is attached to the surface at its middle (position is given in meters and time is given in minutes). The animation shows you both the rod and a magnified view of the right end. What is the coefficient of linear expansion of the rod?

Problem 19.4

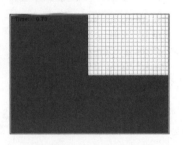

The animation shows a close-up of the bottom-left corner of the square opening in a sheet of material (position is given in millimeters and time is given in minutes).

The initial temperature is 300 K, which changes (increases or decreases) by 200 K over the time of the animation ($t = 2$ minutes). The opening is initially a 20 cm $\times$ 20 cm square.

a. Determine if the temperature is increasing or decreasing.
b. Find the coefficient of linear expansion.

Problem 19.5

A plate with a hole in the middle needs to fit around a ball bearing. In the animation, you are looking at a side view of the plate and ball bearing shown below:

The solid blue lines in the animation are the plate on either side of the bearing, and the dotted blue lines represent the part of the plate going over the ball bearing, which sticks out of the

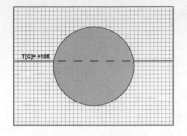

screen. Initially, the plate almost fits around the bearing, but to get it to fit around the middle of the ball, the blue plate and gray ball bearing are heated so that the temperature of the plate changes as indicated (position is given in millimeters). The coefficient of expansion of the ball bearing is much smaller than

the plate so that you can neglect the expansion of the ball bearing in comparison with the plate.

Determine the coefficient of linear expansion of the plate.

Problem 19.6

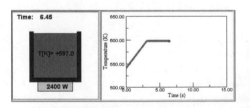

This animation shows a container holding 1 kg of a solid heated by a heater delivering 2400 watts to the material (temperature is given in kelvin and time is given in seconds). Find the specific heat capacity of the material and its latent heat of fusion. Ignore the heat capacity of the container.

Problem 19.7

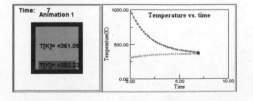

Run the animations and rank the materials in order of their specific heat capacity from smallest to largest (temperature is given in kelvin and time is given in arbitrary units). In each animation the same mass of material starts at the same temperature and is put into a thermally insulated container of water. The graph shows the temperature of the water and the material as a function of time.

Problem 19.8

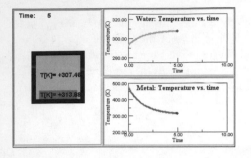

A piece of metal sits in a thermally insulated container of water (position is given in centimeters, temperature is given in kelvin, and time is given in arbitrary units). The dimension of the water you cannot see (into the computer screen) is 25 cm and the dimension of the metal you cannot see is 10 cm. The mass of the metal is 2 kg. The graphs show the temperature of the metal and of the water as a function of time. What is the specific heat capacity of the metal?

Problem 19.9

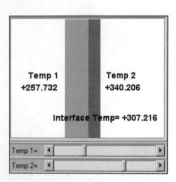

Move the slider to change the temperature on either side of the interface (position is given in centimeters and temperature is given in kelvin). Determine which of the two materials has a higher thermal conductivity.

Problem 19.10

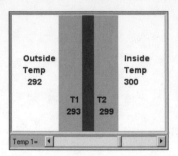

This animation represents a double-pane window with gas between the two panes of glass. Adjust the outside temperature and calculate the ratio of the conductivity of the glass and the conductivity of the gas inside the window (position is given in centimeters and temperature is given in kelvin). Compare (quantitatively) the heat per area conducted across the double-pane window with a window of the same total thickness made completely of glass (of the same conductivity as the glass in this problem).

Problem 19.11

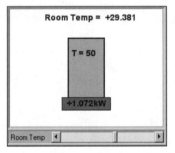

Determine the conductivity of the 10-cm thick insulation around the red water heater shown (this is a cross section of a cylindrical water heater). Assume that all the power from the heater at the bottom of the water heater is delivered to the water to keep it at the same temperature (50°C in this case). Adjust the room temperature to see the change in power required from the heater (position is given in meters and temperature is given in degrees Celsius).

Kinetic Theory and Ideal Gas Law

Topics include Maxwell-Boltzmann distribution, kinetic theory, microscopic and macroscopic connections, kinetic energy and temperature, thermodynamic processes, and the ideal gas law.

Illustration 20.1: Maxwell–Boltzmann Distribution

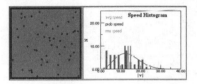

In this animation $N = nR$ (i.e., $k_B = 1$). This, then, gives the ideal gas law as $PV = NT$. The average values shown, $< >$, are calculated over intervals of one time unit.

The particles that make up a gas do not all have the same speed. The temperature of the gas is related to the average speed of the particles, but there is a distribution of particle speeds called the Maxwell-Boltzmann distribution. The smooth black curve on the graph is the Maxwell-Boltzmann distribution for a given temperature. What happens to the distribution as you increase the temperature? The distribution broadens and moves to the right (higher average speed). At a specific temperature, there is a set distribution of speeds. Thus, when we talk about a characteristic speed of a gas particle at a particular temperature we use one of the following (where M is the molar mass, m is the atomic mass):

- Average speed: $(8RT/\pi M)^{1/2} = (8k_B T/\pi m)^{1/2}$
- Most probable speed: $(2RT/M)^{1/2} = (2k_B T/m)^{1/2}$
- Root-mean-square (rms) speed: $(3RT/M)^{1/2} = (3k_B T/m)^{1/2}$

There is not simply one way to describe the speed because it is a speed distribution. This means that as long as you are clear about which one you are using, you can characterize a gas by any of them. The different characteristic speeds are marked on the graph.

Illustration 20.2: Kinetic Theory, Temperature, and Pressure

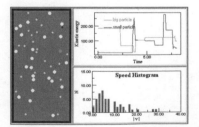

In this animation $N = nR$ (i.e., $k_B = 1$). This, then, gives the ideal gas law as $PV = NT$. The average values shown, $< >$, are calculated over intervals of one time unit.

The equipartition theorem says that the temperature of a gas depends on the internal energy of the gas particles. For monatomic particles, the internal energy of a given particle is its kinetic energy (there is $1/2\ k_B T$ of energy per degree of freedom, and for a monatomic gas there are 3 degrees of freedom). Thus, for an ideal gas made up of particles of different masses, the average kinetic energy is the same for all the particles. In this animation the yellow particles are 10 times more massive than the light blue ones. How does the kinetic energy of the blue particle (representative of the smaller particles) compare with the kinetic energy of the orange particle (which is representative of the yellow particles)? What would you expect in a comparison of the speeds of the two particles? While the average kinetic energy of the particles should be identical, their average speeds should be different, since they have different masses.

Now **triple the temperature**. What happens to the kinetic energy of both the blue particle (representative of the smaller particles) and the orange particle (representative of the yellow particles)? If you triple the temperature, what happens to the kinetic energy? What happens to the speeds of the particles? The average kinetic energy should increase by three, and the speed of the average particle should go up by 1.73, the square root of 3.

Finally, note that $<d\mathbf{p}/dt>$, the average momentum delivered to the walls, is slightly higher than the value of the pressure calculated by the ideal gas law ($P = NT/V$). This is because the ideal gas law assumes that the particles are "point-like" (point particles), while the animation has particles with a definite radius. This means that they interact with the wall sooner (at the edge of a particle instead of the center) and with each other more often. Thus, the average time between collisions with the wall (Δt) is smaller, making $<\Delta\mathbf{p}/\Delta t> = <d\mathbf{p}/dt>$ bigger than predicted by the ideal gas law. In reality, of course, particles are not "point-like," but the size of the particles is typically much smaller in relation to the size of the container (think about the air molecules in a room), so the "point-like" approximation works well. Now **increase the particle size** to see what happens when the particles are quite large.

Illustration 20.3: Thermodynamic Processes

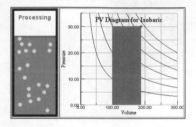

In this animation $N = nR$ (i.e., $k_B = 1$). This, then, gives the ideal gas law as $PV = NT$.

There are many ways for a gas to go from one state (as described by its pressure, volume, number of atoms and temperature) to another. There can be heat flow from the environment to the gas or from the gas to the environment. Work can be done on the gas or by the gas. How the gas goes from one state to another is determined by the heat flow and the work done. However, the change in internal energy depends only on the temperature change (not on how the temperature change occurs). In other words, the change in internal energy is process independent, but the work and heat required are process dependent.

To make things easier, we categorize a set of processes that have names that describe the type of process. We illustrate them for the case when the **number of atoms in the gas remains constant** (there is no requirement that the number of atoms remains constant, but for purposes of this Illustration, we assume a sealed container).

- **Isobaric:** The pressure on the gas remains constant. This means that as the temperature of the gas changes, so will the volume. For example, a balloon that is put in the refrigerator shrinks in size.

- **Isochoric:** The volume of the gas remains constant. This means that any temperature change is accompanied by a change in pressure. For example, the steam in a pressure cooker increases in pressure as the temperature increases.

- **Isothermal:** The temperature of the gas remains constant. As the volume increases, the pressure decreases. For example, a balloon in a vacuum chamber increases in volume as the pressure in the chamber decreases.

- **Adiabatic:** The volume, temperature, and pressure all change. This is a rapid process in which no heat is exchanged with the environment, for example, compressing a piston on a bicycle pump or syringe quickly.

One way to describe the state of a gas is with a PV diagram. You could just as easily use a PT or VT diagram, but we use a PV diagram because it is easy to see the work done by the gas. The work is simply the area under the curve (the red region in the graphs). If you know some calculus, this is because work is given by $W = \int P\, dV$ (and an integral is a calculation of the area under the curve).

A gas does not have to follow any of these special, "named" methods of changing state (**Unknown Process**). For an ideal gas, as long as $PV = NT$ remains true through the process, everything is fine. It is simply a bit harder to mathematically describe the process and, therefore, harder to calculate the work and heat.

When you get a good-looking graph, right-click on it to clone the graph and resize it for a better view.

Illustration 20.4: Evaporative Cooling

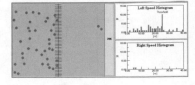

For a gas at a given temperature, there are particles with different speeds, and the particles' speed distribution follows the Maxwell-Boltzmann distribution. In this animation the two containers are separated by a "membrane" in the middle. Initially, no particles can cross the membrane. Once the particles are fairly evenly distributed in the left chamber, you are ready to allow for evaporation—that is, the fastest molecules can escape the left chamber and enter the right chamber. What is the approximate temperature of both sides initially? The light blue wall of the right chamber is at a constant chilly temperature of 20 K so that as particles hit it, they cool (slow down).

Try **letting particles through the membrane**. This animation only allows particles that hit the membrane at a speed of 25 or higher to pass through. This threshold is shown on the speed histogram. After some time passes, particles are no longer passing through the membrane as much. Notice what has happened to the speed distribution in the left chamber. (There may still be particles with speeds greater than the threshold because the speed distribution still follows a Maxwell-Boltzmann distribution.) What has happened to the temperature in the left chamber? This is what happens with evaporation: The fastest particles leave and so the temperature of what remains behind is cooler. This is why sweating cools you off—as the sweat evaporates off your skin, you are cooled down. Thus, evaporation is a cooling process.

Exploration 20.1: Kinetic Theory, Microscopic and Macroscopic Connections

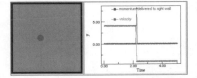

In this animation $N = nR$ (i.e., $k_B = 1$). This, then, gives the ideal gas law as $PV = NT$. The average values shown, $<\ >$, are calculated over intervals of one time unit. Using the ideal gas law, we can make a connection between the macroscopic quantities of temperature (T) and pressure (P) and the individual microscopic properties of a particle of momentum ($\mathbf{p} = m\mathbf{v}$) and kinetic energy ($1/2\ mv^2$).

Let's begin with **one particle** in an enclosed box and bouncing between two walls.

a. What is the change in momentum of the particle (use the velocity vs. time graph) as it hits the wall?

b. What is the average force that the right wall experiences over time? As a reminder, $\mathbf{F}_{avg} = \Delta\mathbf{p}/\Delta t$, so pick a time frame (like 20 or so), multiply the change in momentum each time by the number of collisions with the right wall, and then divide by the total time for this number of collisions.

c. What is the average force on the left wall? The top and bottom walls?

d. Find the pressure on the surface of the box, Force/(area of all the walls). **The dimension of the wall into the screen is 1**.

e. Compare the pressure from (d) with the pressure on the box calculated from the ideal gas law and recorded in the table.

Increase the speed of the particle. The momentum delivered to the wall (and the force it experiences) will increase, thereby increasing the pressure. If the pressure of the gas increases (and if the volume is constant), the temperature of the gas will also increase.

f. What is the new speed of the particle?

g. What is the new pressure?

h. What is the new temperature?

By the same reasoning, increasing the mass of the particle will also increase the pressure, so the temperature should be connected to the mass of the particle as well. **Increase the mass** of the particle.

i.   What is the new mass?

j.   What is the new pressure?

k.   What is the new temperature? The relation between the temperature and increased speed and mass is that the temperature is proportional to the kinetic energy.

One particle in an enclosed box is not very realistic, so let's add a **second particle** (of the same mass) with a different speed. This time, however, we'll plot the kinetic energy of each particle as a function of time and the change in momentum at any wall (an average of this will give us the pressure). The table now shows the average momentum change at the walls. How does this compare with the pressure you calculate using the ideal gas law?

l.   The collision between the particles is an elastic one. How can you tell?

m.   What is the connection between the temperature and the total kinetic energy?

Now let's **add some more particles** of the same mass and different speeds. The table gives the momentum delivered to the wall as particles collide with the wall ($<d\mathbf{p}/dt>$), as well as the pressure calculated from the ideal gas law. This time we plot a histogram of the speeds of the particles. Stop the animation at some time and calculate the total kinetic energy of the ensemble of particles. This, divided by the number of particles, should be the same as the temperature of the system. This is the equipartition of energy theorem: The internal energy of a gas (the sum of the energy of all particles) is equal to $(f/2)k_{\mathrm{B}}NT$, where f is the number of degrees of freedom for the atoms or molecules in a gas. In this case, the particles have 2 degrees of freedom; they can move in the x direction and y direction and thus $f = 2$. Because we are treating the gas particles as "hard spheres" (one of our assumptions in the ideal gas model), the internal energy of the gas is due to the kinetic energy of the particles and is equal to $k_{\mathrm{B}}NT$ and for this animation, $k_{\mathrm{B}} = 1$.

Exploration 20.2: Partial Pressure of Gases

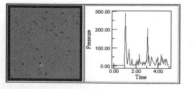

In this animation $N = nR$ (i.e., $k_{\mathrm{B}} = 1$). This, then, gives the ideal gas law as $PV = NT$. Two particles of different masses are in the same container. The total pressure on the container is due to the collisions of both types of particles on the walls. The blue particles are more massive than the red ones (10 times more massive). How can you tell this by watching the animation? (Hint: Temperature is proportional to the average kinetic energy.)

a.   What is the average pressure on the walls? Note: You need to watch the $<P>$ number and wait until it stays around the same number (is not increasing or decreasing, but oscillating about the same number) and then estimate an approximate value. From this pressure and the temperature, use the ideal gas law (in the form $PV = NT$) and calculate the volume of the box the particles are in. If the dimension of the box into the computer screen is 1, what is the length of one wall? Measure the size of the container to verify your calculation.

Run the same animation with only the **red particles**.

b.   What is the pressure on the walls? This is the partial pressure of the red particles.

Run the animation again, but this time with only the **blue particles**.

c.   What is the pressure on the walls? This is the partial pressure of blue particles.

d.   Compare the total pressure with the sum of the partial pressures.

e.   The sum of the partial pressures and the total pressure should be equal. Why?

Now, **start the second animation** when the red and blue particles are in a container with a movable piston between them. The piston generally stays in a position where the pressures on both sides are essentially equal.

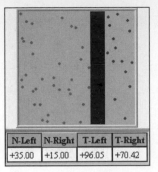

N-Left	N-Right	T-Left	T-Right
+35.00	+15.00	+96.05	+70.42

f. Where does the piston generally stay (right, left, or in the middle)?

g. Why?

h. Remembering that the blue particles have a mass 10 times the red particles, predict what the partial pressure of the red and the blue will be (in the first animation) if there are the same number of red and blue particles (25 each for a total of 50 particles).

Try running the first animation with an **equal number of red and blue** particles (after clicking, scroll back up to the first animation). Run with only **red particles**. Run with only **blue particles**.

i. Was your prediction right? Explain.

j. Predict where you expect the piston will be in the second animation if there are 25 particles on either side of the partition.

k. Try the **second animation with 25 particles** on either side of the partition. Was your prediction right? Specifically, when the temperature is the same on both sides, where is the partition on average?

Exploration 20.3: Ideal Gas Law

The relationship between the number of particles in a gas, the volume of the container holding the gas, the pressure of the gas, and the temperature of the gas is described by the ideal gas law: $PV = nRT$. **In this animation, $N = nR$** (i.e., $k_B = 1$). This, then, gives the ideal gas law as $PV = NT$.

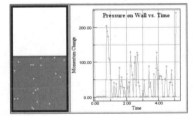

Notice what happens as you change the number of particles, the temperature, and the volume. The pressure is due to collisions with the walls of the container. The graph shows the instantaneous "pressure" (the change in momentum of the particles as they hit the wall and thus exert a force) as a function of time, while the table shows both NT/V (equal to the pressure for an ideal gas) and the average of the instantaneous pressure.

a. Keep the number of particles and the volume constant. What happens to the speed of the particles as the temperature changes? What happens to the pressure (NT/V) if you increase the temperature? (this is known as Gay-Lussac's Law: $P/T = $ constant)

b. If you double the volume (while keeping the same number of particles and the same temperature), what happens to the pressure (and force on the wall)? Why? (This is known as Boyle's law: $PV = $ constant.)

c. If the number of particles is increased (and the temperature and volume stay the same), what happens to the pressure (and the force on the wall)? Why?

d. If you double the volume and halve the temperature (while keeping the number of particles constant), what happens to the pressure? (This is known as Charles's Law: $V/T = $ constant.)

Note that all of these "named" gas laws are included in the ideal gas law: $PV = nRT$. You can also drag the top of the piston to change the volume. In this process both the temperature and pressure can change.

e. Start with a volume of 100. Drag the piston up. What changes and why?

f. Once the particles have spread out into the entire volume available, drag the piston back down. Notice that the particles move fast and the temperature and pressure change dramatically. This is because as the piston comes down and particles hit it, the downward moving piston transfers some of its momentum to these particles and so they speed up. Faster moving particles mean a higher temperature. In a real system, you would not normally see this effect because the particles are moving much faster than any piston being compressed (but if they moved that fast in the animation, you wouldn't see individual particles).

g. How would you need to drag the piston to minimize the change in temperature? Start with a volume of 100 and temperature of 100 again and try to minimize the increase in temperature as you compress the piston.

When you get a good-looking graph, right-click on it to clone the graph and resize it for a better view.

Exploration 20.4: Equipartition Theorem

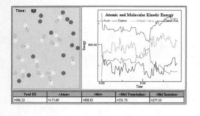

The kinetic energy of a particle can be due to motion in the x, y, and z directions, as well as to rotations. The equiparition of energy theorem says that the kinetic energy of an atom or particle is, on average, equally distributed between the different modes (different degrees of freedom) available. In a monatomic gas, an individual atom has three degrees of freedom because it can move in the x, y and z directions. The energy per particle has an average value of $(f/2)k_B T$, where f is the number of degrees of freedom, k_B is the Boltzmann constant, and T is the temperature.

a. In **this animation of a monatomic gas in a box**, why do the particles only have 2 degrees of freedom? The table shows the total kinetic energy of all particles in the box, as well as the average kinetic energies of particles in the box (the animation averages over a 10-s period, so you need to wait 10 s to read the averages).

b. Record the total energy.

c. What is the energy per particle?

d. If the energy is given in joules/k_B, what is the temperature inside the box?

Try this animation of a diatomic gas with 20 particles. Notice that the graph shows the total kinetic energy of the diatomic particles and the kinetic energies of translation (motion in x and y directions) and rotation.

e. Why is the translational kinetic energy, on average, about two times the rotational kinetic energy? (The animation averages over a 10-s interval, so you need to wait for the animation to run for at least 10 s to read the average values of kinetic energy).

f. From the total energy, what is the energy per particle?

g. If the energy is given in joules/k_B, what is the temperature in the box? (Remember that <energy>/particle = $(f/2)k_B T$ and in this case, f = 3 (Why?).)

Now, **try a mixture of 20 monatomic particles and 20 diatomic particles**.

h. Why is the temperature of the gas in the box a single value (not one value for atoms and another for molecules)? Hint: Think about the air surrounding you at essentially a constant temperature, unless the heater or air conditioner just turned on and made one section of the air a different temperature. Air is made up of monatomic particles (helium) and diatomic particles (water, oxygen, and nitrogen).

i. After waiting at least 10 s, compare the average values of the kinetic energies. What value is the average monatomic kinetic energy close to?

j. Why should those two values (the two averages that you found in part [i]), averaged over a long period of time, be equal to each other and greater than the rotational kinetic energy of the diatomic particles?

k. Explain why the total energy should be equal to $(2/2)20k_B T + (3/2)20k_B T$.

l. From the total energy (given in joules/k_B), what is the temperature?

m. In this animation, if a mixture has 15 monatomic particles, how many diatomic particles should it have so that the average kinetic energies of both particles are the same? Try setting the number of monatomic particles and diatomic particles to check your answer.

Exploration 20.5: PV Diagrams and Work

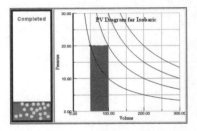

Ideal gas law: $PV = nRT$. **In this animation $N = nR$** (i.e., $k_B = 1$). This, then, gives the ideal gas law as $PV = NT$. The work done during a thermodynamic process depends on the type of process (and can be positive, negative, or zero). Work is given by the equation

$$W = \int P \, dV,$$

so that on a pressure-volume diagram, the area under the curve is the work done **by the gas** during the expansion. In order to analytically solve for the work, you need to know how pressure depends on volume (is pressure constant, changing linearly with volume, etc.?). How pressure varies with volume depends on the type of process (isothermal, isobaric, isochoric, adiabatic).

The three animations show three different processes that start at common temperature and end at common temperature.

a. What is the change in internal energy (ΔU) for these processes (remember that $\Delta U = (3/2)nR\Delta T = (3/2)N\Delta T$ for an ideal monatomic gas)?

b. Estimate the area under the curve (count the blocks on the graph) when the system goes from one temperature to another (from one isotherm on the graph to another). This is the value of the work done since work is $W = \int P dV$. Which process does positive work? Which process does negative work? Which process does zero work?

c. The first law of thermodynamics, $\Delta U = Q - W$, when written as $Q = W + \Delta U$, says that the heat into a system can be used to do work and/or increase the internal energy. Therefore, which process requires the most heat?

d. Compare the area under the curve that you estimated in (b) with the value you calculate using the equations below (found by using calculus and solving the integral):

- Constant pressure: $W = P(V_f - V_i)$
- Adiabatic: $W = (P_f V_f - P_i V_i)/(1 - \gamma)$, where γ (the ratio of C_P/C_V, specific heat at a constant pressure divided by the specific heat at a constant volume) for an ideal monatomic gas is 5/3.

When you get a good-looking graph, right-click on it to clone the graph and resize it for a better view.

Exploration 20.6: Specific Heat at Constant Pressure and Constant Volume

In this animation $N = nR$ (i.e., $k_B = 1$). This, then, gives the ideal gas law as $PV = NT$.

For an ideal monatomic gas, the change in internal energy depends only on temperature, $\Delta U = (3/2)nR\Delta T = (3/2)N\Delta T$.

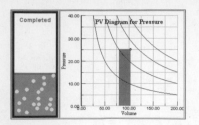

a. Calculate the change in internal energy for the three cases.

b. What is the work done in each case? As a reminder, $W = \int P \, dV$, and pressure can (and does in many instances) depend on volume. Calculate the work done in each case using the following two methods, then compare your answers.

- Graphically: To find the work done is to determine the area under the curve ("area" of the red region on the graph). After estimating the area by counting the grid blocks, click the checkbox above to show a calculation of the area (by numerical integration) on the simulation. Explain any significant differences between your estimation and the numerical integration.

- Analytic solution (a little bit of calculus required): When heat is added at constant pressure (isobaric process), then P, pressure, in the above equation for work is simply a constant of integration. When heat is added at a constant volume (isochoric process), the work done is zero. Why? When heat is added at a constant temperature (isothermal), use the ideal gas law ($PV = NT$) and write the pressure as a function of volume: NT/V (where N and T are constant) and then you can integrate (the answer involves a natural logarithm).

c. Using the first law of thermodynamics, $Q = W + \Delta U$, calculate the heat input and show that it is the same for all three cases.

The specific heat capacity of a material is a measure of the quantity of heat needed to raise a gram (or given quantity) of a material 1°C. For a gas, it requires a different amount of heat to raise the same amount of gas to the same temperature depending on the circumstances under which the heat is added. If the same amount of heat is added, the final temperature of the constant pressure and constant volume expansions are quite different (and, for a constant temperature, heat is added but the temperature does not change!).

d. In which case does the heat input raise the temperature the most? Why?

So, if the specific heat capacity of an ideal gas is to have any meaning at all, it must be defined in terms of the process: specific heat at a constant volume or specific heat at a constant pressure.

e. Go back to your calculation of heat in (c). Calculate the constant of proportionality between heat input and the change in temperature for the constant volume and constant pressure cases: $Q = (\text{Constant}) N \Delta T$.

f. What is the constant in each case? Why is the constant for an expansion at constant pressure greater? (Hint: Think about whether the heat is used only to change temperature or to change temperature and do work.)

Generally, we write the heat capacity as a molar heat capacity (where n is the number of moles) and find that for constant pressure $Q = C_P n \Delta T$ and $C_P = (5/2)R$, and for constant volume expansion $Q = C_V n \Delta T$ and $C_V = (3/2)R$.

We began this discussion by noting that for an ideal monatomic gas, the average internal energy is $(3/2)T$. This comes from kinetic theory and the equipartition of energy, where the 3 comes from 3 degrees of freedom. For a diatomic gas, the average internal energy is $(5/2)T$ because there are two more degrees of freedom (rotation).

g. How are the heat capacity at a constant pressure and heat capacity at a constant volume different for a diatomic gas?

When you get a good-looking graph, right-click on it to clone the graph and resize it for a better view.

Problems

Problem 20.1

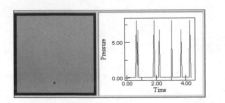

In this animation $N = nR$ (i.e., $k_B = 1$). This, then, gives the ideal gas law as $PV = NT$. The average values shown, < >, are calculated over intervals of five time units. When one particle is present, you see the temperature and average pressure. **Add 24 more particles** with the identical initial speed. What should the missing values in the table be? Explain.

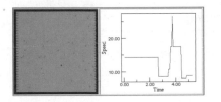

Problem 20.2

The right and left walls of the box are at different temperatures. The graph shows the speed as a function of time for the blue particle (which is identical to the red particles). Which wall is at a higher temperature? Explain how you arrived at your answer.

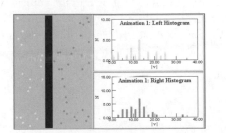

Problem 20.3

In this animation $N = nR$ (i.e., $k_B = 1$). This, then, gives the ideal gas law as $PV = NT$.

Two different groups of particles, blue and another color (depending on the animation), are in a container with a piston between them. In each of the animations the blue particles have the same mass, but the mass of the other particles is different. Rank the nonblue particles from most massive to least massive.

Problem 20.4

The gas in the animation has 20 monatomic particles and 10 diatomic particles. Which of the lines on the graph corresponds

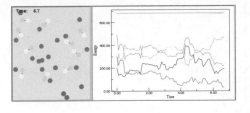

to the total kinetic energy, the total kinetic energy of the monatomic particles, the total kinetic energy of the diatomic particles, and the translational and rotational kinetic energy of the diatomic particles? Explain.

Problem 20.5

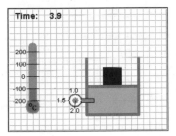

If the gas inside the container being compressed is an ideal gas, what is wrong with this animation? The pressure is shown on the gauge (in atmospheres) and is caused by the black block sitting on the piston (position is given in centimeters).

Problem 20.6

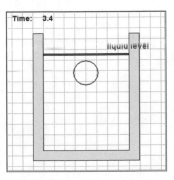

A large balloon filled with an ideal gas is initially held at the bottom of a tank full of a liquid. It is released and floats to the top as shown in the animation. If you click on the graph, you can see the x and y positions of the cursor (position is given in meters). The top of the liquid (indicated by the blue line) is open to air at atmospheric pressure.

Find the density of the liquid (the liquid is at a constant temperature).

Note: You will need to recall what you learned about fluids (and the change in pressure as a function of depth in a fluid).

Problem 20.7

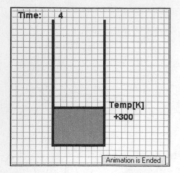

When heated, materials expand in all three dimensions (position is given in meters). The equation for the volume expansion is as follows:

$$\Delta V = V_o \beta \, \Delta T,$$

where the change in volume (ΔV) is equal to the initial volume (V_o) multiplied by the coefficient of volume expansion, β, and by the temperature increase. Note that this equation is similar to the equation for linear expansion (see **Exploration 19.2**), and that for solids the coefficient of expansion, β, is approximately equal to 3α. For a gas we need to be careful because a gas can expand even without a change in temperature (if the pressure decreases).

a. Assuming constant pressure, find the volume expansion coefficient of the gas at this initial temperature (100 K).

b. Assuming an ideal gas, use $PV = nRT$ to find (i.e., derive an expression for) the volume expansion coefficient and see that it varies with temperature (is not a constant).

c. Pick a new initial temperature, use the animation, and verify that the results match the expression you derived.

Problem 20.8

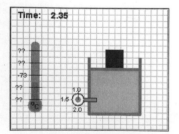

A mass is pushed downward on top of a container (the dimension of the container into the screen is 30 cm) so that an ideal gas undergoes a temperature change as shown in the animation (pressure is given in atmospheres, position is given in centimeters, and time is given in seconds).

a. What is the work done on or by the gas?

b. Sketch a PV diagram of the process.

c. If the initial temperature of the gas is $-73°C$, what is the temperature scale (i.e., what are the divisions on the thermometer: 10°C units? 25°C?)?

Problem 20.9

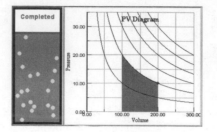

In this animation $N = nR$ (i.e., $k_B = 1$). This, then, gives the ideal gas law as $PV = NT$.

How much heat is added to or removed from the gas during the expansion of the ideal gas?

When you get a good-looking graph, right-click on it to clone the graph and resize it for a better view.

Problem 20.10

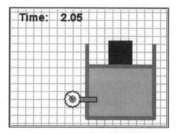

A mass is pushed downward on top of a container with 3 moles of an ideal monatomic gas inside as shown in the animation (position is given in meters and time is given in seconds). The pressure remains constant as indicated by the round pressure gauge.

a. If the initial temperature is 20°C, what is the work done on the gas?

b. How much heat is absorbed or released in the process?

c. Draw a PV diagram.

Problem 20.11

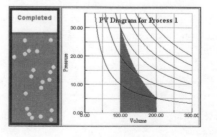

In this animation $N = nR$ (i.e., $k_B = 1$). This, then, gives the ideal gas law as $PV = NT$.

a. Rank the expansions by the work done by the gas (from negative to positive).

b. Rank the expansions by change in internal energy (from negative to positive).

c. From these rankings, if possible, rank the expansions according to the heat required (again from negative to positive).

When you get a good-looking graph, right-click on it to clone the graph and resize it for a better view.

Problem 20.12

In this animation $N = nR$ (i.e., $k_B = 1$). This, then, gives the ideal gas law as $PV = NT$. The same amount of heat is added to an ideal gas in all three processes.

a. What is the value of γ (the ratio of C_P to C_V) for this ideal gas?

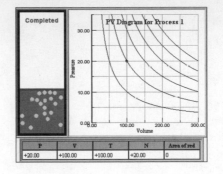

P	V	T	N	Area of red
+20.00	+100.00	+100.00	+20.00	0

b. According to the equipartition of energy theorem, how many degrees of freedom do the particles that make up this gas have?

c. Is the gas monatomic? diatomic? polyatomic?

When you get a good-looking graph, right-click on it to clone the graph and resize it for a better view.

Engines and Entropy

Topics include first and second law of thermodynamics, entropy, PV diagrams, engines (Carnot Cycle, Otto, Brayton, and internal combustion), and refrigerators.

Illustration 21.1: Carnot Engine

In this animation $N = nR$ (i.e., $k_B = 1$). This, then, gives the ideal gas law as $PV = NT$.

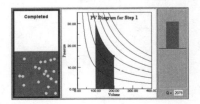

There are four steps to the Carnot cycle: a combination of isothermal and adiabatic expansions and contractions. Which of the steps are isothermal and which are adiabatic? Make sure to step through the animation in order. During step one the gas does a positive amount of work. Step two is adiabatic, with the gas doing a positive amount of work. During step three there is a negative amount of work done by the gas. Finally in step four, which is adiabatic, the work done by the gas is negative. Notice that the total work done (the remaining area) is positive because positive work is done at high temperatures and negative work is done at lower temperatures. During step one heat is absorbed ($Q > 0$) and during step three heat is released ($Q < 0$). More heat is absorbed than is released for the entire cycle. This is the basis of how engines work: Heat (from the hot reservoir) is transferred into mechanical work (piston moving).

Illustration 21.2: Entropy and Reversible/Irreversible Processes

In **Animation 1** you see an ensemble of particles that looks "natural" once the particles spread throughout the box. You would not expect to see the reverse of this process. Why? Consider **Animation 2**. This one looks the same running forward or backward. The first animation is an example of an irreversible process, while the second animation is an example of a reversible process. What differentiates the two processes? The concept of entropy.

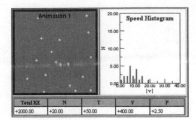

Run both animations again. Look at the total energy (kinetic energy) in both cases. Does the energy change? No. The conservation of energy (stated in thermodynamics as the first law of thermodynamics) does not help us determine which of these animations is more likely (since energy is conserved in both cases).

To determine which animation is a more realistic picture of gas particles in a box, we must apply the second law of thermodynamics and the associated concept of entropy. Entropy is a measure of the disorder of a system. Which animation has greater entropy (disorder)? Why? Clearly Animation 2 is a much more ordered system. Animation 1 starts out ordered, yet ends up disordered.

As you watch the first animation, also notice the many different velocity distributions that are possible that still result in the same overall energy (temperature) and pressure. Statistically speaking, it is much, much, much, much more likely for the speeds of an ensemble of particles to be closer to a Maxwell-Boltzmann distribution than to all have the same speeds.

Entropy and the second law of thermodynamics describe what is more likely to happen. It is more likely for particles to go into states of greater disorder because there are more possible "disorderly states" than ordered ones (and the number of possible states is related to the entropy). For example, there are many more ways for a group of particles to follow a Maxwell-Boltzmann distribution than the distribution having identical speeds for each particle. The second law says that entropy either increases or stays the same; irreversible processes cause an increase in entropy. This is our way of knowing whether or not a movie is running forward or backward—as we move forward in time, entropy increases. If entropy decreases somewhere (when electrons are organized to light up a computer screen in a certain manner so you can read this, for example), energy is required, and the energy needed results in an increase in entropy elsewhere so that globally, entropy increases.

Illustration 21.3: Entropy and Heat Exchange

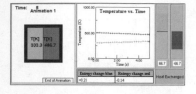

Animation 1 shows two objects of the same size, same mass, and same specific heat ($mc = 2$ for both), initially at different temperatures but in thermal contact with each other (temperature is given in kelvin and heat exchange is given in joules). The color-coded bars indicate the heat exchanged between the red and blue objects.

When two objects are in thermal contact with each other, we expect them to eventually reach the same temperature. However, there is nothing in the first law of thermodynamics that requires this. The only requirement from the first law is that energy be conserved, that the heat from one object goes into the other.

Try **Animation 2**. Is energy conserved? Does the heat from one go into the other? How does the heat exchange in this case compare with Animation 1? What you see in Animation 2, of course, does not happen, even though energy is conserved. The second law of thermodynamics, that entropy (in an isolated system) either increases or stays the same, governs this. The change in entropy, ΔS, is given by $\Delta S = \Delta Q/T$ (for reversible processes at a constant temperature), and since $Q = mc\Delta T$, with a bit of calculus, you get

$$\Delta S = mc \ln (T_f/T_i),$$

where c is the specific heat capacity of a material and m is the mass of the material. What is the change in entropy for each object in the first animation? What is the total change in entropy? What about for the second animation? Notice that the net change in entropy for Animation 1 is positive, and in Animation 2 the net change in entropy is less than zero. According to the second law, processes don't naturally decrease entropy (it requires the input of energy), so what you see in Animation 2 does not occur in an isolated system because that would violate the second law of thermodynamics.

Illustration 21.4: Engines and Entropy

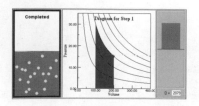

In this animation $N = nR$ (i.e., $k_B = 1$). This, then, gives the ideal gas law as $PV = NT$.

This engine cycle, the Carnot cycle, is considered to be a reversible process because it can run forward or backward. For a reversible process, we define a change in entropy as $dS = dQ/T$ or $\Delta S = \Delta Q/T$. (Note that if T changes, some calculus is involved so that $\Delta S = \int dQ/T$.)

For the Carnot cycle, we can calculate the change in entropy over a cycle by finding the change in entropy for each step. What is the heat input or output divided by the temperature in each step? Note that for the two adiabatic steps, although the temperature changes, the heat input is zero and the process is reversible, so you do not need to use calculus, and $\Delta Q = 0$. Adding the two nonzero terms, you should find that the entropy change is zero for this cycle. The second law of thermodynamics says that to keep the entropy change at zero is the best you can do (the entropy of a cyclic process either increases or remains zero).

Entropy is important for calculations regarding engines because it tells you the best efficiency you can hope for. The efficiency of any engine is the (work out)/(heat in) = $|W|/|Q_H|$ where Q_H is the heat input from the high temperature reservoir (from whatever heats the gas: burning gasoline, propane, boiling water, etc.). To calculate the efficiency of this engine, take the work done by the gas (698 total for all steps) and divide by the heat absorbed by the engine (2079 during step 1) and we get 0.33.

For an ideal engine (no frictional losses, reversible processes), $|W| = |Q_H| - |Q_L|$, and the efficiency is $|W|/|Q_H| = 1 - |Q_H|/|Q_L|$, where Q_L is the heat exhausted to the low temperature reservoir. Since the entropy change of this cycle is zero, this means that $Q_H/T_H + Q_L/T_L = 0$. Thus, for an engine operating between two temperature reservoirs, the maximum efficiency is $1 - |T_H|/|T_L|$.

Entropy, S, is a state variable (does not depend on the process) just like pressure, volume, and temperature and unlike work and heat, which do depend on the process. We can thus describe a thermodynamic process by including entropy on a graph. We often describe a process on a TS diagram, because the area under a curve in a TS diagram is the heat. **Click here to change the graph from a PV diagram to a TS diagram**. (Note that the initial entropy is arbitrary—we are simply interested in the change in entropy.)

Exploration 21.1: Engine Efficiency

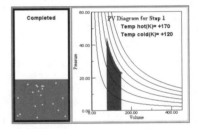

In this animation $N = nR$ (i.e., $k_B = 1$). This, then, gives the ideal gas law as $PV = NT$. Assume an ideal monatomic gas. The efficiency of an engine is defined as $\varepsilon = $ (work out)/(heat in) $= |W|/|Q_H|$.

a. Pick a temperature for the hot reservoir (between 200 K and 150 K) and a lower temperature (between 150 K and 100 K) for the cold reservoir. (Note that new reservoir temperatures only register at the beginning of the engine cycle and that you should run the engine steps in order for the animation to make sense.) Find the work done for each step and the heat absorbed or released (remember that $\Delta U = (3/2)nR\Delta T = (3/2)N\Delta T$).

b. Calculate the efficiency of the engine for these temperatures.

c. Pick a different pair of temperatures for the reservoirs. Is this engine more or less efficient? (Calculate the efficiency for this new engine.)

d. Why is the engine in (c) more or less efficient?

e. What would make the engine even more efficient? Try it and explain.

f. Calculate the difference in temperatures of the reservoir divided by the temperature of the hot reservoir: $(T_H - T_L)/T_H = 1 - T_L/T_H$. Compare this value with the efficiency number for each case above. For a Carnot engine, either calculation gives the efficiency because of the following:

g. For step 1, $W = Q_H = nRT_H \ln(V_1/V_0) = NT_H \ln(V_1/V_0)$, where V_0 is the volume at the beginning of step 1 and V_1 is the volume at the end of step 1. Explain why (do some algebra!) and verify with the animation.

h. Similarly, for step 3, $W = Q_L = nRT_L \ln(V_3/V_2) = NT_L \ln(V_3/V_2)$, so $|Q_L| = NT_L \ln(V_2/V_3)$, where V_2 is the volume at the beginning of step 3 and V_3 is the volume at the end of step 3. Explain why (do some algebra!) and verify with the animation.

i. Steps 2 & 4 are adiabatic, so $Q = 0$. From steps 2 and 4, $P_1 V_1^\gamma = P_2 V_2^\gamma$ and $P_3 V_3^\gamma = P_0 V_0^\gamma$, where γ is the adiabatic coefficient (ratio of the specific heat at constant pressure to specific heat at constant volume). Using these relations and the ideal gas law, show (more algebra) that $(V_1/V_0) = (V_2/V_3)$.

j. Therefore, show for a Carnot engine $|W|/|Q_H| = 1 - |Q_L|/|Q_H| = 1 - T_L/T_H$.

The Carnot engine efficiency of $1 - T_L/T_H$ is the ideal efficiency for any engine operating between two reservoirs of T_H and T_L, because the net change in entropy is zero for a Carnot cycle (see **Illustration 21.4**). Often, you will compare the efficiency of other engines, $|W|/|Q_H|$, to the ideal Carnot engine efficiency. Note that you cannot have a 100% efficient engine because that would require $T_L = 0$ (which is forbidden by the third law of thermodynamics). Another way to think about this is that to get an efficiency of 100% you would need to put the heat released into the low temperature reservoir back into an engine. But that engine would need to run between T_L and a lower temperature, and again you can't get to $T_L = 0$.

Exploration 21.2: Internal Combustion Engine

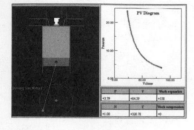

In this animation $N = nR$ (i.e., $k_B = 1$). This, then, gives the ideal gas law as $PV = NT$. We will assume an ideal gas in the engine.

The Otto engine cycle is close to the cycle of an internal combustion engine (and closer to a real engine than the Carnot engine). This cycle consists of adiabatic and isochoric processes plus a cycle of exhausting smoke and taking in new gas. Identify which parts of the engine cycle correspond to which process. No net work is done in the complete process of exhausting smoke and taking in gas. Explain why. Notice that during this part of the cycle, the number of particles changes because the red valves at the top open and close to let gas in and out. Thus, in the release of high temperature particles and intake of low temperature particles, heat is exchanged (released to the environment).

a. For the adiabatic expansion, what are the initial pressure and volume? What are the final pressure and volume? (Remember you can click on the graph to read points from it.) From these values, find the adiabatic constant, γ (since $PV^\gamma = $ constant for an adiabatic expansion).

b. Is the gas monatomic ($\gamma = 1.67$), diatomic ($\gamma = 1.4$), or polyatomic ($\gamma = 1.33$)?

c. What is the net work done during the cycle (the work out)?

d. Neglecting the gas exhaust and intake parts of the cycle, in which part of the cycle is heat absorbed? In which part of the cycle is heat released?

e. Calculate the heat absorbed. Remember that $Q = \Delta U + W$ and that $\Delta U = f/2N\Delta T$, where $f = 3$ for monatomic gases, 5 for diatomic gases, and 6 for polyatomic gases.

f. What is the efficiency of this engine? The efficiency of an engine is $\varepsilon = $ (work out)/(heat in) $= |W|/|Q_H|$.

g. Check that your answer is equal to $1 - (V_{min}/V_{max})^{1-\gamma}$ and is therefore dependent on the ratio of the maximum and minimum volume (called the compression ratio).

Exploration 21.3: Entropy, Probability, and Microstates

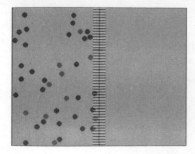

In the animation, two containers are separated by a "membrane." Initially, no particles can cross the membrane. Note that the red and blue particles are identical. They are colored so you can keep track of them. Once the particles are fairly evenly distributed in the left chamber, you are ready to let particles through. Try **letting particles through the membrane**. This animation allows about every other particle that hits the membrane to get through (equally in either direction). When there are about the same number of particles on both the left and right sides, pause the animation and count the number of red particles on each side and the number of blue particles on each side. Let the animation continue and stop it again a few seconds later when there are about the same number of particles on each side. Again, find the number of red and blue particles on each side.

a. Given that there are 30 blue particles and 10 red particles total, if you made many such measurements, what would you expect the average number of red and blue particles to be on each side (when there are a total of 20 particles on each side)?

b. Now, **restart the animation**. Once the particles are fairly evenly distributed in the left chamber, try **letting particles through the membrane a different way**. Again, this animation allows about every other particle through the membrane.

c. When there are about the same number of particles on each side, count the red and blue particles on each side. What is different about the way this membrane is set up?

d. Is it possible for the first membrane to have this outcome?

e. Is this outcome likely?

The reason that the second membrane does not appear "natural" is the second law of thermodynamics. One version of the second law is that the entropy of an isolated system always stays the same or increases (where entropy is defined as a measure of the disorder of the system). In other words, "natural" systems move in the direction of greater disorder. In the animations, the first membrane seems "natural" because it allows for the most disorder—a random distribution of reds and blues on both sides. This is compared to the second membrane that only allows blue particles through, and thus the right side will always have only blue particles in it.

Another way to interpret the second law is in terms of probability. It is possible with the first animation to get 0 red particles in the right chamber, but it is very unlikely (just like it is possible you will win the lottery, but it is very unlikely). It is also possible for the second animation to behave as it does, but again, it is very unlikely. Consider the **animation above with only six particles**: four blue and two red. To keep track of things, we've colored the blue ones different shades of blue and the red ones different shades of red. Run the animation and notice how often there are three blue ones on the right side when there are three particles on each side. What follows will allow you to calculate the probability of this happening and show that when there are three particles on each side there is a 20% chance that there will be three blue ones in the right chamber.

f. Considering the different arrangements of three particles on each side, note that there are four different ways to get three blues on the right and two reds and one blue on the left (list these and **click here** to show them). Similarly, there are the same four ways to get three blue particles in the left chamber.

g. There are six ways to get the light red on the left and the dark red on the right with two blues each (**click here to show these**). Again, there are the same six ways to have the dark red on the left and the light red on the right.

h. This gives a total of how many different arrangements (of three particles on each side)? Since all these states are equally likely, you have only a 20% chance of having three blues in the right chamber.

As we add more particles, it becomes less likely to get all of one color on one side. With 40 particles, 30 blue and 10 red, there is only around a 0.02% chance that when there are 20 particles on each side, there will be 20 blue ones on the left and 10 red and 10 blue on the right. This is not impossible, but not very likely (better odds than your local lottery, where your odds might be around one in a couple of million). A more ordered state (20 blues on the right) is less likely, statistically, than a less ordered state (reds on both sides of the membrane, which is a more even mixing). Entropy is related to the number of available states that correspond to a given arrangement (mathematically, $S = k_B \ln W$, where S is entropy, W is the number of equivalent arrangements or microstates, and k_B is the Boltzmann constant).

Going back to our six-particle example, there were more states that corresponded to one red in each chamber than two reds in one chamber, so one red in each chamber is a more likely state for the system to be in. Most of the time, however, we are dealing with more than six particles (usually around Avogadro's number), so the very ordered state is even less likely to happen. Connecting entropy to probability, then, gives a version of the second law that does not forbid the system from being a highly ordered state; it simply says that it is highly unlikely.

Exploration 21.4: Entropy of Expanding Ideal Gas

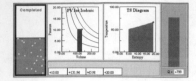

In this animation $N = nR$ (i.e., $k_B = 1$). This, then, gives the ideal gas law as $PV = NT$.

In thermodynamic processes the entropy depends not on the path taken but on the end points. It is a "state function" (in contrast to heat and work, which depend on the process). Since $Q = \Delta U + W$ and $\Delta U = (3/2)nR\Delta T$ (for a monatomic gas),

$$\Delta S = \int dQ/T = \int (3/2)nR dT/T + \int P dV/T = nR[(3/2)\ln(T_f/T_i) + \ln(V_f/V_i)].$$

Thus, $\Delta S = (3/2)N \ln(T_f/T_i) + N\ln(V_f/V_i)$ for an ideal monatomic gas (note that ln represents the natural log, base e).

In the animations, note that the area under the PV diagram is equal to the work.

a. What is the work done in each case?

b. What is the heat absorbed or released in each case?

c. What is the area under the associated TS diagram? (Note that the choice of the initial entropy is arbitrary.)

d. How does the change in entropy compare for the three processes?

e. Compare your measurements from the graphs to the calculated values found using the equation above for an ideal monatomic gas.

Another way to measure the change in entropy is to use $Q = mc\Delta T$ or, for a gas, $Q = CN\Delta T$. In this case,

$$\Delta S = CN \ln(T_f/T_i).$$

f. Show that for the isobaric expansion where $C = C_P = (5/2)$, you get this change in entropy.

Problems

Problem 21.1

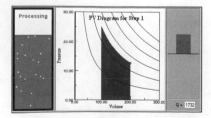

In this animation $N = nR$ (i.e., $k_B = 1$). This, then, gives the ideal gas law as $PV = NT$.

a. Describe the types of expansions and compressions that are a part of this engine cycle.

b. For this ideal gas, what is γ (the adiabatic constant, or the ratio of the specific heat at constant pressure to the specific heat at constant volume)?

c. Find the work done in each step.

d. What is the net work done?

Problem 21.2

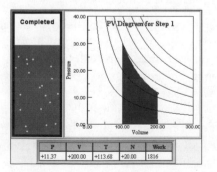

P	V	T	N	Work
+11.37	+200.00	+113.68	+20.00	1816

In this animation $N = nR$ (i.e., $k_B = 1$). This, then, gives the ideal gas law as $PV = NT$. To run this Otto engine (cycles of adiabatic and isochoric expansions and contractions), you must go through the steps in order. The work shown in the data table is the work done during each step.

a. Is the gas a monatomic or diatomic ideal gas?

b. What is the net work and what is the heat absorbed?

c. Find the efficiency of this Otto engine.

(Note that the high temperature and low temperature reservoirs do not stay at constant temperatures).

Problem 21.3

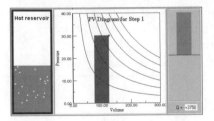

In this animation $N = nR$ (i.e., $k_B = 1$). This, then, gives the ideal gas law as $PV = NT$. To run this Brayton cycle, go through the steps in order.

a. Describe the types of expansions and compressions that are a part of this engine cycle.

b. For this *monatomic* gas, find the work done in each step.

c. What is the net work done?

d. What is the efficiency of this cycle?

(Note that the high temperature and low temperature reservoirs do not stay at constant temperatures).

Problem 21.4

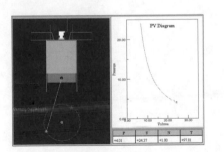

In this animation, $N = nR$ (i.e., $k_B = 1$). This, then, gives the ideal gas law as $PV = NT$. We will assume an ideal *diatomic* gas in the engine.

The Otto engine cycle is close to the cycle of an internal combustion engine and consists of adiabatic and isochoric processes.

a. Identify which parts of the engine cycle correspond to which processes.

b. Explain why, during the gas intake and smoke exhaust process, there is no net work done. Also explain how heat is released.

c. Calculate the work done during the adiabatic expansion and compression.

d. Find the heat exchanged during the isochoric processes.

e. Determine the efficiency of this engine.

Problem 21.5

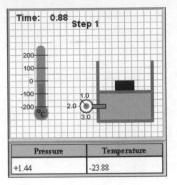

A mass is pushed downward on top of a container (the dimension of the container into the computer screen is 30 cm) filled with an ideal monatomic gas that undergoes a thermodynamic process as shown in the animation (pressure is given in atmospheres, position is given in centimeters, temperature is given in degrees Celsius, and time is given in seconds). Both the dial and the relative height of the black box on the piston represent the changing pressure.

a. Determine how many moles of gas are in the container.

b. Find the following:

- The work done by or on the gas in each step.
- The heat gained or lost by the gas in each step.

c. Draw a PV diagram for this engine cycle.

d. Calculate the efficiency of this engine.

Problem 21.6

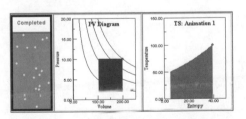

In this animation $N = nR$ (i.e., $k_B = 1$). This, then, gives the ideal gas law as $PV = NT$. Assume an ideal monatomic gas.

Which temperature-entropy graph goes with the animation shown? Explain.

Problem 21.7

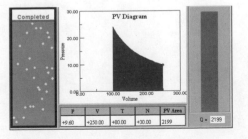

In this animation $N = nR$ (i.e., $k_B = 1$). This gives the ideal gas law as $PV = NT$. Assume an ideal monatomic gas.

a. Identify the thermodynamic process.

b. Draw a TS diagram and calculate the change in entropy.

Problem 21.8

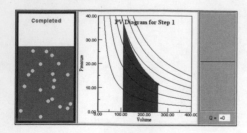

In this animation $N = nR$ (i.e., $k_B = 1$). This, then, gives the ideal gas law as $PV = NT$. Assume an ideal gas.

The reverse of an engine is a refrigerator. An engine uses heat to produce work, while a refrigerator does negative work on the gas (something else does work on the gas) to remove heat. The coefficient of performance is $K = |Q_C|/|W|$ (heat transferred out of the cold reservoir divided by the work required).

a. In which step is the heat removed from the cold reservoir (i.e., heat removed from the refrigerator and absorbed by the gas)?

b. In which step is work done on the gas?

c. For the refrigerator, find the work done during each cycle, the heat transferred from the cold reservoir, and the coefficient of performance.

d. What is the change in entropy for the complete cycle?

Electrostatics

Topics include Coulomb's law, test charges, charge distributions, charging by induction, charge/mass ratios, static electricity, and superposition.

Illustration 22.1: Charge and Coulomb's Law

What is charge? Charge is a property of certain subatomic particles and is **not** a substance that can be transferred from one particle to another. Particles either have charge or they don't. When we say that we are charging an object, what we really mean to say is that we are transferring particles that have charge from one macroscopic object to another macroscopic object.

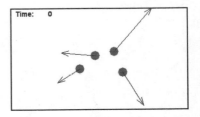

Experiments done over 200 years ago by Benjamin Franklin and others led to the arbitrary assignment of the name "negative" to the property of those particles that are transferred to hard rubber when it is rubbed with wool. Franklin did not, of course, know about elementary particles. We now know that the particles being transferred in rubbing are electrons. We also know that electrons are not the only particles that have the property of charge. You have probably heard about protons, but there are many other particles with nonzero charge. When charging an object, we could say that we are "massing" a ball with electrons rather than "charging" a ball. Such literalism would be correct, but awkward. What we are interested in, after all, is the charge property that is being imparted to the ball by the electrons.

Use the animation to create three like charges at $x = -1$ m, $x = 0$ m, and $x = 1$ m. You can do this by entering the position in the text box and clicking the "add" button (position is given in meters). What is the net force on the middle charge? It is zero because the forces from the other two charges cancel. Now move one of the outer charges around. Is the force on the middle charge still zero? No. The force between two particles always lies on the line between the two particles, is attractive or repulsive depending on the signs of the charges, and varies as 1 over the square of the separation distance $(1/r^2)$.

Add a few charged particles with the same magnitude of charge (and with both positive and negative sign), and move them around by click-dragging them. Use the text box for charge to add the charges. The arrows on the screen show how the particles interact with each other by showing both the magnitude (relative size of the arrow) and direction of the electrostatic force.

Now reset the animation and create two charges with different amounts of charge. Notice the differences and similarities in the force vectors shown on the particles. Also look at the case in which the charges have the same polarity (same sign) and opposite polarity (one positive and one negative). With only two charges, the forces acting on the two objects are always equal and opposite.

The animation allows you to create particles of either polarity and to create particles with any amount of charge. Nature, on the other hand, is constrained. As far as we know, particles can only be created so that the total charge does not change. That is, if a positive particle is created, then another negative particle must also be created. Furthermore, the magnitude of the charge that is created must be an integer multiple of a fundamental unit. Although these restrictions are most apparent in the microscopic world, they manifest themselves in the macroscopic world. For example, a battery requires that equal numbers of charged particles enter and leave the two terminals. (Otherwise the battery would be creating one type of charge.) In the final analysis, all that we can really say is that certain particles have a property called charge that enables them to repel some particles and attract others. If two charged particles repel, then a third charged particle will either repel both particles or attract both particles.

Illustration 22.2: Charge and Mass

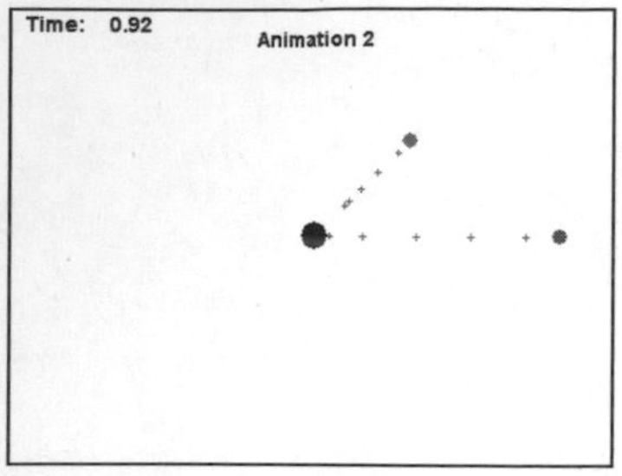

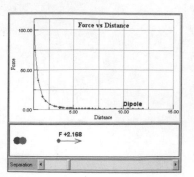

This Illustration shows a fixed charge at the center and one or more test charges (depending on whether you choose **Animation 1** or **Animation 2**) that move under the influence of the fixed charge. Run the animations and observe the motion of the test charges. You can reset the animations and click-drag the test charges. Can you determine the sign of the fixed charge? In other words, is the fixed charge positive or negative? Can you determine the mass of the test charge? What can you say about the force between the fixed charge and the test charge? How is the interaction similar to and how is it different from Newton's universal law of gravitation? These questions are of fundamental importance to physicists who must try to determine the charge, mass, and other physical properties of elementary particles using trajectories produced in experiments at high-energy particle accelerator laboratories throughout the world.

A test charge is defined as a positively charged object whose charge is so small that it does not influence other objects, including other test charges. Therefore, in these animations we assume that the fixed object has much more charge than the test charge(s) so that the motion of the test charges is determined by the Coulomb interaction with the fixed charge. This is similar to having a number of satellites in orbit around Earth. We usually ignore the gravitational attraction between satellites and only consider a satellite's attraction to Earth, and maybe the moon and Sun, when we calculate the satellite's trajectory. In **Animation 2** what are the directions of the forces on the red and green test charges? The directions are radially outward just as it is for only one test charge in **Animation 1**. Remember, we are ignoring the effect of the test charges on each other's motions.

In **Animation 2**, if the test charges have the same mass, can you determine which test charge, red or green, has the bigger charge? If the mass is the same, comparing the accelerations will tell you about the force and, therefore, the charge on the test charge. What if the test objects had different masses as well as different charges? Would this change your answer? The ratio of charge to mass is now proportional to the acceleration, and this ratio affects the motion that you observe. In general, it is not easy to untangle the combination of charge and mass on the motion of charged particles. Early experiments using particle trajectories were performed in 1887 by Joseph J. Thompson and led to the realization that electrons are charged particles, but it took another 24 years before Robert A. Millikan was able to separate the effect of charge from that of mass.

Illustration 22.3: Monopole, Dipole, and Quadrupole

The Coulomb force law predicts that the force of attraction (or repulsion) falls off as $1/r^2$ as the distance between two charges increases. But nature rarely provides us with point charges. Molecules, for example, consist of positive and negative charges bound together by nonclassical forces that can only be explained using quantum mechanics. But electrical forces are still present even if positive and negative charges are bound, and we can develop useful force laws that approximate common charge distributions.

This Illustration allows you to study the force between a movable test charge and orientations of one, two, and four fixed charges. The force between the test charge and a single point charge, known as a monopole, obeys the Coulomb force law. A system consisting of two closely spaced charges of opposite polarity is known as a dipole. Two dipoles placed next to each other form what is called a quadrupole. What can you say about the differences in the force vs. distance graph for the three cases? Does one or more of the plots show a force that decreases at some rate other than $1/r^2$? If so, why isn't this a violation of Coulomb's law? Why does the force drop off more quickly with the addition of more charges?

When you add up the forces due to several charges, the net force experienced by other charges may be different from $1/r^2$ depending on the orientation, magnitude, and sign of the charges. For the dipole, the separation between the positive charge and the test charge is almost the same as the separation between the negative charge and the test charge. If the separations were the same, the two charges would be on top of each other, and the net force on the test charge would be zero. But these separations are not quite identical. When we add up these forces, for the dipole we get a net force that goes as $1/r^3$ and for the quadrupole we get a net force that goes as $1/r^4$.

When you get a good-looking graph, right-click on the graph to get a copy of that graph in order to contrast it with the other animations.

Illustration 22.4: Charging Objects and Static Cling

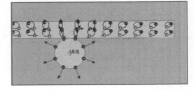

These animations model materials with charges (red = positive and blue = negative). The arrows show the forces between particles.

There are a number of ways to charge objects. You may be familiar with what happens when you rub a balloon against your sweater. Due to the rubbing, the balloon becomes negatively charged. It will stick to the wall or the ceiling, but the wall and ceiling are neutral. Why will it stick to neutral objects? Run the **balloon animation** to see. The model shows the negatively charged balloon near a neutral ceiling. While neutral, there are charges in the ceiling. The ceiling is not chargeless; it just has an equal number of positive and negative charges. What happens to the neutral ceiling? This effect is called polarization (when the charges in an atom get slightly distorted due to other nearby charges).

Another way to charge an object is by induction (to induce it). First, look at the case where the left plate is positively charged and the right plate is neutral (equal number of positive and negative charges) **as in this animation**. Why do the charges separate as they do in the right plate? Charges move according to the forces they experience (like charges attract and opposite charges repel). Suppose we give the charges on the right a place to go (specifically ground) **as shown in this animation**? What happens? Why? Neutral pairs of positive and negative charges separate from each other due to the nearby positive charge. Then the positive charges on the right leak off the grounded plate.

When an object is charged (like a computer screen), other objects can stick to it. We call this "static cling." Consider charged particles near a charged screen as shown **in this animation**. What happens to the positively charged particles? How about negatively charged particles? Now consider neutral particles as **in this** animation. What happens to the neutral particle between two charged screens? It gets polarized and then attracted to the screen. Notice that a charged screen can therefore attract both charged particles and neutral particles. This explains why your computer and television screens (which are charged) collect dust so easily.

Exploration 22.1: Equilibrium

Two fixed charges and a dragable test charge are placed as shown (position is given in meters and force is given in newtons). The blue arrow represents the force on the red test charge. The forces on the fixed charges are not shown. You can examine the forces with the **first** fixed charge, with the **second** fixed charge, or with **both** fixed charges in place. Notice that the net force on the test charge is displayed in the yellow message box when you have only one charge present but not when you have both charges present.

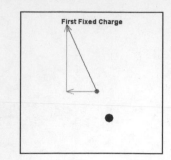

Answer the following questions with both fixed charges in place.

a. Determine the net force on the test charge at the point (3 m, 4 m).

b. Determine the net force on the test charge at a point midway between the two charges.

c. Is (are) there any point(s) where the net force on the test charge is zero? If so, find those points.

d. What is the ratio of the charges?

Exploration 22.2: Explore the Effect of Multiple Charges

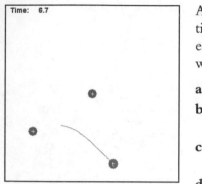

A positive test charge is shown in the animation. You can add positive and/or negative charges. All charges are added to the middle of the animation, so you must drag each newly added charge to a new location. When you push "play," the test charge will move under the influence of the forces from the other charges.

a. Add one positive charge. Describe and explain the motion of the test charge.

b. How can you tell from its motion that the test charge experiences a force, but that the force decreases as the test charge moves away from the positive charge?

c. What do you predict the motion will be if the positive charge is replaced by a negative charge?

d. Clear the screen and try it. Was your prediction correct?

e. How can you configure two charges of the same sign and keep the test charge stationary? Describe your configuration.

f. What happens if you move one of the charges slightly? This is a demonstration of an unstable equilibrium point (like a gymnast on a balance beam; nudge her one way or the other and she will fall).

g. Design and describe a configuration in which the test charge will oscillate back and forth.

h. Explain why (in terms of the forces) the test charge oscillates in your configuration.

i. Clear the charges and add one negative charge. Let the test charge start moving (so it has an initial velocity) and then move the negative charge around so that the test charge orbits the negative charge. Explain (in terms of forces) why it orbits.

Exploration 22.3: Electrostatic Ranking Task

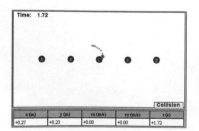

Study the motion of a positively charged test object under the action of five fixed charges. Run the animation a number of times starting with the test object in different positions. Before you move the particle, you must push "reset" (otherwise the particle will simply continue along the same trajectory as before). Note that the motion can be quite complicated depending on your choice of initial position (position is given in meters and time is given in seconds). Rank the fixed charges from most negative to most positive.

a. To begin with, start the test charge close to each individual charge separately to determine the signs of the different charges. Which ones are positive, negative, or neutral?

You will need to use a systematic approach to rank the negative and positive charges. For the positive charges, you might put the test charge fairly close and then watch the motion (and the trail).

b. Which positive charge provides a bigger force to the test charge?

   For the negative charges, put the charge a little way above the unknown negative charge and watch it approach.

c. Which one of the negative charges provides a bigger force?

d. How do you know?

e. Using this approach, rank the charges.

Exploration 22.4: Dipole Symmetry

Each animation shows a positive charge (red) along with two unknown charges (blue). The electrical force on the positive charge is represented with a force vector. You can drag the red charge along a portion of the x axis (position is given in meters and force is given in [newtons/k], where k is the constant in Coulomb's law).

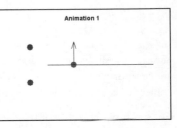

a. In which animation are the two unknown charges a positive and negative charge of equal magnitude?

b. Qualitatively speaking, what charge configuration would produce the results in the other two animations?

c. For the animation with a positive and a negative charge of equal magnitude, what is the value of the magnitude of the two blue charges if the red charge is 2.5 coulombs?

Exploration 22.5: Pendulum Electroscope

Two identical balls are hung pendulum-like in a laboratory as shown (position is given in meters and time is given in seconds). The charge on each ball, in mC, can be varied by using the slider. Position can be measured by click-dragging.

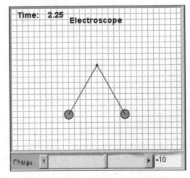

a. Is there any difference in behavior if the charge on both balls is changed from negative to positive?

b. Notice that you can zero the velocity. Can you find a spot where the balls are in equilibrium? (You may need to set the velocity to zero several times to get the balls in equilibrium.)

c. What is the mass of each ball? (Assume that the charge on the two balls is uniformly distributed.)

d. How large a charge is required for the angle (as measured from the pivot) between the two balls in equilibrium to be 90 degrees? How large a charge for 180 degrees?

Exploration 22.6: Run Coulomb's Gauntlet

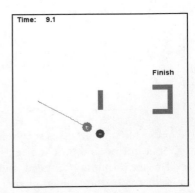

A positive test charge is shown in the animation. You can add positive and/or negative charges. All charges are added to the middle of the animation, so you must drag each newly added charge to a new location. When you push "play," the test charge will move under the influence of the forces from the other charges.

   Move the charges so that the test charge can make it from its starting place to the finish line without hitting a wall.

a. Describe your technique.

b. What is it about the Coulomb force that makes this so difficult? Explain.

Problems

Problem 22.1

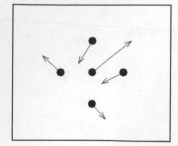

Five charged objects are shown on the screen along with vectors representing the forces on each object. You can click-drag on any object to change its position (position is given in meters).

a. Find two objects that repel each other and place them aside. Can you find any object that is attracted to one of these objects and repelled by the other? What does this observation tell you?

b. How many charges are alike?

Problem 22.2

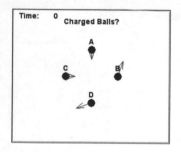

Four charged objects are shown on the screen along with vectors representing the forces on each object. You can click-drag on any object to change its position (position is given in meters).

What, if anything, is wrong with the animation?

Problem 22.3

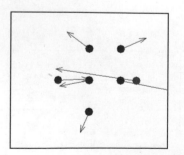

A projectile is launched as shown in the animation (position is given in meters and time is given in seconds). Where does the ball reach its minimum speed, and what is its speed when it gets there? In the animation the electrostatic force on each charge is indicated by an arrow (position is given in meters).

The magnitude of each charge is 1 C, and the blue charge is negative. What is the net charge shown in the animation?

Problem 22.4

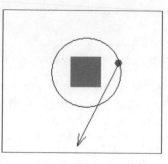

A positive charge (red) is shown along with a region blocked by a gray curtain (position is given in meters). The electric force on the positive charge is represented with a force vector. The region behind the gray curtain contains two charges of equal magnitude but unknown polarity (unknown sign). You can drag the red charge along the black circle.

Draw a picture showing the sign and location of each charge within the covered area.

Problem 22.5

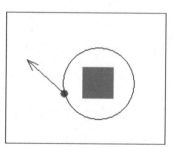

A positive charge (red) is shown along with a region blocked by a gray curtain (position is given in meters). The electrostatic force on the positive charge is represented by a force vector. The region behind the gray curtain contains three charges of equal magnitude but unknown polarity (unknown sign). The charges are arranged in an equilateral triangle located around the center of the square. You can drag the red charge along the black circle.

Draw a picture indicating the sign and location of each charge within the covered arca.

Problem 22.6

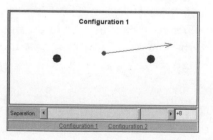

In the animation there are two black fixed charges and a blue movable positive test charge (position is given in meters and force is given in newtons). The test charge has an attached arrow that indicates the direction and relative magnitude of the net electric force. The charge on the left is 25 times that of the right-hand charge. The initial separation is $d = 8.0$ m. Drag the test charge and note that the force is displayed in the yellow message box.

For each configuration,

a. There is a single point at which the electric force on the test charge is zero. Find that point at which the separation of the two charges is 6 m.

b. Does the animation provide sufficient information to calculate the charge on the two black objects? If so, calculate the charge in coulombs.

c. Find an expression for the position of zero force for any value of the charge separation. Show that your result agrees with the animation.

Problem 22.7

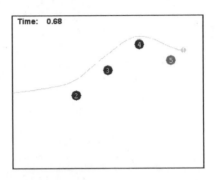

An unknown green charge (with a #1 on it) is shot into a region containing four fixed charges (numbered 2–5), one of which is known to be positive (the red one). You can measure position using a mouse down (position is given in meters and time is given in seconds).

Determine the signs of the unknown charges. You may consider neutral as a possible answer. Justify your response.

Problem 22.8

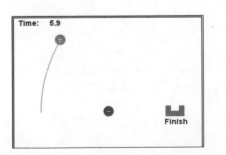

A positive test charge of 1×10^{-5} C with a mass of 0.9 kg is shown near a variable charge with a fixed position (at the origin). You may change the charge of the central charge and the initial velocity of the test charge (position is given in meters and time is given in seconds).

Set the charge of the central charge to -20×10^{-5} C.

a. What initial velocity must you give the test charge so that the test charge can make it from its starting place to the finish line in a circular path?

b. For an arbitrary negative central charge, Q, what initial velocity must you give the test charge so that the test charge can make it from its starting place to the finish line in a circular path? Your answer should be a formula for v in terms of Q. When you have an answer, test it with $Q = -10 \times 10^{-5}$ C and $Q = -30 \times 10^{-5}$ C.

Problem 22.9

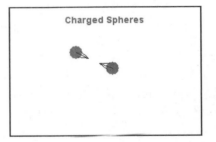

Two conducting spheres of the same mass and volume are shown (position is given in meters and force is given in newtons). The magnitude of the force on a sphere is shown when you drag it.

Drag the spheres so that they make contact. What was the initial charge on each sphere?

Problem 22.10

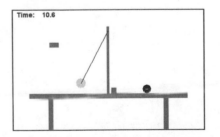

A charged ball with a mass of 30 grams is hung pendulum-like as shown (position is given in meters and time is given in seconds). A second ball with a negative charge of 3 mC is placed on a wood table. Play the animation and move the ball on the table toward and away from the pendulum. What is the charge on the pendulum? (Assume that the charge on the two balls is uniformly distributed.)

Problem 22.11

A 20-gram charged ball is connected to a spring and placed on a frictionless tabletop as shown (position is given in meters and time is given in seconds).

a. Find the equilibrium position.

b. Find the spring constant.

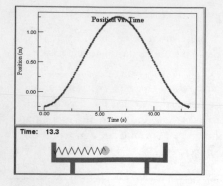

Time: 13.3

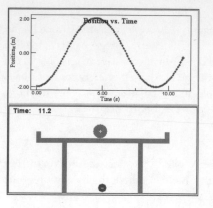

Time: 11.2

Add a 2-μC charge on the right-hand side of the table.

c. Has the spring constant changed? Explain.

d. Find the new equilibrium position.

e. What is the charge on the ball?

Problem 22.12

A charged ball of mass 100 grams is placed on a frictionless table-top with a second charge fixed underneath. Play the animation and observe the motion. You may drag the charged ball to any position before beginning the animation.

If the charge of the fixed charge (beneath the table) is a negative 2 μC, determine the value of the charge on the ball. (**Show hint**.)

Electric Fields

Topics include vector fields, field lines, electric fields, electric force, and charge distributions.

Illustration 23.1: What is a Vector Field?

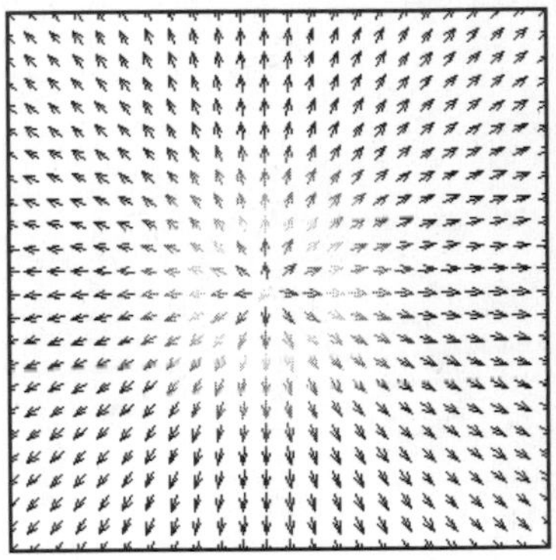

This animation plots a vector field when you enter values for the x component and y component of the field. You should try several values to get a sense of what a vector field is.

Begin by creating a simple uniform vector field by entering 5 N/C for E_x and updating the field. Notice that the animation displays a grid of arrows pointing to the right. If you enter -5 N/C for E_x, the field arrows will point in the opposite direction. Enter 3 N/C for E_x and 4 N/C for E_y and update the field again. The arrows now point at an angle of 37 degrees with respect to the x axis. Now, if you enter 2 N/C for E_x, what do you see? How is it different from 5 N/C for E_x? What does the color of the vector show? Why do you think we do not represent the magnitude of the vector field by the length of the vector?

Now, build a field that you are familiar with (whether you know it or not) by putting in 0 N/C for E_x and -4.9 N/C for E_y. This is a representation of the vector force field for a 0.5-kg mass close to Earth's surface. Why? What would the vector force field be for a 3-kg mass close to Earth's surface? (If it is far away from Earth's surface, like a satellite in orbit, you need to take into account the decrease in gravitational attraction as a function of distance squared).

The values of the field components do not need to be constants. Try $2 * x$ for E_x and $2 * y$ for E_y. What do you see? In this case, the vectors show you a field that changes in both magnitude and direction with position. For $x = 0$ m, $y = 2$ m, what are the values of E_x and E_y? Does the arrow on the screen point in the correct direction at that point? Repeat this exercise with $2 * y$ for E_x and $2 * x$ for E_y.

Try some other set of (nonconstant) values for E_x and E_y. Specifically, try $E_x = x/(x * x + y * y)^\wedge 3/2$ and $E_y = y/(x * x + y * y)^\wedge 3/2$. What does this vector field look like?

Illustration 23.2: Electric Fields from Point Charges

This Illustration allows you to add charges by clicking on the appropriate link. *All charges are added at the center of the animation; you must drag charges from the origin to see the effect of subsequent charges.*

First, examine the field around one charge. What does the field look like for a 1-C charge? Clear the charge and add a 2-C charge. How is that field configuration different? Clear the charges and then add a 2-C charge. What is the difference? Notice that the strength of the field is represented by the color of the field vectors. White is the smallest magnitude (zero) and black is the greatest, with blue, green, and red in between. When the charge is negative, the electric field vectors change direction and now point in the opposite direction. Positive charges have field vectors that point radially outward, and negative charges have field vectors that point radially inward.

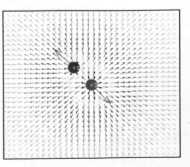

Clear the charges and add two positive charges of the same magnitude. Notice that since the charges are added at the center, you must drag a charge away to see the one underneath. How is the field different with two charges compared with one? Move one of the charges closer and farther away from the other one. When the charges are sitting on top of each other, what does the field look like? When you move them far apart, what does it look like? Notice that the fields add together (it is nothing more than vector addition). The fact that the electric field at any point is the vector sum of the electric fields due to the surrounding charges is simply the principle of superposition. You have seen this in the previous chapter: The force on a charge is due to the sum of the Coulomb forces from the surrounding charges. Notice that the force vector of an individual charge points in the direction of the electric field due to the other charge. It does not, however, point in the direction of the electric field due to both charges. The field configuration shown would be the field experienced by a third particle (not the force experienced by either particle).

What do you predict the field will look like with two negative charges (of equal magnitude)? Try it. What are the similarities and differences between the two positive and two negative charge distributions? The field vectors point in the opposite direction.

What about a dipole, one positive and one negative charge? How is it the same or different from two charges of the same sign? What is the direction of the field at the midpoint between the charges? The vector field can be described in terms of the vector sum of the field from the two particles.

Try two charges of different magnitude. What does the field look like? Notice that there is a point at which the electric field is zero directly in between the two charges. If you added a third charge at that spot, what do you predict the force on it would be? Try it. Notice that the force on the third charge is simply due to the electric field from the other two charges (multiplied by the charge of the third charge).

Add three or four charges and look at the field. Pick one point of the electric field and explain why it points in the direction it does. How can you tell, simply by looking at the field (and not the labels on the charges), which ones are positive and which ones are negative? How can you tell which ones have more charge?

Illustration 23.3: Field–Line Representation of Vector Fields

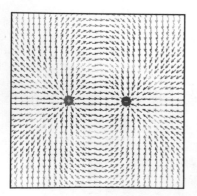

There are different ways to represent the electric field created by a charge distribution. One way is to use field vectors (as you've already seen), but you may find it a bit tedious (and difficult unless you carry around a colored pencil set) to draw that on your paper. Many books use electric field lines as an alternate representation of field vectors.

Switch between the field-vector and the field-line representation for Configuration A. What is the difference between the two representations? In a field-line drawing, the line density is often used to represent, at least qualitatively, field strength (more field lines in an area indicate a larger electric field). The arrows represent the direction of the electric field. Now move the charges around in **Configuration A**. How does the field-line representation reflect the change? Pick a point on a field line. Switch to the vector field representation. What does the field vector look like at that point? Notice that the field vector points in a direction tangent to the field line at any point.

Now, consider **Configuration B** and look at the two representations. Can you tell if the net charge distribution is positive, negative, or zero? Move the charges to check your answer (you can put them all on top of each other).

Illustration 23.4: Practical Uses of Charges and Electric Fields

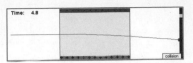

When you first play the animation, there is no electric field present (position is given in centimeters and time is given in seconds). Input values for the electric field in the *y* direction and run the animation again. You will notice that the charge is deflected from its original straight-line path. Consider what happens when you

- increase the charge.
- increase or decrease the initial velocity.
- change the sign of the electric field.

Can you make the charge hit the green target? What happens if an object of zero charge is shot into a region where there is an electric field? The simple ideas demonstrated in the animation have been used in several real world applications.

Ever wonder how your computer and television monitor (those that are not flat-panel LCD or plasma) work? The image you see on the screen is produced using a cathode ray tube (CRT). The CRT uses a heated filament (not unlike that found in a light bulb) to produce electrons traveling at high velocities. The electron is shot into a region of constant electric field. Since the electric field exerts a force on charges, the electron is deflected from its path. The amount of deflection depends on the strength of the electric field and the velocity of the electron before it encountered the field. By controlling either the initial velocity or the field strength, the spot where the charge hits the screen is controlled. In the case of the CRT, the charge is not varied since it is a stream of electrons that is produced. The screen is coated with a substance (phosphor) that will emit light when struck by an electron. The rate at which electrons are striking the screen is very fast, allowing a complete picture to be built up. The applet shows electrons being deflected up and down. Notice that another set of plates could be added to control the right and left movement of the electron. Most new CRTs use magnetic fields to control the electron, but the basic idea is the same. Since the location where the electron hits the screen is directly related to the strength of the electric field, a CRT can be used to measure the electric field. Oscilloscopes use this idea to measure voltages (by measuring the electric fields produced), and display time-dependent voltage information on a screen.

Years ago, most printing was accomplished by impacting something mechanical against the paper. Typewriters and dot-matrix printers both used this concept. Today, high-quality, versatile printing is often done using an ink-jet printer. If you have a printer connected to your home computer, it is probably an ink-jet. The ink-jet printer works by squirting tiny drops of ink on the paper. These drops are very small, with a diameter less than a human hair. The number of dots a printer can place in an inch is specified by the dpi (dots per inch) number and is often around 1200 or more in the horizontal direction. There are two technologies used to drop the ink on the paper: continuous-ink printing and forced-ink printing. Forced-ink printing is the most common today and utilizes some method of forcing the ink to drop when it is directly over the desired location. From a physics point of view, the continuous-ink printing method is more interesting because it controls the location of the ink drop with an electric field.

When an ink drop is ejected from the ink cartridge, it is given a computer-controlled charge by the printer and then passed through an electric field. The location where the drop lands on the paper is determined by the charge on the drop. The same basic idea that is used to light up a particular portion of your television or computer screen can also be used to place ink on a particular location of your paper.

Exploration 23.1: Fields and Test Charges

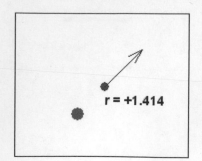

r = +1.414

If interactions between charges can be described by the forces on them, and the electric field is related to the electric force, why talk about electric fields at all? We will consider this issue in the following Exploration.

The animation shows a positive charge of 4 C located at (0 m, 0 m) and a test charge of 1 C located at (1 m, 1 m). The distance between the test charge and the charge at the origin is shown along with a vector representing the force on the test charge (position is given in meters and charge is given in coulombs). The slider allows you to change the charge on the test charge.

a. Drag the test charge to $x = 2$ m and $y = 2$ m. Relative to the force on the charge in its initial position, is the force now greater, less, or the same? How will the force on the test charge compare to its initial force when it is at $(-1$ m, 1 m)? What about when it is at $(0.5$ m, -1 m)? What determines the relative magnitude of the force on the test charge?

b. Now consider leaving the test charge at its initial location but increasing the positive charge to 2 C. How does the force on the test charge change? How will the force on the test charge compare to its initial force when the test charge is 0.5 C? What about when it is -1 C? What determines the relative magnitude of the force on the test charge?

Now suppose you wanted to describe the force that would be felt due to the charge at the center. Can you describe the force in general terms, without specifying the value of the charge to be acted upon? Your answer to (a) and (b) should have shown you that you must know both where the charge is located and its value in order to talk about the force it would feel. If you could talk about the force in terms of the force on a test charge, you will run into difficulty. The force is entirely dependent on the test charge! Change the value of the test charge and you will find a different force. We need a description of the field that is independent of our detecting device.

c. Can you think of a way of quantitatively describing the region around the center charge that does not depend on the properties of a test charge? Write down at least one proposal BEFORE going to (d).

d. The physicist's answer to this dilemma is the creation of the concept of the electric field. The electric field is defined to be the force on a test charge divided by the value of the test charge, $E = F/q$. For a single charge like the one at the center of the animation, $F = (kQq)/r^2$, so the electric field becomes $E = kQ/r^2$.

e. Repeat (a), (b), and (c) considering the electric field instead of the force.

f. Does the force depend on the value of the test charge? Does the electric field depend on the value of the test charge? Test your predictions by changing the value of the test charge and watching the effect on the force vector and the electric field vectors. (Turn the electric field vectors on using the link.)

g. In your own words, why is it useful to define the electric field in the way it is defined? Why not just talk about forces?

Exploration 23.2: Field Lines and Trajectories

The animation shows two fixed charges and a test charge (position is given in meters and time is given in seconds). The electric field lines due to the fixed charges and the force vector on the test charge are shown. The test charge will move under the action of the electric field when the animation is played.

a. Using **Configuration A**, drag the test charge to the approximate position of (−0.8 m, 0 m). Write down a prediction for the path the charge will follow after being released at this point. **After** you have made your prediction, play the animation. Was your prediction correct? If not, what caused your error?

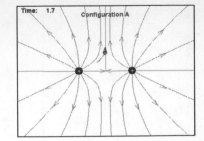

b. Reset the applet and then drag the test charge to the approximate position of (1 m, 0.35 m). As before, write down a prediction for the path the charge will follow after being released. If your prediction was incorrect, explain the flaw in your reasoning.

c. Repeat using **Configuration B** with the charge being released from the point (−0.5 m, 0.5 m).

d. Repeat using **Configuration B** with the charge being released from the point (0 m, 1.3 m).

Exploration 23.3: Adding Fields

A charged bead is placed on a circular wire frame as shown. The center of the circle is at the point (0 m, 1m). In addition to gravity, you can add a uniform electric field in the x direction (position is given in meters, time is given in seconds, and the electric field strength is given in newtons/coulomb). The force field is shown using arrows as in **Illustration 23.1**.

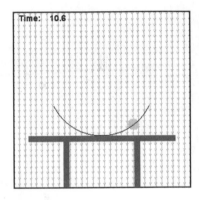

Enter a value for the electric field and click the "set value and play" button to begin the animation. The bead will move unless it is at an equilibrium position. You can set the instantaneous velocity to zero, but the bead will again begin to move unless you happen to damp it at an equilibrium point. Pause the animation, zero the velocity, drag the bead, and play the animation as many times as you like. If the electric field is small enough so it is similar in size to the gravitational field, you can see the field vector at an angle with the horizontal because it is the vector sum of the gravitational force ($m\mathbf{g}$) and the electric force ($q\mathbf{E}$). Determine the charge on the bead if the mass is 10 grams.

a. Notice that the bead oscillates about an equilibrium position. Find a value of the electric field that gives you an equilibrium position somewhere on the wire. Zero the velocity to get the bead to stop at that equilibrium position.

b. Draw a force diagram and show that the y component of the normal force of the wire on the bead must be equal to the weight of the bead, while the x component of the normal force must be equal to the force due to the electric field ($q\mathbf{E}$).

c. Since the normal force is perpendicular to the wire (and therefore points to the center of the wire circle), find the angle that the normal makes with either the horizontal or vertical and then show that the ratio of the gravitational force to the electric force is simply the tangent of the angle that the normal force makes with the horizontal. Therefore, you can find the electrical force ($q\mathbf{E}$) required to keep the bead at equilibrium.

Problems

Problem 23.1

A fixed particle with an unknown charge is shown along with a dragable test charge. The vectors shown point in the direction of the electric field, and the color of the vectors represent the field's magnitude (position is given in centimeters and electric field strength is given in newtons/coulomb).

a. Determine the unknown charge in coulombs.

b. Is the charge due to an excess of electrons or a deficit of electrons?

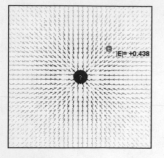

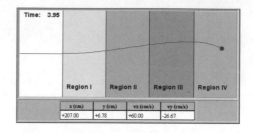

c. Calculate the number of electrons (excess or deficit) that are needed to produce the observed field.

configuration shown below? Red represents a charge of $+Q$ and blue represents a charge of $-Q$.

Problem 23.4

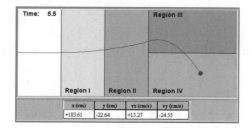

Problem 23.2

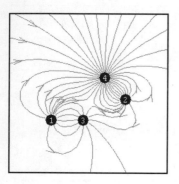

An electron is shot through four regions of constant electric field (position is given in centimeters and time is given in seconds).

a. What is the direction of the electric field in each region?

b. Rank the magnitude of the electric fields of the four regions, smallest to greatest.

Problem 23.5

The electric field lines, due to four charged particles, are shown in the applet (position is given in meters). You can click-drag any of these particles. If you overlap two charges, their charge values will add.

Is the net charge of the distribution positive, negative, or zero?

An electron is shot through four regions of constant electric field (position is given in centimeters and time is given in seconds).

a. What is the direction of the electric field in each region?

b. Rank the magnitude of the electric fields of the four regions, smallest to greatest.

Problem 23.3

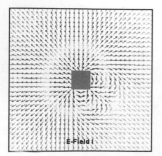

Problem 23.6

An electron is shot into a region of constant electric field shown in green (position is given in centimeters and time is

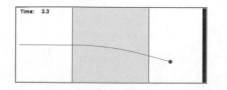

Five animations show the electric field produced by a configuration of hidden charges. The arrows represent the direction of the electric field, and the color represents the intensity of the field. Which electric field would be produced by the charge

given in μ sec, 10^{-6} s). What are the magnitude and direction of the electric field?

Problem 23.7

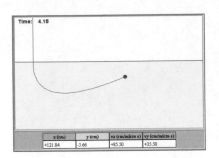

An electron is shot through a region of constant electric field (time is given in μ sec [10^{-6} s] and position is given in centimeters).

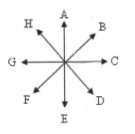

a. Which vector most closely shows the direction of the electric field?

b. What is the magnitude of the electric field?

Problem 23.8

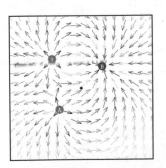

The charge on A is 1 nC and the charge on B is −3 nC (position is given in meters and electric field strength is given in newtons/coulomb).

a. Where does the unknown charge (movable) need to be placed for the electric field at the origin (where the black dot is) to be zero?

b. What is the charge on the unknown charge?

Problem 23.9

Four point charges of equal strength but two different polarities are arranged on the corners of a square as shown. The strength of each charge is measured in nC and can be varied using the slider. You can mouse-down to read the magnitude of the field at any point in the animation (position is given

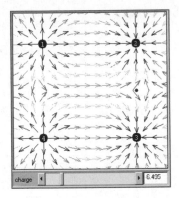

in meters and electric field strength is given in newtons/coulomb).

Find a formula for the electric field midway between charge 2 and 3 (indicated by the small black dot) as a function of the strength of the charges. Include a vector diagram that also shows the direction of the resultant field.

Problem 23.10

Derive an expression for the electric field as a function of position along the line shown (position is given in meters and force is given in newtons). If the purple test charge has a

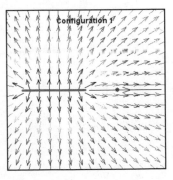

charge of 1 μC, what is the total charge on the red rod in the animation?

a. <u>**Configuration I:**</u> Field along the axis of the red rod.

b. <u>**Configuration II:**</u> Field perpendicular to the red rod.

Remember that a test charge is defined as a charge that is so small that it does not influence other charges.

Gauss's Law

Topics include Gauss's law, symmetry, flux, and charge distributions.

Illustration 24.1: Flux and Gaussian Surfaces

The bar graph shows the flux, Φ, through four Gaussian surfaces: green, red, orange and blue (position is given in meters, electric field strength is given in newtons/coulomb, and flux is given in $N \cdot m^2/C$). Note that this animation shows only two dimensions of a three-dimensional world. You will need to imagine that the circles you see are spheres and that the squares you see are actually boxes. Flux, Φ, is a measure of the amount of electric field through a surface. Gauss's law relates the flux to the charge enclosed ($q_{enclosed}$) in a Gaussian surface:

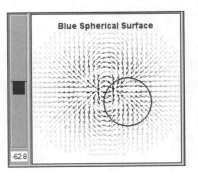

$$\Phi = q_{enclosed}/\varepsilon_0 \quad \text{and} \quad \Phi = \int \mathbf{E} \cdot d\mathbf{A} = \int E \cos\theta \, dA$$

where ε_0 is the permittivity of free space ($8.85 \times 10^{-12} \, C^2/N \cdot m^2$), $\mathbf{E}$ is the electric field, $d\mathbf{A}$ is the unit area normal to the surface, and θ is the angle between the electric field vector and the surface normal.

Begin by moving the green Gaussian surface around. What is the flux when the surface encloses the point charge? What is the flux when the point charge is not inside the surface? What about the red surface? Since the flux is the electric field times the surface area, why doesn't the size of the surface matter? As long as the point charge is enclosed, the flux is the same and is equal to $q_{enclosed}/\varepsilon_0$. When the charge is not enclosed, the flux is always zero. Notice that both the green and red Gaussian surfaces can be moved to either enclose or not enclose the charge. Therefore, the two fluxes should, and do, agree. However, only when these surfaces are centered on the charge can you use them to determine the electric field.

The orange surface has a different symmetry from the point charge (and its electric field). With the orange surface, why doesn't the shape matter in finding the flux? Again, what matters is whether the charge is enclosed or not. Move the surface to a point where the flux is zero. Is the electric field zero at the surface of the box? If the electric field is not zero, why is the flux zero? If you think about flux as a flow of electric field through an area (a bit like fluid flow), which was the early analogy for electric field and flux, then when there is no charge inside, the electric field that comes into the box must also leave. There is no source of electric field inside the box. However, the cubical box no longer has the same symmetry (a spherical symmetry) of the point charge. While the flux is zero in these scenarios, the value of the flux cannot be used to determine the electric field. The integral $\int E \cos\theta \, dA$ is not equal to the integral $E \int \cos\theta \, dA$ because E is not uniform across the Gaussian surface.

Finally, try two charges using the blue surface. What happens when the blue surface encloses just one charge? What happens when it encloses both charges?

Illustration 24.2: Near and Far View of a Filament

The different configurations of the animation show different views of the same charged filament: an intermediate view, a close view, and a view of the filament

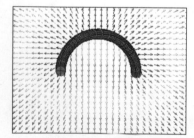

from far away (position is given in meters, electric field strength is given in newtons/coulomb, and flux is given in N·m²/C). A dragable Gaussian "surface" (you must imagine each of these to be three dimensional; boxes are cubes, circles are spheres) mimics the symmetry of the charge distribution as closely as possible.

Compare the fluxes through the Gaussian surfaces in both the intermediate and far views when the Gaussian surface encloses the entire filament. Why are they the same? In the far view, why is the flux the same even if the detector is not centered on the charge? The same amount of charge is enclosed by the Gaussian surface in both cases; therefore, the flux is the same.

For each view, what do you think the electric field lines will look like? Check your answers using the "show E-field" links. Why have we chosen the Gaussian surfaces to have different shapes for the near and far views? In the **intermediate view** the electric field does not have a symmetry, while in the **near view** the electric field has an approximate rectangular symmetry (if you are close enough to the charge distribution), and in the **far view** the electric field has an approximate spherical symmetry (if you are far enough away from the charge distribution). This means that you can only use Gauss's law for the **near view** and the **far view**. Given that there are two different symmetries, the electric field you find using Gauss's law will be different for the two cases. That is okay, because Gauss's law allows you to calculate the electric field on the Gaussian surface (not anywhere else).

Move the surfaces for the "show E-field" links so that the electric field vectors are either perpendicular to or parallel to the surfaces for the near and far views. Can you do the same for the intermediate view? No. This lets you know that the symmetry for the intermediate case does not allow you to use Gauss's law to calculate the electric field at the surface. Gauss's law still holds for all three configurations. You can still calculate the flux through the surface (it is proportional to the charge enclosed). However, for the intermediate view, because the electric field varies at different spots on the surface (points in different directions relative to the surface), you cannot use Gauss's law to calculate the electric field at the surface (you must use Coulomb's law). Thus, although true, Gauss's law is only useful for calculating the electric field for certain symmetrical charge distributions (spherical, cylindrical, and planar).

Illustration 24.3: A Cylinder of Charge

In this animation each charge is a line or rod of charge into and out of the screen. You can create charge distributions and see the electric field that results (position is given in meters and electric field strength is given in newtons/coulomb).

Begin by adding one line charge. Notice the field lines. Imagine that this charge extends both into and out of the computer screen. Clearly, there is cylindrical symmetry. You could imagine putting a tube centered on this charge. How could you convince another student that the magnitude of the electric field at any point on the tube would be the same as any other point on the tube? Well, there is nothing special about the placement of your tube in the direction out of the screen. There is nothing special about how long that tube is.

Now, add another line charge. What happens to the symmetry? Far away from the two charges, there would still be cylindrical symmetry. Why? If you are far enough away it looks (approximately) like a single rod. Look at what happens to the field as you add 10 more line charges and then half a cylinder. When you create a full cylinder, what is the symmetry? The field everywhere inside the cylinder is zero. What Gaussian surface could you use to find the electric field inside the cylinder or outside the cylinder? The full cylinder has cylindrical symmetry about the middle of the cylindrical shell of line charge. Since there is a symmetry, we can use Gauss's law to calculate the electric field. Since there is no charge enclosed by a Gaussian surface of radius 1.65 m, the flux is zero, and because of the symmetry we can say that E is zero inside the cylindrical shell. We could also use a cylindrical Gaussian surface to calculate the electric field outside of the cylindrical shell of line charge.

As you work problems using Gauss's law, you will need to be able to identify symmetry (look at the symmetry of the charge configuration), as well as recognize the direction of the field and where it cancels out. Be sure that you can do that for this configuration.

Exploration 24.1: Flux and Gauss's Law

In this Exploration we will calculate the flux, Φ, through three Gaussian surfaces: green, red, and blue (position is given in meters and electric field strength is given in newtons/coulomb). Note that this animation shows only two dimensions of a three-dimensional world. You will need to imagine that the circles you see are spheres.

Flux is a measure of the electric field through a surface. It is given by the following equation:

$$\Phi = \int_{\text{surface}} \mathbf{E} \cdot d\mathbf{A} = \int_{\text{surface}} E \cos\theta \, dA,$$

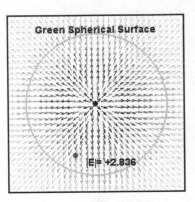

where $\mathbf{E}$ is the electric field, $d\mathbf{A}$ is the unit area normal to the surface, and θ is the angle between the electric field vector and the surface normal.

Move the test charge along one of the Gaussian surfaces (you must imagine that it is a sphere even though you can only see a cross section of it).

a. What is the magnitude of the electric field along the surface?

b. In what direction does it point?

c. What direction is normal to the Gaussian surface?

If the electric field, $\mathbf{E}$, and the normal to the Gaussian surface, $\mathbf{A}$, always point in the same direction relative to each other, and the electric field is constant, then the equation for flux becomes: $\Phi = E \cos\theta \int dA = EA \cos\theta$.

d. In the case of the point charge in (a)–(c), what is the angle between the electric field and the normal to the surface?

This means that $\cos\theta = 1$. Therefore, for this case, $\Phi = EA$.

e. Calculate the flux for the surface you've chosen (remember that the surface area of a sphere is $4\pi R^2$).

f. Calculate the flux for the other two surfaces.

Because the electric field decreases as $1/r^2$, but the area increases as r^2, the flux is the same for all three cases. This is the basis of Gauss's law: The flux through a Gaussian surface is proportional to the charge within the surface. With twice as much charge, there is twice as much flux. Gauss's law says that $\Phi = q_{\text{enclosed}}/\varepsilon_0$.

g. What is the magnitude and sign of the point charge?

Exploration 24.2: Symmetry and Using Gauss's Law

Gauss's Law is always true: $\Phi = \int_{\text{surface}} \mathbf{E} \cdot d\mathbf{A} = q_{\text{enclosed}}/\varepsilon_0$, but it isn't always useful for finding the electric field, which is what we are usually interested in. This should not be too surprising, because to find $\mathbf{E}$, using an equation like $\int_{\text{surface}} \mathbf{E} \cdot d\mathbf{A} = q_{\text{enclosed}}/\varepsilon_0$, $\mathbf{E}$ has to be able to come out of the integral, and for that to happen, $\mathbf{E}$ needs to be constant on a surface. This is where symmetry comes in. Gauss's law is only useful for calculating electric fields when the symmetry is such that you can construct a Gaussian surface so that the electric field is constant over the surface, and the angle between the electric field and the normal to the

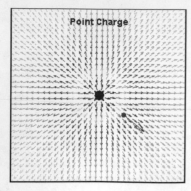

Gaussian surface does not vary over the surface (position is given in meters and electric field strength is given in newtons/coulomb). In practice, this means that you pick a Gaussian surface with the same symmetry as the charge distribution.

Consider a **sphere around a point charge**. The blue test charge shows the direction of the electric field. There is also a vector pointing in the direction of the surface normal to the sphere.

a. By moving the surface normal vector on the sphere and putting the test charge at three different points on the surface, find the value of $\mathbf{E} \cdot d\mathbf{A} = E\, dA \cos\theta$ (set $dA = 1$) at these three points (read the electric field values in the yellow text box). Are they the same? Why or why not?

Now, **put a box around the same point charge**. The test charge now shows the direction of the electric field, and the smallest angle between the vector and a vertical axis is shown (in degrees). The red vector points in the direction of the surface normal to the box (two sides show).

b. By moving the surface normal vectors on the box and putting the test charge at three different points on the top surface, find the value of $\mathbf{E} \cdot d\mathbf{A} = E\, dA \cos\theta$ (set $dA = 1$) at these three points. Are they the same? Why or why not?

c. In the context of your answers above, why is the sphere a better choice for using Gauss's law than the box?

Let's try another charge configuration. **Put a sphere around part of a charged plate** (assume the gray circles you see are long rods of charge that extend into and out of the screen to create a charged plate that you see in cross section).

d. Would the value of $\mathbf{E} \cdot d\mathbf{A} = E\, dA \cos\theta$ be the same at any three points on the Gaussian surface?

e. Explain, then, why you would not want to use a sphere for this configuration.

Now, **put a box around part of a charged plate** (assume the points you see are long rods of charge that extend into and out of the screen to create a charged plate that you see in cross section).

f. Find the value of $\mathbf{E} \cdot d\mathbf{A} = E\, dA \cos\theta$ at three points on the top. Are they essentially the same?

g. What about $\mathbf{E} \cdot d\mathbf{A} = E\, dA \cos\theta$ on the sides?

For the plate, using a box as a Gaussian surface means that $\mathbf{E} \cdot d\mathbf{A} = E\, dA \cos\theta$ is a constant for each section (top, bottom, and sides) and the electric field is a constant on the surface. This means you can write

$$\int_{\text{surface}} \mathbf{E} \cdot d\mathbf{A} = E \int_{\text{surface}} dA = EA \text{ (for the surfaces where the flux is nonzero).}$$

h. Knowing that the charge per unit area on the big plate is σ, use Gauss's law to show that the expression for the electric field above or below a charged plate is $E = \sigma/2\varepsilon_0$ and the direction of the electric field is away from the plate for a positively charged plate. In your textbook you will probably also see an expression that says that the electric field is σ/ε_0 above or below the charged sheet. This holds true for conductors where σ is the charge/area on the top surface and there is the same amount of charge/area on the bottom surface (there is no net charge inside a conductor).

Exploration 24.3: Conducting and Insulating Sphere

What is the difference between the electric fields inside and outside of a solid insulating sphere (with charge distributed throughout the volume of the sphere) and

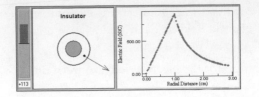

those inside and outside of a conducting sphere? Move the test charge to map out the magnitude of the electric field as a function of distance from the center (position is given in centimeters, electric field strength is given in newtons/coulomb, and flux is given in $N \cdot cm^2/C$).

a. Compare the electric fields inside and outside of the two spheres. What is the same and what is different (same total charge on both spheres)?

b. If a Gaussian surface larger than the two spheres is put around each, how will the flux through each compare? Why?

Try putting a **big Gaussian surface around the insulator.** The bar measures the flux. Now try **around the conductor.**

c. Why is the flux the same?

d. How much charge is on each sphere? How do you know?

e. What do you expect the flux to be through a Gaussian surface inside the conductor? Why? **Try it** and explain the results.

Now try putting the same size **small Gaussian surface inside the insulator**.

f. What flux value do you get?

g. How much charge is enclosed in this smaller surface?

h. What is the ratio of the charge enclosed in the small surface to the total charge on the insulating sphere?

i. What is the ratio of the volume of the small surface compared to the volume of the insulating sphere? Explain why the two ratios in (h) and (i) are the same.

j. Use Gauss's law for the smaller surface to calculate the field at that point inside the sphere. Verify that it agrees with the value on the graph.

As a reminder, Gauss's law relates the flux to the charge enclosed ($q_{enclosed}$) in a Gaussian surface through the following equation:

$$\Phi = q_{enclosed}/\varepsilon_0 \quad \text{(and Flux} = \Phi = \int \mathbf{E} \cdot d\mathbf{A} = \int E \cos\theta \, dA),$$

where ε_0 is the permittivity of free space ($8.85 \times 10^{-12} \, C^2/N \cdot m^2$), $\mathbf{E}$ is the electric field, $d\mathbf{A}$ is the unit normal to the surface, and θ is the angle between the electric field vector and the surface normal. The surface area of a sphere is $4\pi r^2$.

Exploration 24.4: Application of Gauss's Law

A point charge has radial (spherical) symmetry about the center of the charge, while a line charge has cylindrical symmetry about the center of the wire (position is given in meters and electric field strength is given in newtons/coulomb). However, a two-dimensional view of both can look the same.

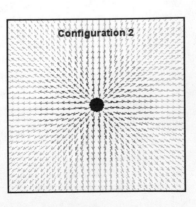

Consider the two configurations. One is a point charge and one is a line of charge (pointing into and out of the screen). Which is which? The electric field is different for the two cases (and you use two different Gaussian surfaces).

a. As a function of the distance away from the charge (as a function of r), what is the electric field of a point charge?

b. Therefore, if you measure the electric field at some point and then measure it twice as far away, how much should the electric field be decreased?

c. Which configuration, then, is a point charge?

d. Use Gauss's law to find an analytic expression for the electric field around a line of charge. You may find the following diagram useful:

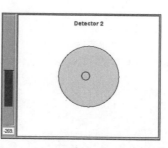

e. If you measure the electric field at some point and then move twice as far away, how should the field drop off from a line of charge?

f. Does the electric field of the other configuration agree with this?

Problems

Problem 24.1

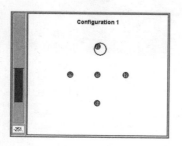

The bar graph displays the electric flux passing through the cylindrical Gaussian surface (position is given in meters and flux is given in N·m²/C). Drag the surface and, from the flux readings, rank the charges from greatest to smallest.

Problem 24.2

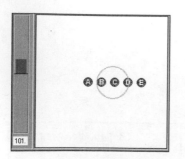

The bar graph displays the electric flux passing through the cylindrical flux detector (position is given in meters and flux is given in N·m²/C). Drag the surface and observe the flux readings. Rank the charges (lines of charge extending into and out of the screen) from most negative to most positive.

Note: The Gaussian surface encloses two or three charges at a time. It never partially encloses a line charge.

Problem 24.3

The bar graph displays the electric flux passing through the cylindrical Gaussian surface (position is given in meters and flux is given in N·m²/C). Drag the surface and, from the flux readings, describe the charge distribution.

Problem 24.4

The bar graph displays the electric flux passing through a cubical flux detector and several spherical flux detectors. Drag each detector and observe the flux readings (position is given in meters and flux is given in N·m²/C). Describe the charge distribution in as much detail as possible. The resolution of the detector is **1 mC**, and the gray sphere is there for your reference only.

Problem 24.5

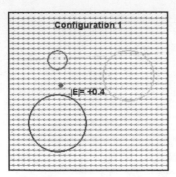

Three spherical shells (red, green, and blue) are located on the screen. You can only see the part of the sphere that is in the plane of the page. A test charge is also shown that measures the electric field at that point (position is given in meters and electric field strength is given in newtons/coulomb). Calculate the flux through each spherical shell. You can click-drag on the test charge to change its position.

Problem 24.6

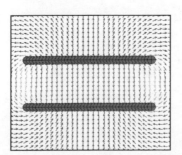

The green square represents a cross section of a cube. Use the test charge to explore the direction of the electric field inside the cube (position is given in meters and electric field strength is given in newtons/coulomb). Click-drag the cursor anywhere inside the cube to measure the magnitude of the electric field. Find the flux through the top, bottom, left, and right sides of the cube.

Problem 24.7

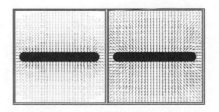

Each "charge" represents a long charged rod that extends into and out of the screen (position is given in meters).

a. Explain the electric field you see when the capacitor (long parallel sheets of charge) is complete.

b. The top and bottom sheets have equal and opposite charge. Which one is positively charged? Which is negatively charged?

c. What type of Gaussian surface would you use to find the field inside the capacitor? What type would you use outside the capacitor?

d. Sketch the Gaussian surface and explain why you chose it.

Problem 24.8

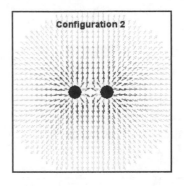

Although electric fields can be quite complicated close to a charge distribution, they often become very simple at large distances (position is given in meters and electric field strength is given in newtons/coulomb).

a. What symmetries does each of these configurations have at large distances?

b. When possible, find an analytical expression for the field at large distances using Gauss's law for each configuration.

Note that each configuration contains either point charges or line charges (into and out of the screen), but not both. You can measure the electric field strength by click-dragging in the animation.

Problem 24.9

Which configuration (when complete) represents a line of charge, and which represents a sheet of charge (where the points you see on the screen extend into and out of the screen)? Explain your reasoning. You can read the value of the electric field at any point by clicking with your mouse into the screen of interest (position is given in meters and flux is given in N·m²/C).

Problem 24.10

In this animation each charge represents a line or rod of charge extending into and out of the computer screen (position is

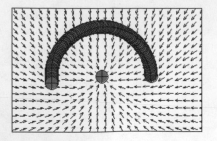

given in centimeters and electric field strength is given in newtons/coulomb).

a. Explain how you know that the charge on the inner rod and the total charge on the outer cylinder are equal and opposite.

b. From the electric field measurements (read values when you drag the mouse on the screen), find the value of the charge per unit length, λ, as well as an expression for the electric field as a function of λ at any point. (The charge per unit length, λ, is the total charge Q divided by the length of the object L.)

Problem 24.11

Each "charge" represents a long charged rod that extends into and out of the screen (position is given in centimeters and electric field strength is given in newtons/coulomb).

a. Add the top plate of the capacitor. Measure the electric field close to the top plate and then use Gauss's law to find the charge/area on the plate.

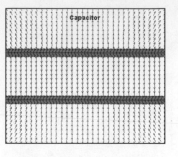

b. Do the same for the bottom plate.

c. What is the field near the center of the completed capacitor?

Problem 24.12

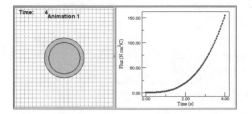

The graph shows the flux through the expanding spherical surface (position is given in centimeters, time is given in seconds, and flux is given in $N \cdot cm^2/C$). Push "play" to start the shell's expansion.

a. Describe the spherical charge distribution.

b. Find the equation of the electric field for the two charge distributions.

Electric Potential

Topics include voltage, work, energy, equipotential surfaces, conductors, charge distributions, and vector fields.

Illustration 25.1: Energy and Voltage

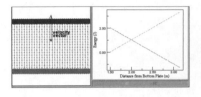

The animation shows a positive test charge in a uniform electric field created by two parallel plates of constant charge. You can drag the test charge between the plates to any spot and then push "set value and play" to see it move. You can also read position, voltage, and magnitude of the electric field. The graph plots the kinetic energy and potential energy as a function of height above the bottom plate.

- Does the charge experience a constant force? Explain.
- As the charge moves, is the work done on it by the electric force positive or negative? Explain.
- As you move the charge to a different starting place, how does the total energy change?
- To what positions can you move the charge to decrease the total energy, to increase it, and to keep it the same?

The charge experiences a constant force since the potential energy curve is linear (as with gravity). Since the kinetic energy increases, the work done by the electrostatic force is positive (and the change in potential energy is negative). Since work here is $\mathbf{F} \cdot \mathbf{\Delta x}$, the larger the displacement, the larger the work done and the larger the change in kinetic energy.

The potential energy divided by the charge of the particle is the electric potential (measured in volts). Triple the charge of the center particle. In this new configuration the potential energy increases, but the electric potential remains the same (for the charge at the same position). Why? What would need to happen to change the electric potential (voltage)?

Illustration 25.2: Work and Equipotentials

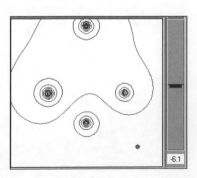

The animation shows the equipotential contours around a charge distribution. The bar shows the work done to move the red test charge (position is given in meters, electric potential is given in volts and work is given in microjoules). Equipotential surfaces are simply surfaces (in this two-dimensional representation they are lines) of constant electric potential.

Equipotential contours are the same as a topographic contour map you might see for mountains (as below). The contours are equally spaced so that each contour represents a given change in voltage (topographic maps have contours equally spaced for certain heights). What is the difference in voltage between contours on the equipotential surface in the animation?

A mountainous region (left) and its contour map (right).
Image credits: United States Geological Survey:
http://interactive2.usgs.gov/learningweb/teachers/mapsshow_act4.htm

- What is the work done on the test charge (by you) as you move it along the equipotential curve it started on?
- What is the work done (by you) as you move it toward another charge?
- The test charge is positive. If the work done by you is positive as you move it toward a charge, what sign does that charge have?

The change in potential energy is proportional to the change in electric potential (with the charge being the proportionality constant). If the change in electric potential is zero, then so is the change in potential energy, and therefore the work done is zero. As you move a positive charge toward another positive charge, you must do a positive amount of work (the charges repel and you must counteract this). As you move a positive charge toward a negative charge, you must do a negative amount of work (the charges attract and you must counteract this).

The electric field at any point is perpendicular to the equipotential contour. You can see this by looking at the **force vector on the test charge** as you move the charge around in this potential field. The direction of the electric field corresponds to the direction of steepest slope on a topographical map. If this were a topographical map, this would be a map of three steep mountains and a steep valley. How many positive and negative charges are on this electrostatic contour map? Note that, since the electric field is perpendicular to the equipotential lines, if we move along an equipotential, no work is done (the electric force and the displacement are perpendicular).

Illustration 25.3: Electric Potential of Charged Spheres

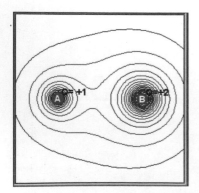

The animation shows the equipotential contours around two charged spheres. You can change the charge of particle A by using the text box. As you click-drag your mouse around the screen, you can see the magnitude of the electric field as well as the electric potential (position is given in meters, electric potential is given in volts, and electric field strength is given in newtons/coulomb). The zero point of electric potential is infinity (far away from the charge distribution).

Change the value of the charge of A so that the charges are equal. Where (if anywhere on the screen) is the electric field zero? Where is the electric potential zero? What happens if the charges have equal and opposite charge? In this case, where is the electric field zero? Where is the electric potential zero?

What happens to the equipotentials if you make charge A more positive? What happens to the equipotentials as you make charge A more negative?

The electric potential due to one point charge is proportional to the charge divided by the distance to the charge ($V = k\,q/r$). When charge A is equal to charge B (in magnitude and sign), where do you need to put a third charge, negative but with the same magnitude of charge, in order for the potential in the middle of A and B [at point $(0, 0)$] to be zero? **Add a charge** and move it to the correct spot to check your answer. Is there more than one place you can put this charge? The electric potential of the original two charges, as measured at the origin, is $V = k\,(2Q)$, since $r = 1$ m. The electric potential of the third charge must be $V = -k\,(2Q)$ to cancel this electric potential at the origin. Therefore, the third charge must be placed at any position that is a distance of $r = 0.5$ m from the origin.

Illustration 25.4: Conservative Forces

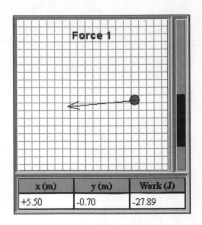

This animation shows the work done in moving a particle in two different force fields. As you move the particle, a vector shows you the direction of the force, and the bar and table show you the total work done as you move the particle around (position is given in meters and work is given in joules). You can zero out the work at any position you want by pushing the "set Work = 0" button.

Simply from the direction of the force (if we assume a positive test charge), if both of these fields were electrostatic, where could charges be located to produce this type of force? As you move the test charge around, notice that the force is biggest at $y = 0$ m on the right edge and points to the left. As you move the charge away from the right edge at $y = 0$, notice that the force decreases quickly and points radially outward from a point near $x = 10$ m, $y = 0$ m. This lets you know that a positive charge could be located near $x = 10$ m and $y = 0$ m to approximately produce these fields.

One of these fields, however, cannot be an electrostatic force field because it is not conservative. In other words, the amount of work done depends on the path taken. If you move the particle to a particular point, it matters if you go straight there or take a circuitous route. Which force field is conservative and which is not? You can mark an initial point and an ending point on the grid, if you want to help keep track of where you've moved the particle, to compare the work done along different paths between the same two points.

Drag the particle away from an initial point and then bring it back to the same spot. How much work is done in the conservative force field? How much is done in the nonconservative force field? For which one of these forces could you get a different answer by moving the particle differently? This means that it takes a different amount of energy to bring the particle to the same spot. Could you uniquely define the potential energy function?

No, you could not. Only conservative forces can have potential energy functions since we can define the potential energy uniquely. Since the electrostatic force is a conservative force, we can develop an associated electric potential that provides easier ways to solve problems, in many instances.

Exploration 25.1: Investigate Equipotential Lines

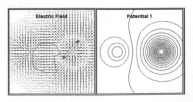

The panel on the left displays an electric field vector plot. The arrows in the field plot represent the direction, and the colors represent the magnitude of the electric field.

a. What is an equipotential line?

b. How can such a line be determined from an electric field representation like that shown in the left panel?

c. Draw the equipotential lines for this field by dragging the pencil (at its tip) after clicking the "draw on" button. Use the print screen function on your computer to print out a copy of your drawing.

d. After you have drawn your lines, determine which potential plot (1, 2, 3, or 4) best represents the region. Explain your choice.

e. Was your drawing a good representation of the actual equipotential lines? Did you have any misunderstandings? Explain.

Need help? **Enable Contour Lines on Double-Click** to see the actual contours in the electric field view by double-clicking in the region.

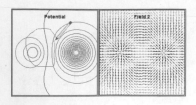

Exploration 25.2: Electric Field Lines and Equipotentials

The panel on the left displays an equipotential plot.

a. What is an equipotential line?

b. How are electric field lines related to equipotential lines?

c. Draw the electric field lines for this potential by dragging the pencil (at its tip) after clicking the "draw on" button. Use the print screen function on your computer to print out a copy of your drawing.

d. After you have made your drawing, determine which electric field (1, 2, 3, or 4) best represents the region. Explain your choice. Note that the arrows in the field plot represent the direction, and the colors represent the magnitude of the electric field.

e. Was your drawing a good representation of the actual field? Did you have any misunderstandings? Explain.

Need help? **Enable Field Lines on Double-Click** to see the field lines in the contour view by double-clicking in the region. Note: There is a short delay between the "click" and the drawing of the field line.

Exploration 25.3: Electric Potential around Conductors

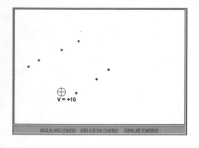

In this animation you can measure electric potential using a probe. You can click-drag to measure position and electric potential (position is given in meters and electric potential is given in volts). Clicking an "add marker" link will add a dot at the current position of the probe. There are two hidden conductors in the animation.

a. How do you know when the cursor moves over a conductor?

b. Make a sketch of the animation. Begin by labeling the position of the hidden conductors on your drawing. As you find the edges of the conductors, you can use the markers to outline them (use one color of marker for the first conductor you find and the other color for the second conductor you find).

c. How could **one** battery be connected to the two conductors to produce the above system? What would the voltage of the battery need to be? (Hint: Remember that the zero point of potential energy is arbitrary. Does either pole of the battery have to be at 0 V? Why or why not?)

d. Draw the battery in your sketch of the conductors from part (a).

e. A conductor is an equipotential surface. Why?

f. Sketch a representative number of equipotential contour lines for this system. Pick a specific voltage and then move the cursor around to find the loop of constant voltage. This maps out one equipotential contour line. As you map out other contour lines, remember that the change in voltage from one contour to the next is constant. Choose values wisely. You probably do not want to map out equipotential surfaces for voltage changes of 0.1 V.

g. Where is the electric field the strongest? Where is it the weakest?

Exploration 25.4: Time–of–Flight Mass Spectrometer

Positively charged particles start in the center of a uniform electric field (created by the charged gray plates; the field is shown, but fringe effects are not). When you push "play," four particles leave the parallel plates and head toward the detector. The graph simply plots the signal at the detector, showing a spike every time a particle hits the detector (position is given in centimeters and time is given in microseconds). This is a time-of-flight mass spectrometer and is used to detect what types of charged particles are in an atomic beam.

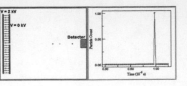

a. Given that the electric field is uniform and the voltage at the left plate is 2000 V and at the right plate is 0 V, explain how you know that the voltage in the middle of the plates (where the particles are) is 1000 V.

b. How much potential energy does each charged particle have if its charge is 1.6×10^{-19} C? (Each need one electron to be neutral again.)

c. After each particle leaves the region with a constant electric field and enters the region without an electric field, what is the value of its potential energy?

d. What then is the value of its kinetic energy?

e. Since the particles do not have the same speeds, rank the masses of the particles from least massive to most massive.

f. By measuring the time it takes the particle to arrive at the detector and the distance the particles travel through the field-free region, determine the speed of each particle.

g. From your calculation of kinetic energy, find the mass of each particle in kilograms and atomic mass units (1 amu $= 1.67 \times 10^{-27}$ kg).

h. Looking on a periodic table, what is the atomic mass of aluminum? It should be essentially the same as the value that you calculated for the mass of the smallest particle, as well as the mass difference between each larger particle. Therefore, this animation represents a particle beam where the first particle to hit the detector is a charged aluminum atom, and the second particle is two aluminum atoms bound together, and so forth. One way to find out what material is in an unknown substance, then, is to do this type of mass spectrometry. (**Illustration 27.3** and **Exploration 27.2** demonstrate other types of mass spectrometry.)

Exploration 25.5: Spherical Conductor and Insulator

How does the electric potential around a charged solid insulating sphere (with charge distributed throughout the volume of the sphere) compare with the electric potential around a charged conducting sphere? Move the test charge to map out the electric potential as a function of distance from the center (position is given in centimeters and electric potential is given in volts).

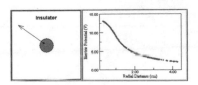

a. Why is the voltage constant inside the conductor?

b. Why is there no electric field and no force on the test charge inside the conductor?

c. Looking at the plots you make of voltage as a function of radial distance (as you move the test charge), what is the same and what is different between the two cases? Given that both spheres have the same total charge, explain the similarities and differences in the plots.

d. The electric field outside both spheres is $Q/4\pi\varepsilon_0 r^2$. Using this and the reference point of $V = 0$ volts at infinity, find an expression for the potential at a point outside the sphere and a distance r from the center of the sphere. $V = -\int \mathbf{E} \cdot d\mathbf{r}$ and integrate from $r =$ infinity (where $V = 0$ volts) to a point r.

e. Measure the voltage at some point outside the sphere and find the charge on both spheres. Verify that the total charge is the same.

Now for the voltage inside the uniformly charged insulator. Here the electric field is $Qr/(4\pi\varepsilon_0 R^3)$, where R is the radius of the sphere itself. In this case, to find the electric potential as a function of r, you again need to integrate $V = -\int \mathbf{E} \cdot d\mathbf{r}$, but this time you must break up the integral and integrate from infinity to R using $E = Q/4\pi\varepsilon_0 r^2$ (to find the electric potential associated with getting all the charges to the surface of the sphere) and then integrate from R to r (an arbitrary point inside the sphere) using the expression for the electric field inside the insulating sphere.

f. Verify that your calculation gives the same results as shown on the graph.

Problems

Problem 25.1

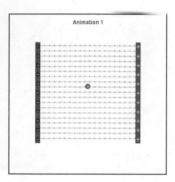

a. For each of the animations, does the potential energy of the particle increase or decrease as it moves?

b. Is the beginning point or ending point at a higher electric potential (higher voltage)?

c. Is an external force required or is the movement the result of the force due to the electric field? If an external force is required, describe it.

Problem 25.2

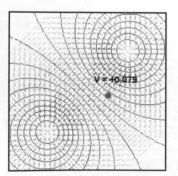

An equipotential plot is shown in the animation (position is given in meters and electric potential is given in volts). The electric potential is shown next to the electron. You can drag the red electron with the mouse. How much work must an external force do in order to move the red electron from $[x, y] = [1.1 \text{ m}, -1.5 \text{ m}]$ to $[x, y] = [-1.6 \text{ m}, 1.2 \text{ m}]$?

Problem 25.3

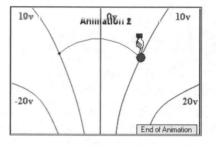

An **electron** is moved through a region with an electric potential defined by the equipotential lines shown (position is given in meters, time is given in seconds, and electric potential is given in volts). Rank the work done by the external force (smallest to greatest) for Animations 1–5. Explain the basis for your ranking.

Problem 25.4

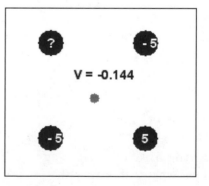

The charges shown are given in units of 10^{-10} C (position is given in meters and electric potential is given in volts). You can measure the electric potential by dragging the test charge around.

Determine the charge of the unknown charge for the following distributions:

a. **Two charges**.

b. **Three charges**.

c. **Four charges**.

Problem 25.5

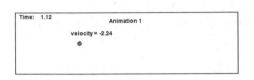

In the animation the electric potential changes from $0\,V$ to $1\,V$ as shown by the equipotential lines (position is given in meters, time is given in seconds, velocity is given in meters/second, and electric potential is given in volts). Click-drag to place the $1\text{-}\mu C$ test charge anywhere in the animation before you press "play." What is the mass of the particle?

Problem 25.6

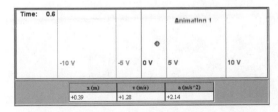

Each animation shows a negative test charge moving under the influence of an electric potential (position is given in meters, time is given in seconds, and velocity is given in meters/second).

For each animation,

a. Sketch a graph representing the electric potential versus x for the region shown in the animation.

b. Sketch a graph representing the electric field versus x for the region shown in the animation.

Problem 25.7

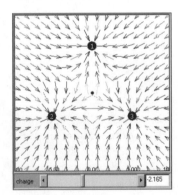

The animations show a *positive* test charge in a region where there is an electric potential, as shown by the equipotential lines (position is given in meters, time is given in seconds, and electric potential is given in volts). The positive test charge is placed at $x = 0\,m$ with no initial velocity. Which animation most closely represents the charge's subsequent motion?

Problem 25.8

The animations show four charged conductors (position is given in meters and electric potential is given in volts).

a. Configuration 1: Rank the magnitude of the electric fields in the three regions between the conductors from

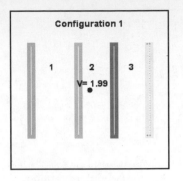

largest to smallest. You can move the black circle around to show the voltage at any point.

b. Configuration 2: Rank the magnitude of the electric fields in the three regions between these conductors from largest to smallest.

c. Configuration 3: Rank the voltages on the four plates from largest to smallest. As you move the *positive* test charge around, the vector shows the strength of the force on the test charge.

Problem 25.9

Three point charges of equal charge are arranged on the corners of an equilateral triangle, as shown in the animation. Each charge is measured in nC (nanocoulombs or 10^{-9} C) and can be varied using the slider. You can click-drag to read the value of the voltage at any point in the animation (position is given in meters, voltage is given in volts, and charge is given in nanocoulombs).

a. Find a formula for the voltage midway between the charges (indicated by the small black dot) as a function of the charges.

b. Looking at the electric field vectors, where is the electric field zero? Explain.

c. Is the voltage at that spot also zero? If not, explain why you can have a zero electric field, but a nonzero voltage.

Problem 25.10

A charged ball with a mass of 30 grams is suspended from a string as shown. A second ball with a negative charge of 3 mC

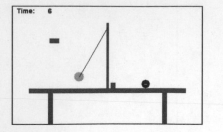

is placed on a wood table. Play the animation and move the ball on the table toward and away from the pendulum (position is given in meters and time is given in seconds).

a. Find the equilibrium point of the pendulum for two different positions of the ball on the table.

b. Write an expression for the change in potential energy (both gravitational and electrostatic) from one equilibrium position to the other.

c. What is the charge on the pendulum? Assume that the charge on the two balls is uniformly distributed.

Problem 25.11

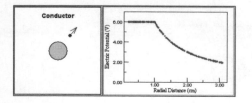

What is the total charge on the conducting sphere in this animation? Move the test charge to map out the magnitude of the electric potential as a function of distance from the center of the conducting sphere (position is given in centimeters, time is given in seconds, and electric potential is given in volts). The electric potential at infinity is 0 volts.

Problem 25.12

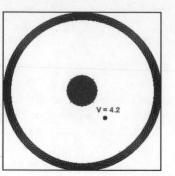

The animation shows either concentric cylinders or concentric spheres (position is given in centimeters and electric potential is given in volts). You can move the test charge around to measure the electric potential. Develop an equation for the voltage as a function of position for concentric cylinders and concentric spheres. The electric field outside a charged sphere $= kQ/r^2$. Outside a charged rod it is $2k\lambda/r$, where $k = 9 \times 10^9$ Nm^2 and $\lambda =$ charge/length.

a. If this animation represented concentric spheres where the inner sphere has a voltage of 10 V and the outer sphere has a voltage of 0 V, how much charge would be on the center sphere? What, then, would be the equation for the voltage as a function of position between the spheres?

b. If this animation represented concentric cylinders where the inner cylinder has a voltage of 10 V and the outer cylinder has a voltage of 0 V, what would be the charge per unit length be on the inner cylinder? What, then, would be the equation for the voltage as a function of position between the cylinders?

c. Does the animation represent concentric cylinders or concentric spheres?

Capacitance and Dielectrics

Topics include capacitance, electric field, equipotential surfaces, voltage, dielectrics, equivalent capacitance, energy, and charge distributions.

Illustration 26.1: Microscopic View of a Capacitor

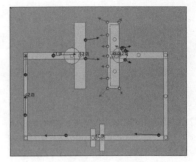

The animation shows a parallel-plate capacitor (at the top) connected to a battery (at the bottom). This Illustration shows you what happens when the battery is connected and the blue electrons are separated from the positively charged atoms. This animation only shows what happens for ten charge pairs.

As the electrons pile up on the left plate, what is the direction of the electric field between the plates? The electric field always points from positive charges and toward negative charges; therefore, the electric field points to the left. Charges move until the electric potential between the two plates matches the electric potential of the battery.

Now change the configuration to **add a thin dielectric between the capacitor**. What happens to the atoms on the dielectric material between the plates (represented by the circles)? Due to the electric field created by the charges on the capacitor plates, the charges in the dielectric are polarized. Since positive and negative charges experience forces in opposite directions in the same electric field, the electrons move to the right and the positively charged atoms move to the left. Note that these charges are not free to completely separate and move like the charges in the plates and wires. They only polarize. Because these charges are still bound together, we call them bound charges.

What is the direction of the electric field due to the separation of charge (the bound charge) in the dielectric? It is in the opposite direction from the initial electric field. Thus, the overall electric field between the plates (for the same number of charges) is smaller. Since the potential across the plates matches the potential of the battery, is this battery bigger or smaller than the first one? If the charge on the capacitor is the same in both animations, and the capacitance is increased with the inclusion of the dielectric ($C = k\,\varepsilon_0 A/d$), then $\Delta V = Q/C$ shows us that the electric potential difference is reduced. In addition, since the electric field is reduced, the electric potential is similarly reduced (for a constant electric field $\Delta V = -Ed$).

If the battery was the same in the two animations (ΔV would be the same between the plates), the set of plates with a dielectric between them would be able to hold more charge. Using $\Delta V = -Ed$, the electric field between the plates would have to be the same in the two animations. Because the bound charge of the dielectric reduces the electric field between the plates, there would have to be more charge on the plates with the dielectric present than without. This explains why the capacitance of the capacitor is greater with a dielectric.

Illustration 26.2: A Capacitor Connected to a Battery

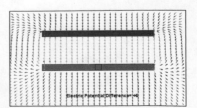

The animation represents a parallel-plate capacitor connected to a battery that is not shown. The red and blue circles on the plates represent the charge build up on the plates (position is given in meters, electric field strength is given in newtons/coulomb, and electric potential is given in volts). The plates are connected to a battery that maintains a constant potential difference between the plates.

How do the amount of accumulated charge and the magnitude of the electric field between the plates depend on the separation distance between the plates? Make a prediction and then test it by dragging the bottom (red) plate closer and farther from the top plate. (You must drag in the center of the plate.)

Notice that the battery maintains the electric potential difference between the plates. If we ignore the fringing effects of the electric field near the edges of the capacitor plates, the electric field is constant between the plates (from Gauss's law). Given this, the electric potential difference between the plates is related to the electric field by $\Delta V = -Ed$, where d is the distance between the plates. Because of this relationship, a larger separation between the plates, for the same electric potential difference, means a smaller electric field between the plates.

How do the amount of accumulated charge and the magnitude of the electric field between the plates depend on the voltage difference between the plates? Make a prediction and then test it by changing the voltage difference. As stated above, the larger the potential difference, for the same separation between the plates, the larger the electric field, and therefore the larger the charge accumulation on the plates.

Illustration 26.3: Capacitor with a Dielectric

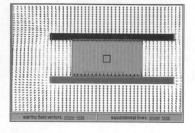

This animation shows a parallel-plate capacitor connected to a battery that is not shown. The battery maintains a constant electric potential difference between the plates even when you move the dielectric. The red and blue circles on the plates and on the dielectric represent the charge on the plates and the dielectric (position is given in meters, electric field strength is given in newtons/coulomb, and electric potential is given in volts). A dragable dielectric is outside the plates (drag at the center of the dielectric).

What effect does the dielectric have on the electric field between the plates of the capacitor and the charge accumulated on the plates? Make a prediction and test it by dragging the dielectric into the region between the plates.

What did you find? You should have found that, as you drag the dielectric into the capacitor, the electric field in between the plates decreases. Why does it decrease? Think about that question while you answer the following question. Make sure that the dielectric is in the capacitor. How does the electric field between the plates and the charge accumulated change when the dielectric constant of the material is increased or decreased? Make a prediction and test it by changing the dielectric constant.

What did you find? You should have found that, as you increase the dielectric constant of the dielectric, the amount of charge induced on the plates and the dielectric is increased. In response to the initial electric field between the two plates, bound charges are created inside the dielectric. While charge is not free to move inside a dielectric as charge is in a conductor, the charges polarize. This means that neutral atoms become little dipoles in response to the electric field: The electrons are one pole and the nucleus becomes the other pole. This polarization is due to the initial electric field between the plates, which points upward. As a consequence, the positive charges in the dielectric experience an upward force, and the electrons in the dielectric experience a downward force. Once the charges in the dielectric polarize, the net effect is for bound charge to accumulate on the top and bottom of the dielectric. There is no net effect from the dipoles in the middle of the dielectric since the effect of neighboring dipoles cancel. This bound charge creates its own electric field that reduces the initial electric field between the plates, since it points in the opposite direction from the original electric field. The larger the dielectric constant, the larger the bound charge and the more the electric field between the plates gets reduced.

What would happen if the dielectric constant could get really, really big? The bound charge would get bigger and bigger until there was no electric field in between the plates. Such a material is called a conductor.

Illustration 26.4: Microscopic View of Capacitors in Series and Parallel

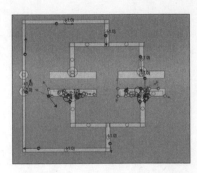

These animations model the charging of parallel-plate capacitors in different configurations as the blue electrons are separated from the positively charged atoms due to the electric potential difference.

Consider two capacitors in parallel connected to a battery. When you push the "play" button, the battery is connected to the capacitors and the electrons slowly begin to leave the vicinity of the positive charges. The charge pairs separate. Charges pile up until the electric potential between the two plates of the capacitor matches the electric potential of the battery. Note that the number of charges on the top plates of the two capacitors is essentially the same. They may differ by a few charges. This is because the two capacitors are identical and therefore have the same capacitance. Since they are in parallel, they have the same electric potential difference between their plates. Therefore, the number of positive and negative charges on each of the pairs of plates must be the same because capacitance is defined as $C = Q/\Delta V$. Note that, if the capacitors were not identical, they would still have the same electric potential difference across their plates, but their charges would be different.

Contrast this with **two capacitors in series connected to a battery**. Again, push the start button to connect the battery to the capacitors. Notice that the negative charges that end up on the left capacitor come from the right capacitor. Also note that the negative charges that end up on the right capacitor come from the left capacitor. Therefore, the charge on one pair of capacitor plates is the same as on the other pair of capacitor plates. In this case, for a series configuration, the sum of the electric potential differences across each of the individual capacitors equals the electric potential difference of the battery.

Instead of connecting to a battery, now let's assume that there is a reservoir of excess positive charge on one end of the configuration and a connection to ground (a reservoir of negative charges) at the other end of the configuration of capacitors. Consider **two capacitors of different capacitance in parallel**. What happens? Which capacitor holds more charge? Why? Since the capacitors are in parallel, the electric potential difference across each of the capacitors must be the same. As a consequence, since the capacitance must be different for each capacitor, the amount of charge on each capacitor must be different. The capacitor on the right has the larger capacitance ($C = \varepsilon_0 A/d$) and therefore has the larger charge.

Finally, consider **two different capacitors connected in series**. As in the series case with the battery, both capacitors contain the same charge. Which one has the greater potential difference across it? Since $\Delta V = Q/C$, the capacitor on the left (with the larger capacitance) has the larger electric potential difference across it.

Exploration 26.1: Energy

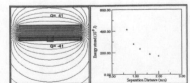

Wait for the calculation to finish. As you move the bottom capacitor (and allow the applet to finish calculating after each move), the graph shows the stored energy as a function of separation distance between the plates (position is given in millimeters, charge is given in nanocoulombs, and energy is given in nanojoules).

a. Given that the stored energy (the potential energy) is $QV/2$, what is the voltage difference between the plates?

b. Does the voltage difference between the plates change?

c. How does the capacitance change as you move the plates?

d. What is the area of the plates for this capacitor?

e. Why does the charge change as you move the plates?

f. As you move the plates closer together, does the stored energy increase or decrease?

g. Does that mean you would need to do positive work to push the plates together or pull them apart? Explain.

h. Since potential energy $U = QV/2$, if V is kept constant, what is U (potential energy) as a function of the separation distance? Verify that this is the relationship shown on the plot.

Exploration 26.2: Capacitors, Charge, and Electric Potential

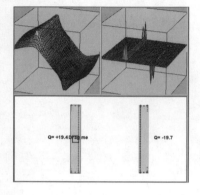

This animation shows a parallel-plate capacitor and the charges on the plates, the total charge, and the electric potential difference between the plates. You can move the left plate by click-dragging the middle of the plate (at the "Drag Me" label). The plots show you the electric potential and charge as a function of (x, y) position (position is given in centimeters, charge is given in coulombs, electric field strength is given in newtons/coulomb, and electric potential is given in volts). You can click-drag in a graph to rotate the plot and see it from a different angle.

a. Which plot corresponds to electric potential as a function of position? Which one is charge as a function of position? Explain how you know.

b. From the charge plot, where is there the most charge on the plates? Why?

Consider the configuration with a constant electric potential difference between the plates.

c. How does the charge change as you move the left capacitor plate? Explain.

d. Does this result correspond to a capacitor connected to a battery or a charged capacitor disconnected from a battery? Explain.

Now consider the configuration with a constant charge on the plates.

e. How does the voltage difference between the plates change as you move the left capacitor plate? Explain. (Note: If the animation says "Failed to converge" after you move the plates, simply click on the plate again to have the animation recalculate. The charge will stay in about the same range.)

f. Does this correspond to a capacitor connected to a battery or a charged capacitor disconnected from a battery? Explain.

Exploration 26.3: Conductors and Dielectrics

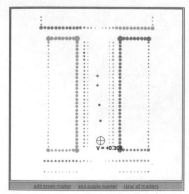

Wait for the calculation to finish. There are hidden conductors and dielectrics in the animation. The light red circles represent positive charges, and the light blue circles represent negative charges. This charge can be either bound or unbound (free); in other words it could be the charge on a dielectric or on a conductor. You can measure electric potential using the probe and you can click-drag to measure position and electric potential (position is given in meters and electric potential is given in volts). Clicking an "add marker" link will add a circular marker at the current position of the probe.

a. Sketch and label your best guess as to the configuration of the hidden conductors and dielectrics. You may want to use the markers to outline conductors and/or dielectrics.

b. What is the minimum number of external batteries needed to produce the system? Show the voltage values of these batteries and how they should be connected to the system.

c. Where is the electric field strongest? Where is it weakest?

d. Sketch the electric field using electric field lines; that is, draw a representative number of field lines.

e. Sketch the equipotential lines.

Exploration 26.4: Equivalent Capacitance

This animation contrasts two configurations of capacitors and a battery (capacitance is given in farads). The table shows the voltage across the battery as well as across each capacitor.

First, consider **capacitors in series**. Pick a value of the capacitance for capacitor A that is bigger than capacitor B.

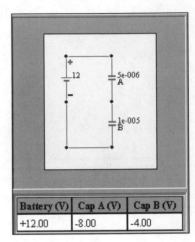

a. What is the charge on each capacitor (use $Q = CV$)?

b. Why are the charges equal (explain in terms of where the charge originates to charge the plates)?

This is the total charge stored in this circuit. If the battery were removed from this circuit and we wanted to use the stored charge for an electrical appliance, notice that the charges stored on the two sides of the capacitors connected to each other (the bottom plate of A and the top plate of B) would not be available to another circuit. Thus, the total charge stored is the charge stored on either capacitor.

Battery (V)	Cap A (V)	Cap B (V)
+12.00	-8.00	-4.00

c. If you wanted to replace the two capacitors in this circuit with one capacitor that stored the same amount of charge at the same total voltage, what would the value of that capacitance be?

d. Verify that the equivalent capacitance is equal to $(1/C_A + 1/C_B)^{-1}$.

Now consider **capacitors in parallel**. Pick a value of the capacitance for capacitor A that is different from capacitor B.

e. What is the same for the two capacitors? What is different?

This time, the charge stored on each capacitor would be available to an electrical appliance if the battery were removed; thus, the total charge is the sum of the charge stored individually on each plate.

f. What is the equivalent capacitance for these two capacitors (i.e., what size capacitor would store the same total charge at this voltage?)?

g. Show that it is equal to $(C_A + C_B)$.

Exploration 26.5: Capacitance of Concentric Cylinders

Wait for the calculation to finish. This animation shows a coaxial capacitor with cylindrical geometry: a very long cylinder (extending into and out of the page) in the center surrounded by a very long cylindrical shell (position is given in centimeters, electric field strength is given in newtons/coulomb, and electric potential is given in volts). The outside shell is grounded, while the inside shell is at 10 V. You can click-drag to measure the voltage at any position.

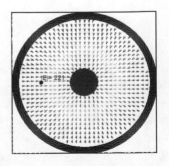

a. Use Gauss's law to show that the magnitude of the radial electric field between the two conductors for a cylindrical coaxial capacitor of length L is $E = Q/2\pi r L\varepsilon_0 = 2kQ/(rL)$, where Q is the total charge on the inside (or outside) conductor and r is the distance from the center.

b. If L = 1 m, measure the electric field in the region between the two conductors and determine the charge on the inside (and outside) conductor.

c. Use $V = -\int \mathbf{E} \cdot d\mathbf{r}$ to show that the potential at any point between the two conductors is $V = (Q/2\pi L\varepsilon_0) \ln(b/r) = (2kQ/L) \ln(b/r)$, where b is the radius of the outer conductor.

d. Given that the potential difference between the two cylinders is 10 V, verify your answer to (b) and find the charge on each conductor.

e. Given, then, that the potential difference between the two conductors is $V = (Q/2\pi L\varepsilon_0) \ln(b/a) = (2Qk/L) \ln(b/a)$—$b$ is the radius of the outer shell and a is the radius of the inner cylinder—show that the capacitance of this capacitor is $(2\pi L\varepsilon_0)/\ln(b/a) = (L/2k)/\ln(b/a)$.

f. What is the capacitance (numerical value) of this capacitor?

Problems

Problem 26.1

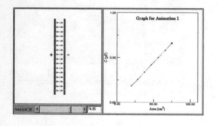

As you move the slider, you change either the area or the separation of a capacitor (position is given in centimeters and capacitance is given in picofarads). Which of the graph(s) is (are) correct? Explain.

Problem 26.2

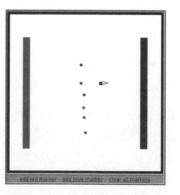

A parallel–plate capacitor is shown. A battery connects the two plates together but is not shown. Position and electric field strength are shown when you drag the mouse (position is given in meters and electric field strength is given in newtons/coulomb).

a. Is the green plate positively charged, negatively charged, or neutral? Explain.

b. Is the purple plate positively charged, negatively charged, or neutral? Explain.

c. Sketch a diagram of the charge distribution on the plates.

d. What is the voltage of the battery that connects the two plates?

Problem 26.3

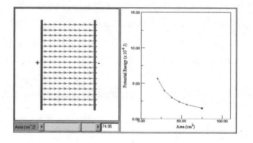

Move the slider to change the area of the two capacitor plates (position is given in centimeters, area is given in centimeters squared, and energy is given in microjoules). Is this a capacitor connected to a battery or a charged capacitor that is not connected to a battery? Explain.

Problem 26.4

The animation shows a parallel-plate capacitor with a hidden dielectric between the two plates. You can click–drag to measure position and electric potential (position is given in meters and electric potential is given in volts). The small black circle in the center is a dragable test charge that can be used to show the direction and magnitude of the electric field. Click on the add marker links to add a blue or red marker at the location of the test charge to outline the edges of the dielectric.

a. Sketch and label an estimate of the location of the dielectric.

b. Sketch a representative number of equipotential lines.

Problem 26.5

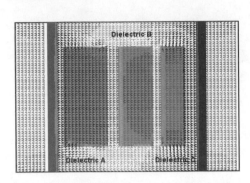

A parallel-plate capacitor has three dielectrics inserted between the plates. The battery connecting the two plates is off screen. The blue and red circles on the plates represent the accumulated charge. Position and electric field strength are shown when dragging the mouse (position is given in meters and electric field is given in newtons/coulomb). The arrows represent the electric field vectors.

Rank the dielectrics based on their dielectric constants from smallest to greatest. Explain your ranking.

Problem 26.6

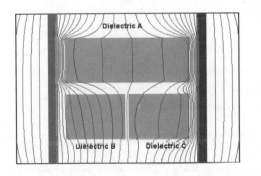

A parallel-plate capacitor has three dielectrics inserted between the plates (position is given in meters and electric potential is given in volts). A battery that connects the two plates together is off screen. The blue and red circles on the plates represent the accumulated charge on the plates. The lines represent equipotential surfaces.

Rank the dielectrics based on their dielectric constants from smallest to greatest. Explain your ranking.

Problem 26.7

The animations show a parallel-plate capacitor with a movable dielectric, while the plots show the electric potential (left plot) and charge (right plot) as a function of the (*x, y*) position (position is given in meters and the electric field strength is given in newtons/coulomb). You can click-drag your mouse in a graph to rotate a plot and see it from a different angle. You can move the dielectric into and out of the

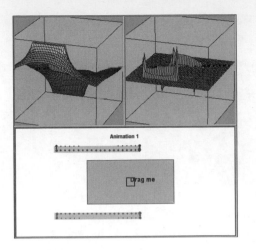

region between the two plates by clicking and dragging the middle of the dielectric.

a. In one of the animations a capacitor is connected to a battery, and in the other one a charged capacitor is not connected to the battery. Which is which?

b. Describe the changes you see on the plots in the two animations and explain what is happening as the dielectric is dragged between the plates (i.e., why do you see the increases or decreases in potential or charge?).

Note: If the animation says "Did not converge" after you move the dielectric (especially if you move it from one end to another), simply click on the dielectric again to have the animation recalculate.

Problem 26.8

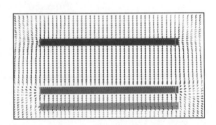

The animation shows a parallel-plate capacitor that has been charged and then disconnected from a battery so that the charge on the plate remains constant. The red and blue circles inside the plates represent the charge buildup on the plate, and the electric field vectors are also shown (position is given in meters, electric field strength is given in newtons/coulomb and electric potential is given in volts). A dragable dielectric (drag at the center) is below the capacitor. What is the value of the dielectric constant of the dielectric?

Problem 26.9

The animation shows two parallel-plate capacitors connected by conducting wires to a battery (the two circles to the left). When you push the "play" button, the battery is connected to the capacitors and electrons are pulled away from the associated positive charges (charge separation occurs).

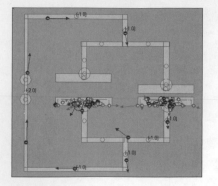

a. Which capacitor has the larger capacitance?

b. Which has the larger electric potential difference between the plates?

c. Which has the greater electric field between the plates? Explain.

Problem 26.10

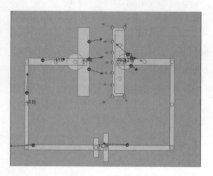

The animations show parallel–plate capacitors connected by conducting wires to a battery. If all the capacitors are identical, how do the batteries in the two cases compare?

Problem 26.11

In the animation, you can close and open switches and read the voltage across capacitors A, B, and C. The capacitance of

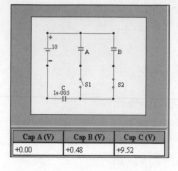

Cap A (V)	Cap B (V)	Cap C (V)
+0.00	+0.48	+9.52

capacitor C is 1×10^{-5} F, and the voltage of the battery is 10 V. What, if anything, is wrong with the animation?

Problem 26.12

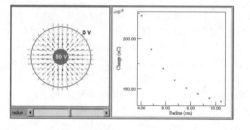

The animation shows a spherical conductor surrounded by a conducting spherical shell. The two conductors have equal but opposite charge, and the voltage difference between the two is 50 V (position is given in centimeters and charge is given in nanocoulombs). The field vectors for the electric field (kQ/r^2) are shown in the region between the shell and the inner sphere and are zero everywhere else.

a. Develop an expression for the charge as a function of the radius of the outer sphere and verify it with the animation.

b. Develop an expression for the capacitance as a function of radius.

c. Does the capacitance increase or decrease with increasing shell size? Explain.

Magnetic Fields and Forces

Topics include magnets, magnetic fields, Lorentz force, and current.

Illustration 27.1: Magnets and Compass Needles

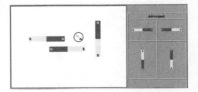

This Illustration allows you to consider the magnetic field around a bar magnet. By default, the page will load with a magnet in the center of the animation. Use the compass to explore the magnetic field around a bar magnet by dragging the compass around the magnet. A compass utilizes a small permanent magnet; its arrow points toward the north pole of its magnet. Make a diagram showing the direction the compass needle points at various locations. Include enough points to establish a pattern.

Now that you have completed this diagram, turn the magnetic field vectors on to see this representation of the magnetic field. Does your diagram look like that of the animation's field vectors? It should. The magnetic field vectors are like little compass needles placed at various points around space. The color of the arrow of the magnetic field vector represents the strength of the field at the point, while the arrow shows the direction of the field.

You can also double click at any point in the animation to draw a magnetic field line through that point, which yields another representation. There will be a small delay after double clicking before the line appears (the line needs to be calculated). Double click at enough points to get an accurate picture of the magnetic field lines around the magnet. What is the difference between the field vectors and the field lines? Notice that in the field-line representation the field lines are tangent to the field vectors and also tangent to the direction of a compass needle placed at that point. In the field-line representation the field lines are drawn with the same color. In the field-vector representation the strength of the field is depicted in the color of the field vectors. How do we represent the magnitude of the magnetic field in the field line representation? The density of field lines (lines per square length unit) is greater where the field is stronger.

Clear the screen. Place two magnets beside each other with the north pole of one lined up with the south pole of the other:

Display the field vectors and/or double click to display the magnetic field lines. How does the magnetic field of the two magnets compare to the field of one magnet? What do your observations suggest about how a bar magnet would behave when broken in half? The magnetic field vectors and lines for the above configuration look just like those for one large magnet. In fact, if you broke a bar magnet in half, what you would get would be two bar magnets, each with their own north and south pole. This is because there are no magnetic monopoles in classical magnetism. There are electric monopoles, which we call electric charges.

Predict what the magnetic field will be if you place a north-south magnet directly on top of a south-north magnet. Try it.

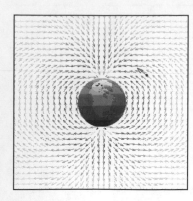

Illustration 27.2: Earth's Magnetic Field

This Illustration demonstrates the magnetic field of Earth. We describe the magnetic field by mapping out magnetic field vectors and/or magnetic field lines. This Illustration allows you to try both representations. What is the difference between the two representations?

Before adding either field vectors or field lines, add a compass and move it around. A compass utilizes a small permanent magnet; its arrow points toward the north pole of its magnet. Now show the **field vectors**. Notice that the compass arrow lines up with the field vectors. The field vectors essentially tell you where little compass needles would point at different places. Now show the **field lines**. On the field-line representation, notice that the compass needle is always tangent to the field line.

Now show the **geographic poles** of Earth. Move the compass around (again the compass is a little magnet with the arrow on its north pole). Is the geographic north pole also the magnetic north pole? Check your answer by showing the **magnetic poles**.

Why do you think we call the pole in the Arctic the North Pole? This is because the north pole of a compass points there (even though by this definition the North Pole is a south pole). Although we know where the poles are on Earth at the present time, over thousands of years the magnetic north and south poles have flipped back and forth, and we do not yet have a satisfactory theory that explains what causes Earth's magnetic field.

Illustration 27.3: A Mass Spectrometer

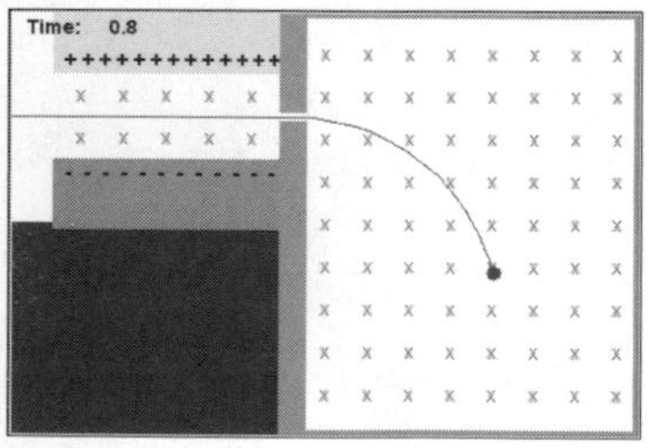

Begin this animation by selecting the **Multiple Masses Demonstration**. This shows five particles passing through a model of a mass spectrometer. The particles have different masses but are otherwise identical. Notice how the particles are separated based on their mass.

You can enter values for the initial conditions and then press the "register values and play" button to see a single particle pass through the mass spectrometer. The particle initially enters a region with an electric field directed downward and a magnetic field directed into the screen. Since the particle is negatively charged, the electric field exerts an upward force ($\mathbf{F} = q\,\mathbf{E}$; see, for example, **Illustration 23.4**) and the magnetic field initially exerts a downward force ($\mathbf{F} = q\,\mathbf{v} \times \mathbf{B}$). Try setting the magnetic or the electric field to zero to see the effect of just one field on the particle. For certain values of the magnetic and electric fields, the magnetic and electric forces on the particle will exactly cancel and the particle will pass through the first region. This region is called a velocity selector, since only particles with a certain initial velocity will pass through for given values of the magnetic and electric fields. Other Explorations and Problems from this chapter will require you to formulate a mathematical relationship between the initial values for particles that pass through the velocity selector.

If a particle is able to pass through the first region, it enters a region where only the magnetic field is present. Since the magnetic field exerts a force perpendicular to the direction of the velocity ($\mathbf{F} = q\,\mathbf{v} \times \mathbf{B}$), the particle follows a circular path (since v and B are constant and $\mathbf{v}$ and $\mathbf{B}$ are perpendicular). The radius of this path depends on the mass. From Newton's second law for uniform circular motion, $|\mathbf{F}| = mv^2/R = q\,|\mathbf{v} \times \mathbf{B}| = qvB$, since $\mathbf{v}$ and $\mathbf{B}$ are perpendicular. By measuring where the particle strikes one of the walls, you can determine the mass of the particle.

Illustration 27.4: Magnetic Forces on Currents

This Illustration shows current flowing through a wire. A current consists of charges moving through a conducting wire (1 coulomb of charge per second = 1 ampere).

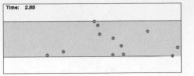

This can occur in a conductor since charges in a conductor are free to move in response to forces.

In the animation there is initially no magnetic field in the region shown. The electrons (the charge carriers that are free to move in conductors) travel in one direction, but the direction of the current is in the opposite direction (it is the direction positive charges would flow). There is nothing strange about this; it is just a convention. A positive current in a particular direction means that it is as if positive charges are flowing in that direction (this is equivalent to negatively charged electrons flowing in the opposite direction). What is the direction of the current in this animation? As there are negative charges moving to the left, there is a positive current to the right.

Turn on the **uniform magnetic field pointing into the screen**. Notice that we represent the field by a series of circles with "x"s inside them. Since we represent vectors with arrows, the "x" is supposed to represent what it would look like if you saw an arrow moving away from you. You would see the back view of an arrow and the "x" made by the feathered-end of the arrow. Similarly, for a magnetic field pointing out of the page, we use a series of circles with dots inside them (the view of an arrow point coming right at you). From where the electrons are located, what direction is the force on the electrons? Switch the direction of the **magnetic field (out of the page)**. Now what is the direction of the force on the electrons? Notice that the electrons are moving to the left. We can use $\mathbf{F} = q\mathbf{v} \times \mathbf{B}$ to determine the force. For the **uniform magnetic field pointing into the screen**, q is negative, $\mathbf{v}$ is to the right, and $\mathbf{B}$ is out of the page. Since $\mathbf{v}$ and $\mathbf{B}$ are perpendicular, we have that $F = |\mathbf{F}| = |q|vB$. The right-hand rule (point your fingers towards v, curl them towards B, the direction your thumb points is the direction of $\mathbf{v} \times \mathbf{B}$) gives downward for the force direction, but we must factor in the fact that we have negative charge. Thus, the force on the electrons for this case is upward. When we go through the same process for **magnetic field (out of the page)**, we get that the force on the electrons is downward.

Change the **direction the electrons are moving**. Turn on the **magnetic field into the page**. What direction is the force now? If you followed the arguments of the previous paragraph, this should be easy.

With the **electrons moving in the original direction**, try a **magnetic field pointing to the right**. What is the direction of the force? What do you expect for a magnetic field pointing left? **Try it**. What can you conclude about the force on a moving charge (and, therefore, the force on a wire) in a magnetic field?

The force on a moving charged particle is perpendicular to both the velocity and the magnetic field (and the velocity and magnetic field cannot point in the same direction) and is described by the vector product or cross product $F = q\mathbf{v} \times \mathbf{B}$. Remember that the charge on the electron is negative, so the force points in the opposite direction from $\mathbf{v} \times \mathbf{B}$.

Illustration 27.5: Permanent Magnets and Ferromagnetism

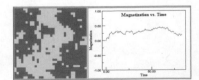

This is a simplified model of permanent magnets called the Ising model. In this Illustration you can change the temperature and background magnetic field to see how these variables affect the production of permanent magnets.

To turn an ordinary nail into a magnet, you can put it in a magnetic field. The iron will be magnetized and it will retain its magnetization even when the field is removed. This model illustrates how that is possible. The red and green represent regions within a material with magnetic moments that are lined up in one direction (red) and the other direction (green). After you push "play," notice that, to begin with, there are essentially equal areas of red and green regions. This is recorded on the graph as a magnetization of about 0. This means that inside our iron there is no organization of the magnetic moments. The thermal energy available in the material allows for the changing between red and green that you see.

Put this material in a magnetic field (push the "$B > 0$" button). What happens? Now the magnetic moments (little magnets inside the material) are lined up with the applied field. What do you expect will happen if you push the "$B < 0$" button? Why? Try it.

Now, what do you expect will happen if you push the "$B = 0$" button? Try that. What happens? Does the magnetization go to zero right away? Even when the magnetic field is no longer in place, the magnets want to stay lined up. It takes energy to disorganize them again. Over a long period of time, they can get randomly oriented again, but they will line back up quickly with an external field. Verify this.

Another way to get the magnets randomly oriented again is to increase the temperature (give them enough thermal energy to destroy the order). To simulate this, first magnetize the material (either red or green), set the field back to zero, and then push the "increase temperature" button.

Exploration 27.1: Map Field Lines and Determine Forces

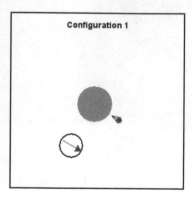

In the animation there is an object underneath the gray circle that creates a magnetic field.

a. Use the compass to determine the direction of the magnetic field. Sketch the vector field and the field lines for each configuration.

b. Check your field-line diagrams by double-clicking on the animation to show a field line at the location of the mouse.

Add a wire with current coming out of the screen (electrons moving into the screen). Click-drag the wire to move it around. The arrow shows the direction of the force on the wire.

c. Explain why the force vector points the way it does at two different locations of the wire for each configuration.

d. If the current was in the other direction, what direction would the force be at the two locations you chose? Explain.

e. Check your answer by adding a wire with current going into the screen.

Exploration 27.2: Velocity Selector

A mass spectrometer measures the mass of particles. The first step in the operation of the mass spectrometer is to select particles of a particular velocity. As you work through this exploration you will see how a velocity selector operates. The animation shows a positively charged particle entering a constant magnetic field directed into the screen.

a. BEFORE you play the animation, PREDICT the path the charge will follow. **I have already made my prediction; let me see the path.** Were you correct? If not, what caused your error?

Now, suppose a constant electric field is added to the region with the magnetic field.

b. In which direction (right, left, up, down, into screen, or out of screen) should the electric field be oriented such that it could possibly cancel the force due to the magnetic field?

c. In order to create the electric field, two charged plates are used. Which plate should be positively charged and which should be negatively charged to create the desired field?

d. I have made a prediction; let me check my thinking. Were you correct? If not, what misunderstanding caused your error?

e. Derive a mathematical relationship between the electric field, the magnetic field, and the velocity a particle must have to pass through the region undeflected.

f. In the animation the electric field produced by the plates is 6000 N/C and the magnetic field is 0.3 T. Use your mathematical relationship to find the velocity the particles have in order to pass straight through the two fields. Once you have calculated your answer, put it in the box and press "play" to see if you were correct.

Exploration 27.3: Mass Spectrometer

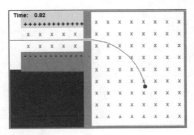

A negatively charged particle enters a region with a constant magnetic field directed into the screen and a constant electric field produced by two charged plates. If the particle is able to pass through the first region, it enters a region where only the magnetic field is present.

The Exploration demonstrates how a mass spectrometer works (See **Illustration 23.4** and **Exploration 25.4** for related examples). Many particles might be injected into the first region. For certain values of electric and magnetic fields, only particles with a particular velocity will pass through undeflected. By subjecting the particles to the velocity selector, we know the velocity of the particle when it enters the second region.

a. If the initial velocity is 50 m/s, the magnetic field is 0.5 T, the mass is 0.3 gram, and the charge is -1×10^{-3} coulombs, what must the electric field be in order to "select" the 50 m/s particle? Calculate your answer first and then test it using the animation.

b. If you change the value of the magnetic field, is the 50 m/s particle still "selected"?

c. What if you change the mass or the charge? Explain.

d. Once you are able to select the 50 m/s particle and it passes into a region where only the magnetic field is present, it follows a circular path. Why?

Now change the mass from 0.3 gram to 0.1 gram. Notice that the curved path of the charge changes. For every mass, the curved path will be slightly different. This allows you to measure the mass of an individual particle. This is very useful, especially when the mass is too small to easily measure using other methods.

e. By considering the magnetic force in the second region, develop a mathematical expression that relates the mass of the particle to the other variables. Do not include the velocity in your expression. You can use the condition that the particle passed through the region of electric and magnetic fields undeflected to eliminate velocity from your expression. Your expression will also contain the radius of the circular path.

You can measure this radius in the applet using a mouse-down (position is given in meters and time is given in seconds). In a real mass spectrometer the radius is often measured by putting a photographic plate on the wall where the particle hits. When the particle hits the plate it leaves a mark, allowing the experimenter to determine the value of the radius.

f. Check the expression you derived. When you put in the values from above, do you get a mass of 0.1 gram as you should?

Problems

Problem 27.1

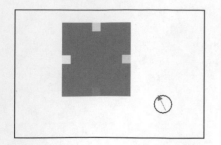

You can move the compass around. You can also move the gray magnet by dragging it at the pink square (position is given in centimeters). Determine the poles of the magnet. The colored blocks are there for your reference.

Problem 27.2

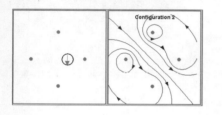

The animation shows four wires carrying current (each wire carries current either into or out of the screen). Drag the compass around to map out the magnetic field in the left-hand panel.

Which of the three configurations (in the right-hand panel) shows the correct field lines for the original animation? Click inside the configuration (in the right-hand panel) to draw magnetic field lines.

Problem 27.3

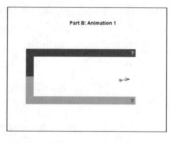

A positively charged particle moves through a uniform magnetic field. The animations show the field vectors (position is given in millimeters and time is given in seconds). Which animation is correct? Explain.

Problem 27.4

The animation illustrates charged particles flowing through a wire (position is given in millimeters and time is given in seconds). A current consists of charges moving ($1 \text{ C/s} = 1$ ampere) through a conducting wire. Remember that in a conductor,

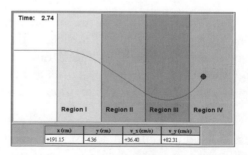

charges are free to move in response to forces. Initially, there is no magnetic field in the region shown. The particles travel in one direction.

a. What is the direction of the current in each case?

b. What is the direction of the magnetic field in each case?

Problem 27.5

This problem has two parts. You must first answer (a) before you can go on to answer (b).

a. Determine the poles of the horseshoe magnet. You can move the compass around.

b. After completing part (a), determine which animation shows the correct force on a wire carrying current out of the computer screen. You can drag the wire around in the animations.

Problem 27.6

An electron is shot through four regions of constant magnetic field (position is given in centimeters and time is given in seconds).

a. In which direction is the magnetic field in each region?

b. Rank the magnitude of the magnetic fields of the four regions, from smallest to greatest.

c. How would the path change if we inverted the direction of the magnetic field in every region? Draw the new path of the electron.

d. If you wanted to ensure that the particle did not enter Region II, would you increase or decrease the speed of the electron entering Region I? Give a mathematical proof of your answer.

Problem 27.7

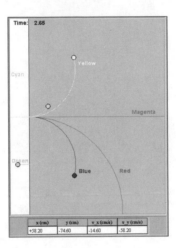

x (cm)	y (cm)	v_x (cm/s)	v_y (cm/s)
+58.20	-74.60	-14.60	-58.20

Six objects (all with a mass of 5 grams) are shot into a region of constant magnetic field. The blue object has a charge of 6 mC. The position and velocity of the **blue** object are given in the table (position is given in centimeters and time is given in seconds). All objects are located at $x = -50$ cm at time $t = 0$ s.

a. Rank the objects based on their velocity when they first encounter the magnetic field, from lowest to highest.

b. How does the magnitude of the velocity of each object change as it interacts with the magnetic field?

c. Rank the objects based on their charge, from lowest to highest.

d. What are the direction and magnitude of the magnetic field?

Problem 27.8

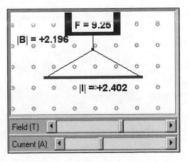

The animation shows a particle passing through a mass spectrometer (position is given in meters and time is given in seconds). There is a constant magnetic field throughout the region directed into the screen. There is a constant electric field in the first region only.

a. Is the particle charged? How do you know? If it is charged, does it have a positive or negative charge? Justify your answer.

b. If the electric field produced by the charged plates in the first region is 80 N/C, what is the value of the magnetic field in that region?

c. What is the charge-to-mass ratio (q/m) of the particle?

For an introduction to the mass spectrometer see **Exploration 27.3**.

Problem 27.9

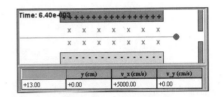

	y (cm)	v_x (cm/s)	v_y (cm/s)
+13.00	+0.00	+5000.00	+0.00

The animation shows a 30-mg charged object entering a region where a magnetic field and/or electric field can be applied (position is given in centimeters and time is given in seconds). The magnitude of the charge is 4×10^{-3} C.

a. Is the object positively or negatively charged? Explain.

b. What is the magnitude of the electric field?

c. What is the magnitude of the magnetic field?

d. What voltage must be applied to the plates to create the electric field?

For an introduction to the velocity selector see **Exploration 27.2**.

Problem 27.10

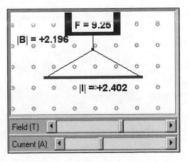

The blue rod has a current flowing through it and sits in a uniform external magnetic field that points out of the page (as represented by the gray circles with white dots). The probe at the top records the force required to support the rod (position is given in centimeters, magnetic field is given in tesla, current is given in amperes, and force is given in newtons).

a. In which direction does the current flow through the rod?

b. What is the mass of the rod?

28

Ampere's Law

Topics include magnetic fields, currents, Lorentz force, symmetry, Ampere's law, and the Biot-Savart law.

Illustration 28.1: Fields from Wires and Loops

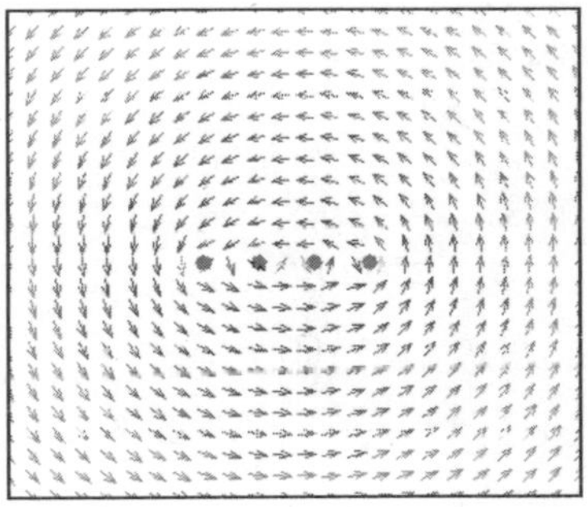

The magnetic field around a long, straight wire carrying current out of the computer screen points in a direction that circles the wire (position is given in meters and magnetic field strength is given in tesla). You may want to use a right-hand rule to determine the field direction. If you point the thumb of your right hand out of the computer screen (as if it pointed in the direction of the current) and close your fingers in a fist, then your fingers will point in the direction of the magnetic field around the wire. Instead of one wire, add four wires (again the current in these wires is coming out of the screen). Notice that the vectors from each wire add up. Double-click inside the animation to draw a field line.

What would you expect the direction of the field and the field lines to look like for many wires all lined up in a horizontal row? Sketch your prediction first. Once you have your prediction, try it by pushing the "plate" button. Explain why the field lines look like they do.

For the plate, you should have predicted that the fields from the individual wires that point in the *y* direction should all cancel. This leaves just a field in the *x* direction. Since the currents in the wires are all out of the screen, the field points to the left above the plate and points to the right below the plate.

Now, let's put a loop perpendicular to the page. In this representation you are looking at the edge of a loop of wire: The wire goes into the screen, circles around, and comes back out. The blue and red dots simply represent a slice of the wire, with red indicating current coming out and blue indicating current going in. For this case, describe how the current travels (Does the current flow into the screen at the top or the bottom of the loop?). The field points to the right along the center axis of the loop and diverges out from there. Adjust the size of the loop by click-dragging either the red or blue dot. Notice that the region near the center of the loop becomes more and more uniform as the loop gets bigger and bigger.

If you place many loops side by side, what do you expect? Try it by pushing the "solenoid" button. What are the similarities and differences between the field inside a solenoid and the field above a plate? Again the magnetic fields in the *y* direction all cancel, leaving the magnetic field in the *x* direction. Given that there is no current enclosed for an Amperian path in the plane with sides outside of the solenoid, the magnetic field is zero there. Inside the solenoid, however, the magnetic field is rather uniform and points to the right.

In order to use Ampere's law, you will need to have a sense of the direction of magnetic fields from a wire or group of wires and then build Amperian loops to match the symmetry of the fields.

Illustration 28.2: Forces Between Wires

The wire in the center has a fixed current. You can change the current in the blue wire by using the slider (position is given in meters, current is given in amperes, and magnetic field strength is given in tesla). The animation shows the magnetic field vectors (you can also double click on the screen to draw the field lines).

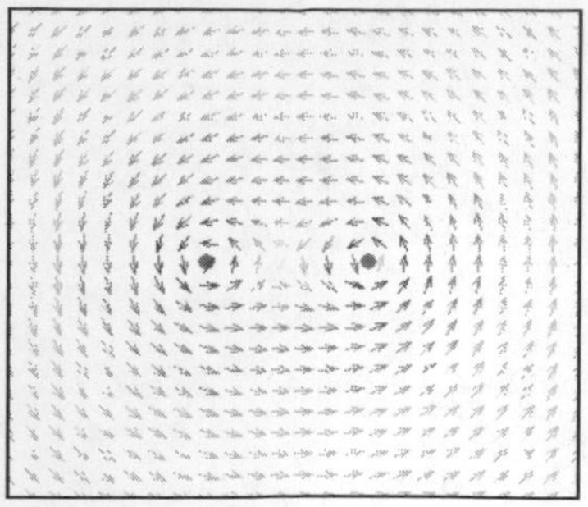

Keep the current in the blue wire at zero. In what direction is the magnetic field at the point where the blue wire is located? When the current is turned on in the blue wire, the current will either come out of the page (positive) or go into the page (negative). If you put positive current in the blue wire, in what direction would the force be on those moving charges (the current in the wire)? This is the Lorentz force on the charge carriers, so you'll need to use the right-hand rule you used in the previous chapter. Use the slider to put positive current through the blue wire. The vector shown is the force on the wire. Using the right-hand rule, the direction of the force is $q\,\mathbf{v}\times\mathbf{B}$. The positive charges are moving out of the screen, and the direction of the magnetic field is in the plane of the screen and perpendicular to the line separating the currents. Using the right-hand rule gives you a direction toward the red wire.

Move the blue wire to a new position. The force points in a different direction at this position, but it still points toward the other wire. What happens if you increase the current? The force gets bigger. What happens if you make the current negative? Now the direction of the current changes, and so will the direction of the force from the right-hand rule. The currents will now repel instead of attract.

Why is there a force vector on the center (red) wire? Well, this wire also experiences a magnetic field due to the blue wire. We get the force on the red wire to be equal and opposite to that of the force on the blue wire from the right-hand rule, but we could have just as easily predicted this result from Newton's third law.

Illustration 28.3: Ampere's Law and Symmetry

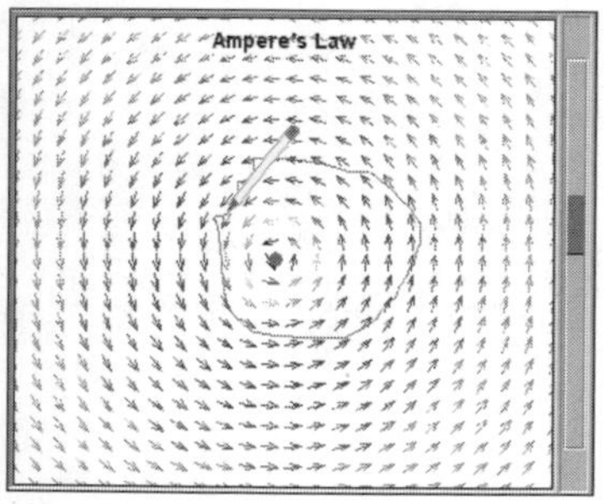

A single wire carrying current in the z direction (out of the computer screen) has radial symmetry about the center of the wire. Two systems that differ only by a rotation about the center of the wire will be indistinguishable. This symmetry is, however, broken if a second wire is added, because the displacement vector from the first to the second wire defines a unique direction. Calculations to determine magnetic field strength that depend on the ability to follow a closed Amperian path are much more difficult because it is not possible to write a simple analytic expression for a path along which $|B|$ is constant.

Look at the magnetic field vectors for the one-wire configuration. Notice how there is a circular symmetry about the center of the wire. Because of this symmetry, we can use Ampere's law to determine the magnetic field. Now look at the configuration with two wires. You can drag the wires either toward or away from each other. Notice that with two wires the magnetic field lines no longer have a circular symmetry. As a consequence, we cannot use Ampere's law to determine the magnetic field. Do not think that Ampere's law is no longer valid. Ampere's law is always true. It is just that in certain cases it is easy to use Ampere's law to calculate the magnetic field, and in others it is too difficult.

What is the analytic expression for the magnetic field on a path that has constant $|B|$ if there is only one wire? Move the wires closer together and farther apart. Under what circumstances can this expression be used as an approximation in the case of two wires? Around one wire, $|B| = \mu_0\,I/2\pi r$ and it points in the direction tangent to a circle centered on the wire. If there are two wires, we may add together the magnetic fields due to each individual wire. Be careful: You must add these fields as vectors, not as just numbers.

The magnetic field produced by two long, straight wires is far from irregular. What types of symmetry does this system still have? There is still a symmetry in the z direction, but this symmetry is not useful for calculations using Ampere's law. Why? How do you use an Amperian loop to calculate the magnetic field? To use this symmetry, the loop or rectangle would have to be centered on the wire and have one side along the z axis and the other side in the xy plane. Using this rectangle, how much current is enclosed? Since the loop is infinitesimally thin, there is no current enclosed, and therefore the result for the magnetic field is zero. We already know that the magnetic field in the z direction and the radial direction are zero. So this symmetry also tells us something about the field.

Illustration 28.4: Path Integral

Ampere's law is $\int \mathbf{B} \cdot d\mathbf{l} = \mu_0 I$, where the integration is over a closed loop (closed path), $d\mathbf{l}$ is an element of the path in the direction of the path, μ_0 is the permeability of free space ($4\pi \times 10^{-7}$ T·m/A), and I is the total current enclosed in the path. When you turn the *integral on*, this animation shows the path integral as you move the pencil around **(results from the path integral are given in 10^{-7} T·m)**.

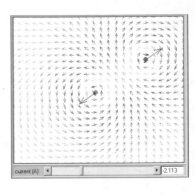

Start with one wire. Notice the magnetic field vectors. Pick a starting point, turn the integral on, drag the pencil around the wire counterclockwise, and come back to the starting point. What is the value of the path integral? Push the "set integral = 0" button to zero out the integral. Turn off the integral and pick a different starting point. After turning the integral back on, go around the wire again, taking a different path (but going the same direction counterclockwise). What is the path integral? Notice that with a different path the value of the magnetic field ($\mathbf{B}$) along the path and the direction of $d\mathbf{l}$ are both different, but by the time you get back to your starting point the sum (integral) of $\mathbf{B} \cdot d\mathbf{l}$ is the same. This is what Ampere's law says: that the integral around the path only depends on the current enclosed (times μ_0).

What do you expect if you go in the other direction (clockwise vs. counterclockwise)? Try it (zero out the integral before each time). Note that $d\mathbf{l}$ now points in the opposite direction. Therefore, the value of the integral is negative. The current flowing through this loop is negative with respect to the normal to the loop (since the normal to this loop is into the screen). This corresponds to a current flow out of the screen. This agrees with the result from the other (counterclockwise) path.

Now try this with two wires. Again notice the magnetic field vectors. Pick a starting point and mark it. Drag the pencil around the red wire. Why is the integral the same as before? Zero the integral and drag the black dot in a circle around both wires. What is the integral? What does that mean about the total current enclosed in the circle you just made? How does the current in the blue wire compare with the current in the red wire? Since the integral is zero, we know that the currents must be equal and opposite.

If the path integral is zero, does that mean that the magnetic field along the path is zero? Why or why not? When the path integral is zero, the total current enclosed is zero, but it is not necessarily true that the magnetic field is zero. There must be a symmetry in the problem in order for Ampere's law to be useful in determining the magnetic field.

Exploration 28.1: A Long Wire with Uniform Current

The gray circle in the center represents a cross section of a wire carrying current coming out of the computer screen. The current is uniformly distributed throughout the wire (position is given in centimeters and magnetic field strength is given in millitesla). The black circle is an Amperian loop with a radius you can change with the slider.

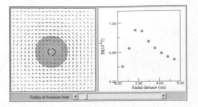

Begin with the Amperian loop with a radius larger than the radius of the wire.

a. What is the radius of the Amperian loop?

b. What is the magnetic field at this radius?

You will use Ampere's law to find the total current in the wire:

$$\int \mathbf{B} \cdot d\mathbf{l} = \mu_0 I, \quad \mu_0 = 4\pi \times 10^{-7} \text{ T·m/A},$$

where the integration is over a closed loop (closed path), $d\mathbf{l}$ is an element of the path in the direction of the path, and I is the total current enclosed in the path. Pick a point on the Amperian loop and draw both the direction of the magnetic field at that point and the direction of $d\mathbf{l}$ (tangent to the path).

c. The magnetic field and $d\mathbf{l}$ should be parallel to each other. What is $\mathbf{B} \cdot d\mathbf{l}$?

Pick another point on the Amperian loop.

d. What is the magnitude of the magnetic field at that point? At any point on the loop?

e. This means you can write $\int \mathbf{B} \cdot d\mathbf{l} = B \int dl$. Why?

$\int d\mathbf{l}$ is simply the length of the Amperian loop (in this case the circumference). Therefore, $B = \mu_0 I/2\pi r$ outside the wire.

f. Calculate the current carried by the wire from your measurement of the magnetic field.

g. Change the radius of the loop (but leave it still bigger than the wire) and predict the magnetic field on that loop. Measure the value to verify your answer.

Make the Amperian loop smaller so it fits inside the wire. This time, the current inside the loop is not equal to the total current, but instead it is equal to the total current times the fraction of the area inside the loop, Ir^2/a^2, where a is the radius of the wire.

h. Why?

i. Use this fraction and Ampere's law to predict the magnetic field inside the loop.

j. Measure the field to verify your answer.

k. Show that the general expression for the magnetic field inside the wire is $\mu_0 I r/2\pi a^2$.

Exploration 28.2: A Plate of Current

Ampere's law states that $\int \mathbf{B} \cdot d\mathbf{l} = \mu_0 I$, where the integration is over a closed loop (closed path), $d\mathbf{l}$ is an element of the path in the direction of the path, μ_0 is the permeability of free space ($4\pi \times 10^{-7}$ T·m/A), and I is the total current enclosed in the path (position is given in millimeters and the magnetic field is given in millitesla 10^{-3} T, so the integral is given in mT·mm $= 10^{-6}$ T·m). To use Ampere's law to calculate the magnetic field, Amperian loops need to mimic the symmetry of the field so that $\mathbf{B} \cdot d\mathbf{l}$ is constant over the loop (or sections of the loop).

a. The blue dots represent wires carrying current into or out of the computer screen. In which direction does the current flow in the wires? Explain.

This animation shows the path integral (the value in table and on the bar graph) as you move the cursor (the circle with crosshair) around, as well as the position of the cursor as you move it. Move the cursor along the top portion of the loop.

b. Is the integral positive or negative? Why? (Hint: $d\mathbf{l}$ points along the path in the direction you move the dot around the path).

Move the cursor to a corner and re-zero the integral (push the "set integral = 0" button). Now move the cursor along one of the vertical sides of the loop.

c. How does the size of this integral compare with the integral along the top part of the loop? Why? (Hint: What is the direction of **B** along the side, and what is the direction of $d\mathbf{l}$? So what then is $\mathbf{B} \cdot d\mathbf{l}$?)

d. Do the complete path integral (take the cursor completely around the loop). What is its value?

e. From this value, if each wire has the same current, what is the current in one of the wires?

f. From the path integral, what is the magnetic field above the series of wires? (Hint: If we neglect edge effects, $\int \mathbf{B} \cdot d\mathbf{l} = BL$ on the top and bottom of the loop, and $\int \mathbf{B} \cdot d\mathbf{l} = 0$ for the sides.)

g. Compare your calculated value (from the path integral) with the value you measure by click-dragging around the animation. Comment on any differences.

h. Show that the general expression for the magnetic field above or below the series of wires is $B = (\mu_0/2)(\text{current/length})$ where the current/length is the current per length across the cross section of the plate (along the x axis in this animation).

i. Verify this expression for this animation.

Exploration 28.3: Wire Configurations for a Net Force of Zero

The purple wire and the green wires have fixed currents and fixed positions. You can change the current in the gray wire by using the slider, and you can also drag the gray wire around to new positions. The animation shows the magnetic field vectors as well as the force on the wires. You can also add field lines by double clicking.

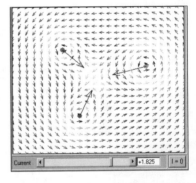

a. What are the directions of the currents in the green and purple wires?

b. Which one carries a larger magnitude of current? Explain.

Keep the current in the gray wire at zero (you can use the "I = 0" button to set the current to zero). Move the gray wire to a spot where the magnetic field is zero. Now increase the current in the gray wire.

c. Why is the force on the gray wire zero?

d. Why isn't the force on the other wires zero?

e. Is the force on the other wires different from the force on those wires before the current was turned on in the gray wire? Explain.

With current in the gray wire, move it to some point where the force is nonzero. The force on the wire is due to the current in the gray wire and the magnetic field it sits in (due to the other wires), $\mathbf{F} = q\mathbf{v} \times \mathbf{B} = I\mathbf{L} \times \mathbf{B}$, where **L** is the length of the wire and points in the direction of the current in the wire. To determine the direction of the force, then, you use the right-hand rule. Turn the current off in the gray wire.

f. What is the direction of the net magnetic field (make a sketch)?

g. Positive current comes straight out of the computer screen (negative current is into the screen). Therefore, in what direction is $I\mathbf{L} \times \mathbf{B}$ for negative current (indicate this on your sketch)? Try it and verify your answer.

h. With a negative current, where does the gray wire need to be located so that the force on the purple wire is zero? So that the force on the green wire is zero? Explain.

i. If you change the current, how does your answer to (h) change? Explain.

Try a **different configuration**.

j. Where will the force be zero on the gray wire when it has a current flowing in it?

k. If the gray current has a current of about $-1A$, where do you have to put it in order to get the force on the green wire to be zero? Where do you have to put it in order to get the force on the purple wire to be zero? Where do you have to put it in order to get the force on the yellow wire to be zero? Explain.

Problems

Problem 28.1

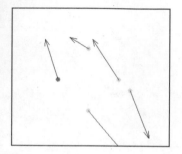

The animation shows wires coming out of the screen. For each configuration, how many wires have current flowing in the same direction as the red wire? You can drag the red wire as well as the other wires. Explain.

Problem 28.2

The current in the two red wires is 3 A coming out of the page (position is given in meters). The black wire has a current of zero right now, but can have current either into or out of the page depending on the value of the slider.

a. Where does the black wire need to be placed (you can drag it around), and what current does it need to have so that the magnetic field will be zero at the origin (0 m, 0 m)?

b. Develop an expression for the position of the black wire as a function of current so that the magnetic field will be zero at the origin. Verify your expression with the animation.

Problem 28.3

Find the current carried by each of the four wires in the animation. Each wire carries current either into or out of the

computer screen (position is given in millimeters and the magnetic field strength is in given in millitesla, 10^{-3} T, so the integral is given in mT·mm $= 10^{-6}$ T·m). You can turn the integral on, and the cursor will change to a pencil and draw the path as it calculates the integral of the magnetic field in the direction of the path you take (path integral of $\mathbf{B} \cdot d\mathbf{l}$.) You can re-zero the integral at any point or turn the integral off and move the cursor to another spot.

Problem 28.4

a. Find $\int \mathbf{B} \cdot d\mathbf{l}$ for each side of the square wire (position is given in millimeters and the magnetic field strength is given in millitesla, 10^{-3} T).

b. What is the total value of the path integral $\int \mathbf{B} \cdot d\mathbf{l}$ for the field shown? Explain.

Problem 28.5

A coaxial cable consists of an outer conductor and an inner conductor separated by an insulating plastic filler. These two conductors usually carry equal currents in opposite directions. You may have seen this type of configuration on cable TV hookups as well as on certain types of computer networks. The animation shows the progression of the magnetic fields as

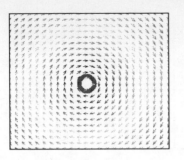

many small wires are added to produce a coaxial configuration (position is given in millimeters and magnetic field strength is given in microtesla, 10^{-6} T).

a. Build the coaxial cable by adding the current-carrying wires. Explain why this type of cable might be preferable to household wiring or lamp cord, which consists of two side-by-side wires carrying currents in opposite directions.

b. Explain why the field is zero outside the cable and within the center (blue) cable.

c. Click-drag to measure the field at any point. Find the current passing through the inner conductor.

Problem 28.6

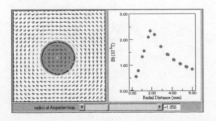

Find the current in the red wire. You can drag the red wire around and read the force/length on it. You can also read the value of the magnetic field at given points in the animation when your mouse is not on a wire (position is given in centimeters, magnetic field strength is given in 10^{-5} tesla, and force/length is given in newtons/meter). You can also turn the current in the red wire on and off by pushing the buttons.

Problem 28.7

The gray circle in the center represents a cross-section of a wire carrying current coming out of the computer screen (position is given in millimeters, 10^{-3} m, and magnetic field

strength is given in millitesla, 10^{-3} T). The current is uniformly distributed through the wire. The black circle is an Amperian loop with a radius you can change with the slider.

a. What is the direction of the current in the wire?

b. What is the current in the wire?

Problem 28.8

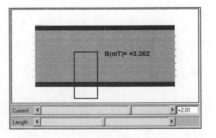

When you push the "add current" button, current-carrying wires are added to a region of space with a uniform magnetic field to form a plate carrying current either out of the screen or into the screen (position is given in centimeters and the magnetic field strength is given in millitesla).

a. Before you add the current, draw an Amperian path that makes use of the symmetry of the field. What is the value of the path integral?

b. Add the current. Draw a path that encloses a number of wires and use this path to find the current per unit length along the x axis (i.e., if you think of the plate as made up of the wires you see in cross section, how much current is carried on the wires from $x = 0$ cm to $x = 1$ cm?). Ignore edge effects.

Problem 28.9

A solenoid is made by wrapping a long wire many times around a cylinder. This animation shows a cross section of the solenoid (cylinder). Each loop of the wire circles behind and in front of the computer screen, so your view of the solenoid is a long tube sliced in half, lying on its side (position is given in centimeters and magnetic field is given in millitesla). Use the slider to change the current through the wire. This solenoid has a fixed number of wire coils, but by using the slider, you can stretch it or compress it (think of a solenoid made from a Slinky® in which the ends of the Slinky® are connected to a current source).

a. The black box is an Amperian loop. For which sides of the box is $\int \mathbf{B} \cdot d\mathbf{l} = 0$? Why? For the other side, show that $\int \mathbf{B} \cdot d\mathbf{l} = BL$, where L is the length of the side and B is the magnitude of the magnetic field at that point. How much current is enclosed in the Amperian loop?

b. Therefore, how many loops/centimeter are there?

c. How many total loops are there in this solenoid (how many coils in the Slinky®)?

d. Develop an expression for the magnetic field as a function of the length of the solenoid (same number of loops as the solenoid is stretched or compressed).

Problem 28.10

The animation shows the cross section of a loop of wire oriented perpendicular to the screen. It carries a current into and out of the screen (position is given in millimeters and magnetic field is given in millitesla). You can move the loop (drag at the middle) or change its radius (drag either the red or blue dot).

a. What is the current in the loop?

b. Develop an expression for the magnetic field in the center of the loop as a function of the radius of the loop. Verify your expression with the animation.

Faraday's Law

Topics include Faraday's law, Lenz's law, emf, flux, inductance, and the Lorentz force.

Illustration 29.1: Varying Field and Varying Area

In this animation we consider Faraday's law, which tells us how a changing magnetic flux creates an electromotive force (an emf), $-d\Phi/dt =$ emf. Magnetic flux, Φ, is a measure of the amount of magnetic field flowing perpendicularly through an area. It is given by $\mathbf{B} \cdot \mathbf{A}$ for uniform magnetic field and constant area (position is given in meters, magnetic field strength is given in millitesla, emf is given in millivolts, and time is given in seconds).

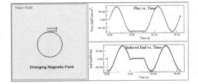

Consider the **Changing Magnetic Field Animation**. A loop is shown in a region in which the magnetic field varies sinusoidally, then is constant, and finally starts to vary sinusoidally again. The graphs on the right show the induced emf in the loop and the magnetic flux through the loop as a function of time. The direction of current in the top of the loop is indicated by the current arrow above the loop. Blue indicates the magnetic field is into the page; red indicates it is out of the page. The intensity of the color is proportional to the magnitude of the magnetic field.

Notice that for the first 1.5 s of the animation there is an increasing flux through the loop due to the magnetic field that is increasing out of the screen. Also notice that there is an emf induced in the wire loop and an induced *clockwise* current. Is the induced current in the loop in the direction you would expect? The induced current may be in the opposite direction as you expected. Because of the minus sign in Faraday's law (Lenz's law), the emf is the negative of the slope of the flux vs. time graph. From $t = 0$ s to $t = 1.5$ s the magnetic field is increasing and therefore the emf is negative. Now watch the animation for the remaining time and see how the emf changes with time.

Consider the **Changing Area Animation**. A loop of changing area is shown in a region where the magnetic field is constant (unlike the previous animation where the magnetic field changed with time) and pointing out of the screen. The graphs on the right again show the induced emf in the loop and the magnetic flux through the loop as a function of time. The direction of current in the top of the loop is indicated by the current arrow above the loop. Blue indicates the magnetic field is into the page; red indicates it is out of the page.

Note that (again) the magnetic flux is increasing for the first 1.5 s of the animation. Again note that the emf is negative during this time interval. In fact, compare the two graphs from the first animation with the pair of graphs from the second animation. What do you notice? Since the emf is related to the changing flux, it does not matter if that changing flux is due to a changing magnetic field or a changing area. In fact, a changing magnetic flux can be due to a changing magnetic field, a changing area, or both.

Illustration 29.2: Loop in a Changing Magnetic Field

A wire loop in an external magnetic field can have an induced emf (and therefore an induced current) if the magnetic flux varies as a function of time. Since the magnetic

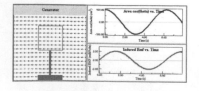

flux is the dot product of the magnetic field and the perpendicular area of the loop ($\mathbf{B} \cdot \mathbf{A}$ for uniform magnetic fields), the flux can change if the magnitude of the magnetic field changes in time and/or if the orientation between the magnetic field and the perpendicular area changes with time (position is given in meters, magnetic field strength is given in millitesla, emf is given in millivolts, and time is given in seconds). The color of the vector indicates field strength, and graphs on the right show the magnetic field in the x direction as well as the induced emf.

In **Animation 1** the loop of wire is perpendicular to the screen, while the magnetic field is to the right. The orientation of the loop and field does not change in time. However, the magnetic field strength changes with time according to the values you select with the slider. You can change the maximum magnitude of the magnetic field as well as the frequency of its oscillation.

In **Animation 2** the loop of wire is again perpendicular to the screen, and now the magnetic field rotates in the plane of the computer screen. The orientation between the loop and field does change in time because the magnetic field changes direction (with respect to the loop) as a function of time. The magnetic field strength in this animation does not change with time. You can set the magnitude of the magnetic field and the frequency of the field's rotation by using the sliders.

What are the differences between the two animations? What are the similarities? In **Animation 1** the magnetic field changes strength as a function of time. In **Animation 2** the magnetic field maintains a constant magnitude, but its direction changes with time. Despite these differences, for the same values of max $|B|$ and frequency, you get the same value for the magnetic field in the x direction as a function of time and the same induced emf. For **Animation 1** the magnetic field changes strength as a function of time according to $\sin(2\pi ft)$. In **Animation 2** the magnetic field maintains a constant magnitude, but its direction changes with time. The component of the field that is in the direction of the area of the loop (a direction normal to the loop or in the x direction) changes as a function of time according to $\sin(2\pi ft)$. As a consequence, $\mathbf{B} \cdot \mathbf{A}$ as a function of time is the same for both animations, as long as you have the same values of max $|B|$ and frequency.

Note that while B changes with time in one animation and changes direction in the other animation, we can still use $\mathbf{B} \cdot \mathbf{A}$ since the magnetic field at every instant in time is uniform across the area of the loop. If it were not uniform over the area of the loop, we would need to use an integral to determine the magnetic flux.

Illustration 29.3: Electric Generator

A wire loop that is rotated by an external motor (or turbine) is shown. The rotating loop of wire is also in a constant magnetic field (created by magnets not shown). There is a current induced in the rotating loop. As the loop rotates around, you see the red (front) side and then the black (back) side of the loop (position is given in centimeters, magnetic field strength is given in tesla, emf is given in millivolts, and time is given in seconds). The green arrow indicates both the direction and magnitude of the induced current.

Consider the **Normal View**. The top graph shows $A \cos(\theta)$, the area of the loop times $\cos(\theta)$, as a function of time, where θ is the angle between the area of the loop and the magnetic field. The bottom graph shows the induced emf in the loop as a function of time.

What is the position of the loop when the magnitude of $A \cos(\theta)$ is a maximum? How about when the magnitude is a minimum? What is the induced emf in the loop? Notice that when the loop is out of the screen (you see only a thin rectangle), $A \cos(\theta)$ is a maximum in magnitude. When the loop points to the left in this animation, $\cos(\theta) = 1$, and when the loop points to the right in the animation,

$\cos(\theta) = -1$. When the loop is completely in the plane of the screen, then $\cos(\theta) = 0$. Notice that the induced emf is related to the negative of the slope of the $A \cos(\theta)$ vs. time graph. Why?

Now consider the **Flux View** in which the graphs on the right show the flux through the loop and the induced emf in the loop as a function of time. What is the position of the loop when the magnitude of the magnetic flux is a maximum? When is the magnitude of the magnetic flux a minimum?

Notice that the flux is the dot product between **B** and **A** or just $BA \cos(\theta)$ for uniform magnetic fields (magnetic fields that are uniform across the area of the loop). If the magnetic field is not uniform we must use an integral. Therefore, when $A \cos(\theta)$ is a maximum [$\cos(\theta) = 1$] or a minimum [$\cos(\theta) = -1$], so is the flux. Similarly, when $A \cos(\theta)$ is zero, so is the flux. Notice that the corresponding induced emf is related to the negative of the slope of the magnetic flux vs. time graph. Since the $A \cos(\theta)$ vs. time graph is proportional to the magnetic flux vs. time graph, with the proportionality constant being the magnetic field strength, this explains the relationship between $A \cos(\theta)$ and the induced emf in the **Normal View**.

In electric power plants turbines generate current based on this principle. Either a wire rotates in a magnetic field (as in this Illustration) or, more commonly, a magnet rotates near stationary coils of wire (changing the magnetic flux through the coils, which induces current in them). For electricity in the United States, turbines make 60 revolutions per second (generating 60 Hz current), while in Europe the turbines make 50 revolutions per second (generating 50 Hz current).

Exploration 29.1: Lenz's Law

Lenz's law is the part of Faraday's law that tells you in which direction the current in a loop will flow. Current flows in such a way as to oppose the change in flux. The magnetic field created by the current in the loop opposes the change in the magnetic flux through the loop's area (position is given in meters, time is given in seconds, and magnetic field strength is given in tesla).

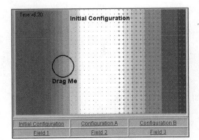

Consider the initial configuration. The center has a field-free region, and the sides have a linearly increasing magnetic field into the computer screen (blue) and out of the computer screen (red). The deeper colors represent a stronger field. Drag the loop from the white (field-free region) into the blue.

a. While you drag it, which way does the current in the loop flow? (right arrow means clockwise current; left arrow means counterclockwise current).

b. Sketch the field that the *current in the loop* generates.

c. In the center of the loop, does this field (created by the induced current) point into or out of the computer screen?

d. So, as you drag the loop to the right, the *external field* in the loop increases in which direction? The field generated by the current points in which direction? According to Lenz's law, these two directions should be opposite.

Now, take the loop over to the far right and then move it slowly to the white region.

e. Explain why the direction of the current points the way it does.

f. What if you take the loop from the center to the left (into the red region)? Explain what you expect to happen and then try it.

g. Can you tell the difference between moving a loop from a blue to a white region and moving from a white region to a red region? Why or why not?

h. Try the two other configurations, Configurations A and B (in which the magnetic field is hidden). Describe the magnetic field as completely as possible.

i. Once you've completed your descriptions, decide which of the magnetic fields (Fields 1, 2, or 3) matches Configuration A and Configuration B.

Check your answers to (i) by adding a loop to a field animation.

Exploration 29.2: Force on a Moving Wire in a Uniform Field

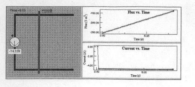

Faraday's Law is a relationship between a time-varying magnetic field flux (Φ) and an induced emf (voltage), emf $= -\Delta\Phi/\Delta t$ (position is given in meters, current is given in amperes, emf is given in volts, and magnetic flux is given in tesla per meter^2). In this animation a wire is pushed by an applied force in a constant magnetic field.

a. What are the fluxes at $t = 1$ s and $t = 3$ s (from the graph)?

b. What is the change in flux/second ($\Delta\Phi/\Delta t$)?

According to Faraday's law, this should be equal to the induced emf.

c. Does your calculated emf agree with the emf reading on the meter connected to the wires?

d. What is the velocity of the sliding rod?

e. What is the change in area/second?

f. Since $\Phi = \int \mathbf{B} \cdot d\mathbf{A}$, which is $\Phi = BA$ for this case (why?), what is the value of the magnetic field the wire slides in?

The sliding wire has a current flowing in it.

g. In what direction is this current and what is the value of the current (read the current value from the graph) at a given time (pick a time)?

h. In what direction is the magnetic force on this current-carrying wire moving in the external magnetic field [the one you found in (f) above]? Remember, $\mathbf{F} = I\mathbf{L} \times \mathbf{B}$.

i. What is the value of the force?

j. Since the wire moves at a constant speed, what must be the direction and magnitude of the applied force? Check your answer by **showing the force on the wire**.

The power dissipated in an electrical circuit is the current times the voltage drop. In this case, it is I times the emf across the rod.

k. What is the power dissipated?

The power delivered by an external force is $\Delta W/\Delta t$, where $W = \mathbf{F} \cdot \mathbf{s}$ is the work done by the applied force, $\mathbf{F}$, and $\mathbf{s}$ is the displacement.

l. Show that the power delivered is also $\mathbf{F} \cdot \mathbf{v}$.

m. What is the power delivered by the external force?

n. Why should this power be equal to the power dissipated by the circuit?

o. Pick a different velocity and calculate the power dissipated by the circuit and the power delivered by the force.

Exploration 29.3: Loop Near a Wire

A loop is near a wire that has a current flowing upward. You can drag the loop (position is given in meters, magnetic field strength is given in millitesla, emf is given in millivolts, and time is given in seconds). The flux through the loop and the induced emf are shown in the graph. The animation will stop after 30 s.

a. How do the emf and the flux through the loop change as you drag the loop toward and away from the wire?

b. How do the emf and the flux through the loop change as you drag the loop parallel to the wire?

c. Are the flux and emf different when the loop is on the left side, instead of the right side, of the current-carrying wire? Explain.

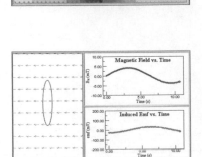

Exploration 29.4: Loop in a Time-Varying Magnetic Field

The animation shows a wire loop in a changing magnetic field. The graphs show the magnetic field in the x direction as a function of time and the induced emf in the loop (position is given in meters, magnetic field strength is given in millitesla, 10^{-3} T, and emf is given in millivolts).

a. The vectors show the field through the loop as a function of time. What do the different colors indicate?

b. What impact does changing the maximum value of the magnetic field have on the induced emf?

c. What impact does changing the frequency of the oscillation of the magnetic field have?

d. Develop an expression to relate the change in the emf to the parameters you can vary.

e. Develop an equation for the magnetic field as a function of time and the parameters you can vary.

f. What is the area of the loop? Therefore, what is the flux through the loop as a function of time?

g. Using Faraday's law, show that the emf should be equal to $|B_{max}|A\omega\cos(\omega t + \varphi)$, where $|B_{max}|$ is the maximum value of the magnetic field in the x direction, A is the area of the loop, ω is the angular frequency of the oscillation, and φ is a phase angle.

h. Verify that this expression matches the graph for the emf vs. time.

Exploration 29.5: Self-Inductance

This animation shows a cross section of a solenoid (think of a long tube cut lengthwise down the cylinder and then looking at the edge) so that the black dots represent the current-carrying wires coming into and out of the screen. The arrows show the direction and magnitude of the magnetic field. You can drag the black dot around to measure the field in different spots (position is given in centimeters, the magnetic field strength is given in millitesla, 10^{-3} T, and current is given in amperes). You can either **change field by varying the current in the wires with the slider** or you can choose to **change the current linearly as a function of time**.

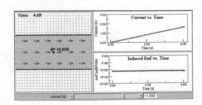

Faraday's law tells us that when a loop is in a changing magnetic field, an induced emf in the loop will result. But what if the loop itself has a changing current? With a changing current, the loop has a changing magnetic field. Wouldn't it make sense, then, for there to be an induced emf and an induced current to oppose the changing flux? The answer is that there are: If the current is changed in a current loop, there is a self-induced back emf. The measure of the back emf produced when a current is changed in a loop is called its self-inductance, or simply inductance, represented by L and measured in henries, H (1 H = 1 T·m^2/A). From Faraday's law, emf = $-d\Phi/dt$, the self-inductance is the back emf = $-L(dI/dt)$.

Run the **change field by varying the current in the wires with the slider**. Instead of considering a loop, we will look at a solenoid (it is easier to calculate the magnetic field inside a long solenoid).

a. For the solenoid above, adjust the current with the slider and determine how the magnetic field varies with current.

b. For this solenoid (given the value of the magnetic field at the current chosen), how many loops per meter are there?

Run **change the current linearly as a function of time**.

c. What is the back emf?

d. Using the equation for the back emf, what is the inductance, L?

e. Using Faraday's law and the equation for the back emf, show that $L = (\Phi/I) N$ for an inductor with N loops.

f. Therefore, show that the inductance, L, of a solenoid is $\mu_0 N^2 A/(length)$, where N is the number of loops, A is the cross-sectional area, and length is the *length* of the solenoid (so that $N/length$ is the number of loops per meter).

g. If this solenoid is 2 m long, calculate the inductance and compare it to your answer in (d) above.

Problems

Problem 29.1

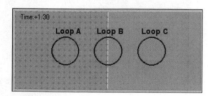

Three loops are shown in a region where the magnetic field is changing (position is given in meters and time is given in seconds). Blue indicates the magnetic field is directed into the screen and red indicates it is directed out of the screen. The intensity of the color represents the magnitude of the field. At any instant of time, the red and blue "fields" have the same magnitude.

For each of the following times, is there a current in each loop (A, B, and C) and, if there is a current, is it flowing clockwise or counterclockwise? Explain.

a. $t = 0.5$ s.

b. $t = 3.1$ s.

c. $t = 4.0$ s.

Problem 29.2

A loop is shown in a region where there is a magnetic field (position is given in meters and time is given in seconds). Blue

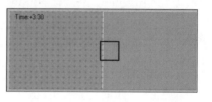

indicates where the magnetic field is directed into the screen and red indicates where it is directed out of the screen.

a. Construct a qualitative graph of current induced in the loop vs. time. Indicate on your graph the time when the current reaches its maximum and minimum values. Consider the current to be positive when it flows clockwise and negative when it flows counterclockwise.

b. How would your graph be different if the region on the left side of the applet had zero magnetic field instead of a field into the screen? Explain.

Problem 29.3

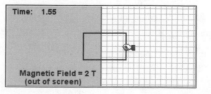

A loop with a resistance of 200 Ω is pulled at constant velocity through a region where there is a magnetic field of 2 T out of the screen and into a region of no magnetic field (position is given in meters and time is given in seconds).

During the time shown in the animation,

a. What are the direction and magnitude of the current in the loop?

b. A force is necessary to pull the loop out of the magnetic field. Why? (Draw a free-body diagram of the loop and explain the origination of each force.)

c. Find the magnitude of the force that was exerted by the hand on the loop in the animation.

d. Describe the subsequent motion of the loop if the same force continues to act on the loop when it is in the region of no magnetic field.

Problem 29.4

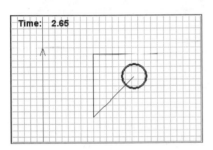

A loop, next to a very long current-carrying wire, moves along the path shown (position is given in meters and time is given in seconds). The trail of the loop is marked. The current in the wire is constant and is directed upward as shown by the arrow.

a. For $t = 0.5$ s, 1.5 s, 2.5 s, 3.5 s, and 4.5 s, is the induced current in the loop clockwise, counterclockwise, or zero?

b. Rank the **magnitudes** of the currents induced in the loop (smallest to greatest, explicitly indicating any ties) at $t = 0.5$ s, 1.5 s, 2.5 s, and 3.5 s.

Problem 29.5

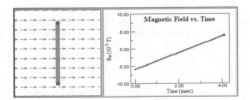

The animation shows the cross section of a wire loop in a changing magnetic field (the wire loops out of and into the screen). The graph shows the magnetic field in the x direction as a function of time (position is given in centimeters, time is given in milliseconds, 10^{-3} s, and magnetic field strength is given in tenths of tesla, 10^{-1} T).

a. Sketch a graph of the induced emf in the loop as a function of time.

b. What is the maximum value of the emf?

c. What is the direction of the current in the loop as a function of time (use the convention that current flowing out of the top of the loop is positive)?

Problem 29.6

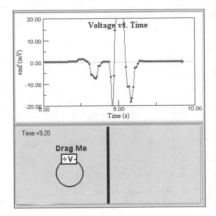

A dragable conducting loop is to the right of a wire as shown in the animation (position is given in meters and time is given in seconds). The graph shows the voltage read by the voltmeter attached to the conducting loop. (The polarity is also shown; when the voltage is positive, it means the voltage is higher at the red terminal than at the blue one, so current flows from the red around the loop counterclockwise to the blue). In which direction is the current flowing in the long straight wire?

Problem 29.7

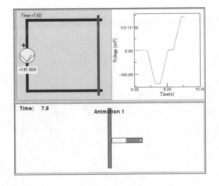

The animation shows a top view of four wires and a galvanometer in the lab. Current flowing into the + terminal, i.e., counterclockwise, will deflect the meter to the right (positive voltage). During the time interval $t = 2$ s to $t = 8$ s a magnet is slowly pushed completely through the rectangle from below to above. The animation starts at $t = 0$ s and stops at $t = 10$ s. Observe the meter reading (shown on the graph) during this simulation. In the animation of moving magnets, you see a side view of the process. Think of the left side of the gray loop as behind the computer screen and the right side as in front of the screen.

Which of the animations of moving magnets best matches what happens to create the graph shown?

Problem 29.8

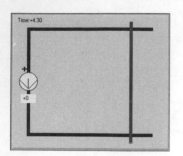

The animation shows a top view of four wires and a galvanometer (position is given in meters and time is given in seconds). There is a constant magnetic field passing through the area enclosed by the wires. Current flowing into the + terminal, i.e., counterclockwise, will deflect the meter to the right. You can drag the black bar on the right. The animation runs from 0 to 10 s and then repeats.

Determine the direction of the magnetic field through the loop.

Problem 29.9

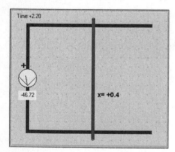

The wire on the right is pulled to the right for 5 s and then to the left for 5 s as shown. Determine the magnitude of the magnetic field passing through the wire rectangle. You may pause the animation and read coordinate values by click-dragging the mouse at any location (position is given in centimeters and emf is given in microvolts, 10^{-6} V).

Problem 29.10

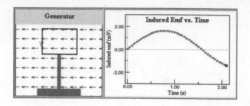

The graph on the right shows the induced emf through the loop as a function of time (position is given in centimeters, time is given in seconds, and emf is given in millivolts). The green arrow shows the direction and magnitude of the induced current.

a. What is the magnitude of the magnetic field?

b. Looking down on this loop from above, in what direction is it rotating (clockwise or counterclockwise)? Explain your answer.

Problem 29.11

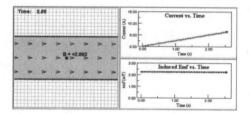

This animation shows a cross section of a solenoid (think of a pipe cut in half lengthwise and looking at the cut edge). The graphs show the current in the solenoid and the induced emf as a function of time (position is given in centimeters, time is given in seconds, magnetic field strength is given in millitesla, current is given in amperes, and induced emf is given in millivolts).

a. What is the inductance of the solenoid shown in this animation?

b. How many turns per meter of current-carrying wire are coiled around this solenoid?

c. How long is this solenoid (it extends beyond the screen)?

30

DC Circuits

Topics include meters, power, resistance, voltage, current, switches, and switched RC.

Illustration 30.1: Complete Circuits

This Illustration examines open and closed circuits by considering a circuit composed of three lightbulbs. These lightbulbs behave very much like resistors and are often represented using the resistance symbol, —⋀⋀⋀—.

Note that even though these lightbulbs are identical, their brightness can vary. The relationship between voltage, current, and brightness is discussed later in this chapter, but you should notice that the brightness of a bulb depends on the voltage across it.

A burned-out filament results in an open circuit in that branch and is equivalent to a very large resistance. First predict what will happen when one or more lightbulbs burn out and then click on the links to test your prediction. The animation will show you both the bulbs (which ones light and which ones do not) and the voltages (voltage is given in volts).

Remember that if there is a voltage drop across a "good" lightbulb (one that is not broken), it should light up. Notice that when bulb 1 is out, none of the lightbulbs light. If only bulb 2 or bulb 3 is out, notice that the other bulbs light. If bulbs 2 and 3 are both out, notice that bulb 1 does not light up. Make sure that you can explain these observations and the associated voltage readings.

By looking at the voltage across each bulb (resistor), you should be able to analyze the circuit. Notice that you cannot determine which lightbulb is broken simply by looking at the bulbs because multiple conditions produce the same visual clues.

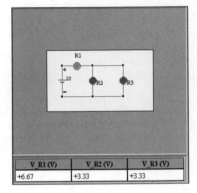

V_R1 (V)	V_R2 (V)	V_R3 (V)
+6.67	+3.33	+3.33

Illustration 30.2: Switches, Voltages, and Complete Circuits

This Illustration allows you to control the current flow through a circuit using switches (voltage is given in volts).

You can open and close switches to turn the lightbulbs on and off. (How is this similar to a bulb breaking in Illustration 30.1?) Notice that when bulb 1 is dark (S1 open and the other switches closed), there is no complete path for the current, so none of the bulbs light. Furthermore, there must be 0 V across bulbs 2 and 3, and there is 10 V across the switch S1. Similarly, when bulb 2 is dark (S2 open and the other switches closed), there is a complete path for current flow through bulbs 1 and 3 (and the voltage across each is 5 V). The same reasoning applies to bulb 3 when switch 3 is open. (When all the switches are open, the voltmeter readings can have any value, as long as the sum adds up to 10 V.)

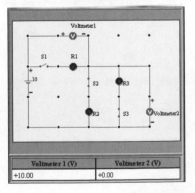

Voltmeter 1 (V)	Voltmeter 2 (V)
+10.00	+0.00

Illustration 30.3: Current and Voltage Dividers

This Illustration shows two different configurations of resistors connected to a battery (voltage is given in volts, current is given in amperes, and resistance is given in ohms).

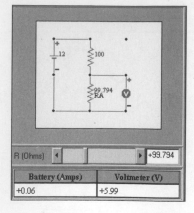

R (Ohms) ◄ ► +99.794

Battery (Amps)	Voltmeter (V)
+0.06	+5.99

Start with the **voltage divider animation.** The circuit shows an ideal battery supplying current to a 100-Ω resistor in series with a variable resistor, R_A. When the resistance of R_A is equal to 100 Ω (the value of the top resistor), the voltage is equally divided across the two resistors. As you increase R_A, what happens? What happens as you decrease R_A? The current from the battery also changes in this process; however, notice that the current through the top resistor and R_A are always equal. This is because the current through the top resistor must also go through R_A.

Now try the **current divider animation.** The 100-Ω resistor is now in parallel with R_A. When the resistance of R_A is equal to 100 Ω (the value of the fixed resistor), the current is equally divided between the two branches of the circuit. As you increase R_A, what happens? What happens as you decrease R_A? The current from the battery also changes in this process, but the voltage across the two resistors is the same because we have assumed that the battery is capable of supplying a large amount of current and because the wires are assumed to have negligible resistance. If you added a third resistor in parallel with the other two, the current from the battery would increase since the battery needs to supply current to that resistor (with the same voltage drop) as well.

Many students think that if you add additional resistors to circuits such as the ones just seen, the current from the battery must decrease (for the voltage to stay the same). Notice, however, that resistors added in parallel increase the total current from the battery, while resistors added in series reduce the total current from the battery.

Illustration 30.4: Batteries and Switches

In this Illustration you can close and open switches to see what happens to the two identical bulbs. The table gives the voltage across the lightbulbs, and the brightness of the bulbs indicates the relative current flow (voltage is given in volts). The batteries are all identical.

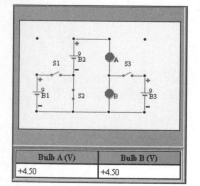

Bulb A (V)	Bulb B (V)
+4.50	+4.50

Notice that S1 and S2 cannot be closed or opened at the same time. What would happen if both S1 and S2 were open? What if both S1 and S2 were closed? (Which situation is bad for battery B1, and why?) If S1 and S2 are both open, no current flows, but if S1 and S2 both are closed, this creates a short circuit. There is essentially no resistance in the path connecting the two terminals of the battery, so as much current as possible flows from the battery (quickly discharging it).

As you switch S1 and S2 together (one open and one closed), what happens to the lightbulbs? Why? (That is, what is the difference between the two circuits?) Notice that when you put another battery in series, the potential difference (voltage) across each bulb increases, as does the current (and the bulbs are brighter).

With S1 closed and S2 open, now close S3. What happens? What is the voltage across each bulb? Why doesn't it change? You have added a battery to the circuit, but nothing happened. Why? Now close S1 and open S2. What happens? Why?

Now look at the voltage across bulb B. When S1 is closed, S2 is open, and S3 is open, the voltage across bulb B is 9 V. When you close S3, the voltage across bulb B is still 9 V. With S3 closed, the voltage across bulb B is always 9 V, no matter what is happening with S1 and S2. When you put batteries in parallel, you don't increase the total voltage. In fact, you need to be careful of adding batteries in parallel because if you have two batteries in parallel with each other at different voltages, you will end up with a great deal of current. This is why you have to be careful using jumper cables: You put two batteries in parallel with each other and even the slightest difference in voltage means a lot of current flows (because the jumper cables don't have much resistance. The current is limited by the internal resistance of the batteries). Of course this also works to your advantage to quickly recharge the dead battery.

Illustration 30.5: Ohm's "Law"

Ohm's law is not a "law" in the sense of "laws of physics" (for example, Newton's laws, conservation of energy, conservation of charge, etc.). Instead, it describes a linear relationship between current and voltage that holds true for some circuit elements, namely resistors. There are, however, other circuit elements that do not follow Ohm's law. Vary the voltage of the source and note the linear current-voltage relationship for the resistor. Compare this to the nonlinear response of a diode and a lightbulbs (voltage is given in volts and current is given in amperes).

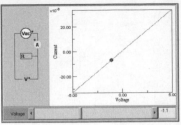

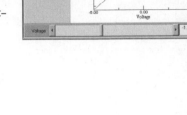

- **Resistor**
- **Diode (voltage scale is tenth of volts)**
- **Real lightbulb**

Illustration 30.6: RC Circuit

In the animation you can close and open switches to see what happens to the lightbulb.

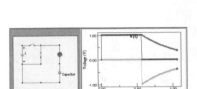

When the animation begins, the capacitor is initially charged. Push the "play" button and then open/close the switches. Watch what happens to the **lightbulb**. When the lightbulb goes out (is dark), throw the switches again. What happens? You should observe that the lightbulb is bright initially and always eventually goes out (even when there is a battery in the circuit). Notice, too, that the lightbulb is the brightest right after you throw the switches. This means that the current is the biggest. When the bulb is out, however, the current is zero. From these observations, what then is the voltage across the bulb as a function of time after the switch is thrown?

Show the graph of voltage vs. time. The voltage across the lightbulb **(green)**, the voltage across the capacitor **(red)** and the total voltage across both **(blue)** are shown. What does the graph look like for the situation when the capacitor is charging (capacitor and resistor in series with the battery)? Notice that the lightbulb voltage plus the capacitor voltage equal the total voltage and that current flows (the lightbulb lights) until the capacitor voltage equals the battery voltage. What does the graph look like when the capacitor is discharging (battery not in the circuit with the capacitor and the resistor)? Notice that the capacitor and resistor voltages are equal and opposite so that the sum of their voltages is zero. A negative voltage across the bulb simply means that current is flowing in the other direction as the capacitor discharges down to 0 V. So as the capacitor charges, the current flows from the battery through the resistor and charge builds up on the capacitor, but as the capacitor discharges, the current flows from the capacitor through the resistor until the capacitor has no charge on its plates.

Illustration 30.7: The Loop Rule

Kirchhoff's loop rule states that the sum of all the potential differences around a closed loop equals zero. In other words, $\Sigma \, \Delta V = 0$ for a complete loop.

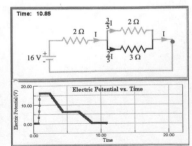

In the **circuit** shown, current from the battery flows through the resistors before returning to the battery. This illustration follows a hypothetical charge as it flows through the upper of the parallel resistors. This is, of course, just a simulation. Current also flows through the lower resistor, and the current through these two resistors is not the same. In fact, an accurate microscopic simulation would need $\sim 10^{20}$ electrons moving counterclockwise around the circuit. Because an electron–flow representation of current would be awkward, we use the standard definition of current and show a hypothetical unit of positive charge flowing out of the positive battery terminal, through the resistors, and into the negative terminal.

In a circuit, there are charges moving through the potential differences from the battery and across the resistors. So another way to state the loop rule is that, when a charge goes around a complete loop and returns to its starting point, its potential energy must be the same. Positive charges gain energy when they go through batteries from the − terminal to the + terminal, and they give up that energy to resistors as they pass through them.

Use the loop rule to determine the current through the battery in a circuit consisting of a 16–volt battery connected to a set of three resistors: a 2-Ω resistor in series with a parallel combination of a 2-Ω resistor and a 3-Ω resistor.

Now consider a **Kirchhoff loop** consisting of the battery and the two 2-Ω resistors. It doesn't matter where we start, as long as we come back to the same spot. Let's go clockwise around the loop starting at the bottom left corner.

$$+16 \text{ V} - (2 \ \Omega) * I - (2 \ \Omega) * 3I/5 = 0$$
$$+16 \text{ V} = (10 \ \Omega) * I/5 + (6 \ \Omega) * I/5$$
$$+16 \text{ V} = (16 \ \Omega) * I/5.$$

This gives $I = 5$ A.

Run the animation and follow the energy of the unit charge as it passes through each circuit element. Each voltage drop represents the amount of energy that is lost or gained when the charge passes through a circuit element. This demonstrates that, as charge flows around a complete loop, the gains in energy are always offset by the losses. The total change in energy is zero.

Exploration 30.1: Circuit Analysis

This Exploration begins with four identical lightbulbs connected to a battery (voltage is given in volts and current is given in amperes). Often, you will be asked to find the current through, the voltage across, the resistance, and/or the power consumed by a given bulb (or group of bulbs). To solve these types of problems you will use Ohm's law, $V = IR$, and an equation for power, $P = VI = I^2R = V^2/R$, where V is the voltage, I is the current, R is the resistance, and P is the power. You will also need to use two rules that are based on conservation laws:

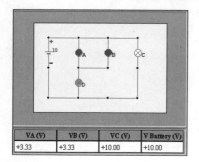

VA (V)	VB (V)	VC (V)	V Battery (V)
+3.33	+3.33	+10.00	+10.00

1. **current in = current out**. Since charge is not created or destroyed (conservation of charge), charge flowing into some point must flow back out unless there is a circuit element that can store charge (a capacitor).

2. **ΔV through a complete loop = 0**. The electric force is a conservative force (which is why we can define an electrostatic potential: V). This means that if you start at one point in a circuit and add up all the potential increases and subtract all the decreases in potential as you trace out a loop, then when you get back to the place you started, the potential must be the same.

Let's use these rules for the initial circuit. The brightness of the bulbs is an indication of the current through the bulbs (brightness actually goes as I^2).

a. Rank the bulbs in order of brightness (and therefore in order of current through them), from highest to lowest.

b. **Show the currents** (in the data table) through the bulbs to check your answer. The arrows indicate the direction of the current through the circuit.

c. Find the current through bulb D by determining how much current must be coming into the node (the dot where the wires come together) above bulb D. The current is coming from bulbs A and B.

d. Check your answer by verifying that the current coming into the node below bulb D (from bulb C and bulb D) is equal to the current going out of the node and into the battery.

Now consider the voltage across various elements. **Show the voltages** (in the data table) across the bulbs.

e. Why is the voltage across bulb C the same as the voltage across the battery (think about tracing a path around the outside "loop" of the circuit)?

f. Why is the voltage across bulb A equal to the voltage across bulb B?

g. Find the voltage across bulb D by picking a complete loop to trace around (battery → bulb A → bulb D → battery) OR (battery → bulb B → bulb D → battery) OR (bulb C → bulb D → bulb A) OR (bulb C → bulb D → bulb B) and finding the value of the voltage for bulb D that makes the change in potential equal to zero.

h. Using $V = IR$, what is the resistance of bulb D? (Check that it is the same as bulbs A, B, and C).

i. What is the power dissipated by bulb D?

Exploration 30.2: Lightbulbs

In this animation you can close and open switches to determine the resistance of each lightbulb. The current and voltage readings show the current through the battery and the voltage across the battery (voltage is given in volts and current is given in amperes). In order to solve such problems, you can always use Kirchhoff loop equations. If it is possible, a faster way to work problems is in terms of the effective resistance of the network of resistors. You consider the resistors that are in parallel and the resistors that are in series. It is, however, worth looking at the circuit in a bit more detail first, before diving into equations, to see if there is a way to understand the problem conceptually and to solve it faster.

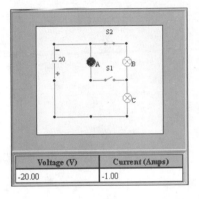

Voltage (V)	Current (Amps)
-20.00	-1.00

Notice that when both switches are closed, bulbs A and B are dimmer than bulb C. This should not be surprising because the current through bulb C is the sum of the currents through A and B. Open one of the switches and leave the other one closed. Now bulb C is in series with one of the bulbs (which one?). Notice that the total current from the battery is less, but that either bulb A or bulb B is brighter than it was before.

a. Why?

Go back to the case in which both switches are closed and notice that bulbs A and B look to be the same brightness. If the brightness were exactly the same, they would have the same resistance.

b. With switch 1 open and switch 2 closed, what is the current?

c. What about with switch 1 closed and switch 2 open?

d. How does this "prove" that bulbs A and B are identical?

e. Furthermore, when one switch is open and one closed, how does the brightness of bulb C compare with the bulb it is in series with?

f. What does that indicate?

Now for some math.

g. Since $R_A = R_B$, explain why, when both switches are closed, the effective resistance of the circuit is $1/2R_A + R_C$. (Hint: with both switches closed, A and B are in parallel with each other).

h. When both switches are closed, use the voltage across the battery and the current through the battery to find the value of the effective resistance.

i. With one switch open and one switch closed, use the voltage across the battery and the current through the battery to find the value of the effective resistance. The effective resistance is equal to $R_A + R_C$ (or $R_B + R_C$).

j. Solving these equations, you should find that all the bulbs are indeed identical (something you surmised from the brightness of the bulbs).

Notice that in this problem, trying to understand conceptually what was happening helped to guide the problem-solving process. Although the Kirchhoff loop rules will work, they are not necessarily the easiest way to solve a problem.

Exploration 30.3: Designing a Voltage Divider

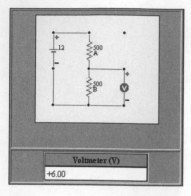

Often with circuits, not only do you want to be able to figure out what a circuit that is already built is doing, you may want to design a circuit for a specific task. In this case our task is to design a circuit that is a voltage divider with a particular output voltage (voltage is given in volts and resistance is given in ohms). You have a 12-V supply that can give you 1 W of power, and you need a 4-V output with as much power as possible. The resistors that you have can dissipate 1 W of power.

To divide the voltage, we can put the power supply in series with two resistors and then use the voltage across one of the resistors to be our 4-V output.

a. What ratio of resistors do you need to divide the supply voltage by one-third? In other words, how many times bigger (or smaller) should resistor A be than resistor B to get an output of 4 V? Try it.

b. Once the ratio is set up, do you have the maximum available power? To determine this, figure out the power used from the voltage source ($P = VI$). To get the maximum power (at a fixed voltage), should you increase or decrease the resistance in the circuit?

c. What is the limit on the total resistance ($R_A + R_B$) and, therefore, the limit on each resistor? Try it.

d. Try using a smaller value of resistance. Does the power supply burn up? (Fortunately, you can simply restart the animation and try again).

e. Double the values of R_A and R_B. How much power does this circuit now draw from the battery?

Now that you have determined convenient values of R_A and R_B that produce a 4-volt output, replace the voltmeter with a **lightbulb**. (Adding a power-consuming circuit element is sometimes referred to as adding a "load.")

f. When this lightbulb is added, what is the voltage across the lightbulb?

g. Why is it less than 4 V?

h. If you increase R_A and R_B more, what happens to the voltage across the lightbulb? Why? This is the reason voltage dividers like this are made from resistors that are as small as possible.

Exploration 30.4: Galvanometers and Ammeters

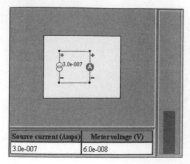

An ammeter measures current through a device and therefore must be in series with whatever element you want to find the current through. In this animation we will contrast the behavior of ideal and real ammeters by exploring the way a basic galvanometer works and finding how you can build an ammeter using a galvanometer (voltage is given in volts and current is given in amperes).

A galvanometer is a very sensitive meter that deflects when a small amount of current passes through it. (The current often goes through a coil that induces a magnetic field that causes an indicator needle to move. We use a red indicator bar instead of a needle.) In the **galvanometer animation** you can enter a source current and push the "galvanometer" button. The current source is shown using two interlocking circles, —⦻— . The indicator bar on the right shows the maximum current that can pass through the galvanometer without damaging the instrument.

a. Change the source current so that the indicator is at 50%. What is the current through the galvanometer and the voltage drop across the galvanometer? (This voltage drop is due to the internal resistance of the coil.)

b. What, then, is the maximum current that should go through the galvanometer to just get the red bar to 100% of the screen?

c. What happens if you exceed the maximum current rating?

We see that a galvanometer is a very sensitive current meter. They are often rated not by the maximum current, but by the internal resistance and the associated voltage drop at the maximum current.

d. Show that the internal resistance of this galvanometer is 0.2 Ω and the voltage drop at maximum current is 0.2 μV.

Suppose we want to use a galvanometer to measure currents up to 1 mA. We know that we want full-scale indication (bar at 100%) at 1 mA and half-scale at 0.5 mA, and so we need for our meter to be made up of the **galvanometer plus a resistor in parallel.** This configuration is called an ammeter. For the galvanometer to just read full scale at 1 mA, only 1 μA of current can go through the galvanometer, and the other 999 μA must go through the parallel resistor.

e. If the voltage drop is 0.2 μV, what value does the resistor in parallel need to have?

f. Try the value (for R_x) and then test by adjusting the power supply to see if you get the appropriate indication over the range of values (e.g., you should get half-scale indication for a source current of 0.5 mA; 80% of the bar would indicate a source current of 0.8 mA, etc.). Here use the "ammeter" button.

g. What would the ideal value of the internal resistance for an ammeter be and why?

Exploration 30.5: Voltmeters

A voltmeter measures the voltage across a circuit element and therefore is put in parallel with that element. We can construct a voltmeter by placing a large resistor in series with the galvanometer, which is indicated by —Ⓐ— (an ammeter symbol) in the circuit because a galvanometer and an ammeter are essentially the same (see **Exploration 30.4**). In this example the galvanometer shows a full-scale indication at a current of 1 μA, and the internal resistance of the galvanometer is 0.2 Ω.

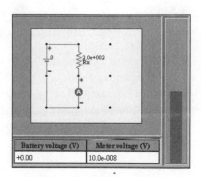

a. What voltage across the galvanometer produces a full-scale reading?

If we want to measure battery voltages of up to 2 V, we'd want the galvanometer needle to give a full-scale indication at this voltage. This means that 0.2 μV must drop across the galvanometer (with a current of 1 μA), while 1.9999998 V must drop across the series resistor.

b. Calculate the value of the series resistor required to produce a full-scale reading when the input voltage is 2 V. Test to see if you get the appropriate reading for a range of battery voltages. (Use the "set values" button.) Specifically, check that, for a battery voltage of 1 V, you get a half-scale reading on the indicator bar.

c. What would the ideal value of internal resistance for a voltmeter be and why?

Exploration 30.6: RC Time Constant

In this animation you can close and open switches to see what happens to the voltage across the capacitor **(red)**, the voltage across the resistor **(green)**, and the total voltage across the capacitor plus resistor **(blue)**. Initially, the capacitor is charged. After pushing "play," you should throw the switches (voltage is given in volts and time is given in seconds).

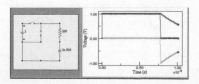

Set the switches so that you can get a good graph of the capacitor discharging and charging.

a. How much time (approximately) does it take the capacitor to charge and discharge?

b. Double the battery voltage. How much time does it take to charge and discharge?

c. Double the capacitance and measure the time to charge and discharge.

d. Double the resistance and measure the time to charge and discharge.

The value of RC (resistance times capacitance) is the RC time constant for the circuit and is a characteristic time. Set the battery voltage to 1 V.

e. When the time equals RC after throwing the switch, what is the capacitor voltage when it is discharging? When it is charging?

f. Compare your measurements to the values found from the equations for a charging or discharging capacitor:

$$\text{Charging: } V = V_0(1 - e^{-t/\text{RC}}) \quad \text{Discharging: } V = V_0\, e^{-t/\text{RC}}$$

Problems

Problem 30.1

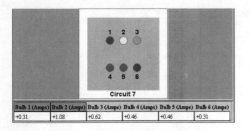

Bulb 1 (Amps)	Bulb 2 (Amps)	Bulb 3 (Amps)	Bulb 4 (Amps)	Bulb 5 (Amps)	Bulb 6 (Amps)
+0.31	+1.08	+0.62	+0.46	+0.46	+0.31

You are given arrangements of identical lightbulbs with the wires connecting them hidden from view. Determine how the lightbulbs are connected by unscrewing and/or screwing, the bulbs. All bulbs are initially screwed in their sockets, and the current through each bulb is given in the table (current is given in amperes).

Draw a schematic diagram representing the hidden circuit for each animation.

a. Circuit 1—Three Bulbs

b. Circuit 2—Three Bulbs

c. Circuit 3—Four Bulbs

d. Circuit 4—Four Bulbs

e. Circuit 5—Five Bulbs

f. Circuit 6—Six Bulbs

g. Circuit 7—Six Bulbs

Problem 30.2

The circuit shown contains three lightbulbs, an ideal battery, and a variable resistor. Answer the following questions by varying the resistance and examining the **current** and **voltage** across each circuit element (voltage is given in volts, current is given in amperes, and resistance is given in ohms).

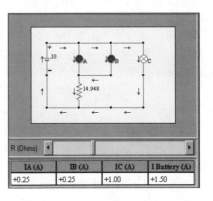

IA (A)	IB (A)	IC (A)	I Battery (A)
+0.25	+0.25	+1.00	+1.50

a. As you change the resistance, what happens to bulbs A, B, and C?

b. Why doesn't the brightness of bulb C change?

c. Why do the currents change in the way that they do?

d. Is the voltage of the battery changing? Why or why not?

e. Why do the voltages across the bulbs change the way that they do?

Problem 30.3

Voltmeter (V)	Ammeter (mA)
+9.00	1.50e+001

Answer the following questions for each circuit in this animation. Assume an ideal battery (no internal resistance) and ideal meters but note that the battery and unknown resistor are different for each circuit (electric potential is given in volts and current is given in milliamperes). Use the slider to change the variable resistor.

a. What is the resistance of the unknown resistance and the voltage of the battery in each circuit?

b. What is the range of power dissipated for each circuit over the full range of variable resistance values?

Problem 30.4

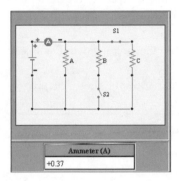

Ammeter (A)
+0.37

Rank the three resistors (from smallest to largest) in each of the two circuits (ammeter current is given in amperes).

a. Circuit 1
b. Circuit 2

Problem 30.5

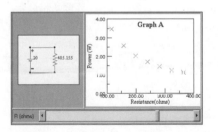

Assume an ideal battery. Vary the resistor and explain which of the graphs are correct and which are incorrect. Pay attention to the labels on the axes (electric potential is given in volts, current is given in amperes, resistance is given in ohms, and power is given in watts).

Problem 30.6

What is wrong with these circuits? Close the switches to see what happens and then explain what is wrong. Note which circuit elements are destroyed (electric potential is given in volts and resistance is given in ohms). Choose a new circuit after a circuit element is "destroyed."

a. Circuit 1
b. Circuit 2

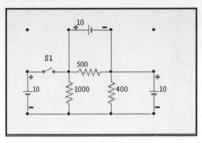

c. Circuit 3 (Hint: What might the power rating be on the resistor?)

Problem 30.7

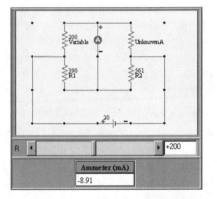

Ammeter (mA)
-8.91

The animation shows a "Wheatstone bridge," which is used to measure unknown resistors. Assume an ideal battery (no internal resistance) and ideal meters (resistance is given in ohms, electric potential is given in volts, and current is given in milliamperes). Begin the **animation (with unknown resistor A)** to read the current on the ammeter. In a Wheatstone bridge you adjust the variable resistor until the ammeter reads 0 and then you can calculate the value of the unknown resistor.

a. What is the unknown resistor in this case?

b. Develop an algebraic expression for the unknown resistance as a function of R1, R2, and the variable resistor.

c. Use your expression to quickly calculate unknown **resistor B** and unknown **resistor C.**

Problem 30.8

Circuits A and B are different configurations of the same circuit elements. Assume the battery is ideal (no internal resistance). Pick an animation to show the voltage and current on the meters (voltage is given in volts and current is given in milliamperes).

Use circuits A and B to determine the internal resistance of the ammeter and the voltmeter that are used in both circuits. You can vary the resistor in circuits A and B (and see the resistor value).

a. Which circuit should you use to find the resistance of the ammeter? Which circuit to find the resistance of the voltmeter? Why?

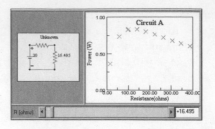

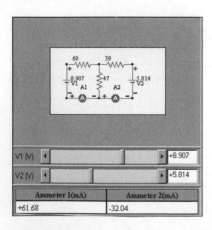

Once you determine which circuit you will use to find the resistance of the ammeter, you should keep in mind the ideal resistance of an ammeter (ideally 0 Ω; why is this the ideal resistance of an ammeter?) and pick your variable resistance appropriately (e.g., if a small resistance is in series with a very large resistor, the voltage drop across the big one will not be measurably different than the voltage drop across both of them, etc.). The same is true for your determination of the resistance of a voltmeter.

b. What is the resistance of both the ammeter and the voltmeter?

c. If you don't know the internal resistance of the meters or the value of the variable resistor (which is often the case), and you simply want to divide the voltmeter reading by the ammeter reading to determine the unknown resistance, which circuit, A or B, is the best for measuring small resistances?

d. Which circuit, A or B, is the best for measuring large resistances? Explain.

Problem 30.9

The batteries shown are not ideal (that is, they have internal resistances) but are otherwise identical. Assume ideal meters (current is given in milliamperes and voltage is given in volts). Vary the battery voltages and find the internal resistances. The internal resistance is the same for both.

Problem 30.10

Assume an ideal battery. The graph shows the power dissipated by the variable resistor as a function of resistance (resistance is given in ohms and power is given in watts).

a. For **circuit A**, what is the value of the unknown resistor?

b. What is the value of the power dissipated by the variable resistor when its resistance is equal to the unknown resistor?

c. Move the slider to vary the resistance. When the variable resistor is equal to the unknown resistor, how does the power dissipated by the variable resistor compare to the power dissipated when the variable resistor is not equal to the unknown resistor?

d. How can this help you find the unknown resistances in **circuit B** and **circuit C**?

e. What are the unknown resistances in those circuits (to within about 10%)?

Problem 30.11

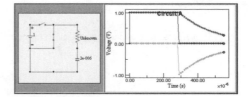

In the animation, you can close and open switches to see what happens to the voltage across the capacitor **(red)**, the voltage across the resistor **(green)**, and the total voltage across the capacitor plus resistor **(blue)**. Initially, the capacitor is charged. After pushing "play," you should throw the switches (voltage is given in volts and time is given in seconds). Rank the resistors in the three circuits from highest to lowest and explain your rankings.

Problem 30.12

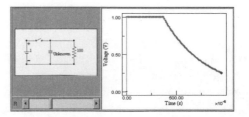

The graph shows the voltage across the capacitor (voltage is given in volts, resistance is given in ohms, and time is given in seconds). Initially the capacitor is charged. Find the capacitance of the capacitor.

Problem 30.13

The graph shows the voltage across capacitor A. Run the **Initial Circuit** to see the voltage when there is only one capacitor (capacitor A) in the circuit. Initially the capacitor is charged, so you

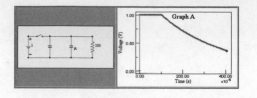

should open the switch (voltage is given in volts, resistance is given in ohms, and time is given in seconds). Which of the graphs, if any, correctly shows the voltage across capacitor A when there are two identical capacitors in parallel in the circuit?

CHAPTER 31

AC Circuits

Topics include impedance, reactance, RC, RL, and RLC circuits, resonance, and filters.

Illustration 31.1: Circuit Builder

Circuit Builder can be used to build and analyze DC and AC electrical circuits. Circuit Builder was written by Toon Van Hoecke at the Universiteit Gent.

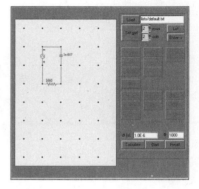

If you want to change the size of the circuit grid, change the number of rows and columns and then press the "Set grid" button. Be aware that all components on the screen will disappear. The "Show −>" button is used to visualize arrows representing the chosen current direction. New components for the circuit are entered by a drag-and-drop method. Press the button of the component you want to add, hold the mouse button, and release it on the position where you want to put this component. Once you drop the component on the circuit grid, a box will appear that will allow you to change the parameters of the component. Any horizontal or vertical position between two black dots can be taken.

The following direction-independent (no positive or negative side) components can be added:

- ─WWW─ Resistor: value entered in ohms.
- ─││─ Capacitor: value entered in farads.
- ⌒⌒⌒⌒⌒ Inductor: value entered in henries.
- ─⊗─ Lightbulb: characterized by voltage (in V) and power (in W). The color varies from black (no current) to white (maximum current).
- ────── Wire: used to close connections.
- ─╱•─ Switch: can be opened and closed.

Some components are polarized and thus have a positive and a negative side. The direction can be set by choosing "+ down/right" or "+ up/left" on the Direction list item. The following direction-dependent components can be added:

- ─│├─ Battery: value entered in V.
- ─Ⓥₛ─ General Voltage source: Its function prescription (in V) can be entered next to the source button and its default value is $\sin(t * 2 * \mathrm{pi} * f)$. Use t as the time variable, f as the frequency variable, and p as the period variable. The frequency can be entered in the text field below.
- ─⊕─ Current Source: value entered in A.
- ─▣∶▮─ Oscilloscope: simulation of a one-beam oscilloscope. A window with the view on the particular oscilloscope can be opened by selecting the "Display Oscilloscope" option on the popup menu that appears when you click the right mouse button on the oscilloscope icon in the circuit.
- ─Ⓥ─ Voltmeter: simulation of a digital voltmeter. Use the "Display Voltmeter" option on the right-mouse-button popup menu.

- —Ⓐ— Ampèremeter: simulation of a digital ampèremeter. Use the "Display Ampèremeter" option on the right-mouse-button popup menu.

The "Calculate" button is used for recalculating data. A number of Step # data points is calculated iteratively with a step size of Step(s). The step size is entered in seconds (default is 1e-6 s).

The "Start/Pause" button and the "Reset" button are used when a real-time clock is necessary. This is in situations with slow varying sources or the displaying of voltage or current graphs. The number of frames per second is $1/(10 * \text{step size})$. This is in real time up to step sizes of 0.01 s.

You can move a component to another position by using drag and drop. Other actions are available as options on the popup menu that appears when you click the right mouse button on the component's icon in the circuit. The possible options are only enabled if they are relevant:

- Delete Component: deletes the selected component.
- Change Value: changes the value or function of the selected component.
- Display Value Knob: pops up a little window with a scroll bar to change the value dynamically (linear steps or logarithmic steps).
- Display Frequency Knob: pops up a little window with a scroll bar to change the frequency of an AC source dynamically. This only works when the f variable is present in the function prescription.
- Show/Hide Value or Function: concerns the display on the circuit grid.
- Set Label: gives a name to the selected component.
- Display Oscilloscope: pops up the oscilloscope window of the selected oscilloscope.
- Display Voltmeter: pops up the digital voltmeter window of the selected voltmeter. The mode can be switched between DC and AC (rms value).
- Display Ampèremeter: pops up the digital ampèremeter window of the selected ampèremeter. The mode can be switched between DC and AC (rms value).
- Display Voltage Graph: pops up a voltage graph for the selected component (use "Start" button).
- Display Current Graph: pops up a voltage graph for the selected component (use "Start" button).
- Change Switch: changes the status of the selected switch, open or closed.
- Change Polarity: switches the + and − signs of components that are polarized.

Illustration 31.2: AC Voltage and Current

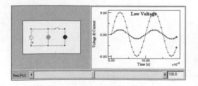

Assume an ideal power supply. The graph shows the voltage **(red)** across and current **(blue)** from the power supply as a function of time. Note the 10^{-3} exponent on the timescale when the animation starts (voltage is given in volts, current is given in hundredths of amperes, and time is given in seconds).

Start the **low voltage** animation. As you change the frequency, describe what happens to the bulb and the graph. Note the factor of 10^{-3} for the timescale when the animation starts. As you close the switch, notice that the voltage does not change, but the current increases. This is because closing the switch adds more resistors in parallel to the power supply.

Since the voltage is positive as much as it is negative, we do not talk about the average voltage (which would be zero) but instead describe the voltage either by the amplitude (the size of the peak voltage. What is it in this case?) or the rms (root-mean-square) voltage ($= V_{\text{peak}}/\sqrt{2}$). For this power supply, the peak voltage is 5 V and the rms voltage is 3.5 V.

Household voltage is 120 V rms. What is the peak voltage? In order to plot the current on the same graph, the current shown is 100 times the actual current. What is the average power of one lightbulb? ($P = I_{rms}V_{rms} = V_{peak}I_{peak}/2$ for a purely resistive load). You should find that these lightbulbs are 60 W bulbs.

With alternating current (AC), fluorescent lights in your room are flickering on and off 120 times/second (frequency in the US is 60 Hz), but you simply don't notice it (just like a movie is made of separate frames, but to you it looks continuous). In Europe the standard frequency is 50 Hz, so the fluorescent lights in Europe go on and off 100 times in one second.

Illustration 31.3: Transformers

A transformer is connected to an outlet. The graph shows the input voltage (the voltage across the primary) and output voltage (the voltage across the secondary) as a function of time.

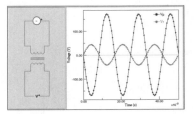

A transformer works by induction. A changing voltage in the primary coil (connected to the outlet) causes a changing current in a primary coil. The changing magnetic flux in the primary coil induces an emf (voltage) in the secondary coil. If you think about a coil of wire, the induced emf depends on the rate of change of the magnetic flux through the coil and the number of windings in the coil. Try changing the number of windings on the primary and secondary. How does the ratio of peak voltages depend on the windings? You should find that the ratio of the voltages is equal to the ratio of the coils. If the number of windings on the primary is greater than on the secondary, it is called a step-down transformer, but if the number of windings on the primary is smaller than on the secondary, it is a step-up transformer.

Both the step-up and step-down transformers conserve energy. For ideal transformers (no heat losses), energy conservation means that the average power ($I_{rms}V_{rms}$) is the same in the primary and secondary. Since the ratio of the number of windings is equal to the ratio of the voltages, for a step-down transformer with 200 turns on the primary and 20 turns on the secondary, 2 A coming into the transformer would yield 20 A out of the secondary. Conversely, for a step-up transformer, less current would be available at the secondary.

The facts that the power is the same in the primary and secondary and that transformers are easy to construct (coils wound around iron cores) are the reasons we use alternating current (instead of DC). Power companies can deliver a large amount of power either with high voltages and low current or with low voltages and high current. For the same amount of power, the lower-current option is preferable because of resistive heat losses on power lines. Consider the following two ways to deliver power over a 10-Ω power line and notice that the power from the plant is the same in both cases.

1. $V = 10,000$ V at the power plant and 2 A through the line. The total power dissipated is given by $I^2R = 40$ W (and the voltage drop between the power plant and the user is 20 V).

2. $V = 1000$ V at the power plant and 20 A through the line. In comparison, the total power dissipated is 4000 W (and the voltage drop is 200 V).

It is clearly better to choose the high voltage, low current route, and so power plants produce electricity at high voltages (around 20 kV). This is stepped up with transformers to a couple of hundred kV (e.g., 300,000 V) for cross-country transmission and then stepped back down in cities and at your house. It is not nearly as easy or as efficient to step up and down DC, which makes AC cheaper to transport over wires.

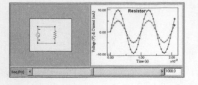

Illustration 31.4: Phase Shifts

Assume an ideal power supply. The graph shows the voltage **(red)** across and current **(blue)** from the power supply as a function of time (voltage is given in volts, current is given in milliamperes, and time is given in seconds).

We start by reviewing the current and voltage relationship for a **resistive load**. As you change the frequency in the animation, what happens (if anything) to the ratio of voltage to current? Notice that the voltage and current are in phase with each other in this circuit.

Try a **capacitive load**. What happens to the amplitude of the current as you increase the frequency? The ratio of V/I is not called a resistance for this type of load; it is called the reactance (or impedance, but impedance includes information about the phase shift between voltage and current). This means that the reactance of a capacitive load changes with frequency. Since the current increases as the frequency increases, the reactance must decrease as the frequency increases.

Notice the phase shift between the current and the voltage. Pause the graph. Which plot (the voltage or the current) is in the "lead?" In other words, if you look at a time in which the current reaches its maximum value, has the voltage already reached its maximum value or will it reach its maximum value at a slightly later time? If the current reaches its maximum first, we describe this as the "current leading the voltage," but if the voltage reaches its maximum first, we call it "current lagging the voltage." Which is the case with a capacitor?

Try an **inductive load**. What happens to the amplitude of the current as you increase the frequency? Does the current lead or lag the voltage in this case? Notice that with a capacitive load the current leads the voltage, while with an inductive load the current lags the voltage.

Because the current and voltage are out of phase when there are capacitive and inductive loads, and the reactance is a function of frequency, the mathematics to calculate the voltage and current is a bit more involved, but Kirchoff's laws still hold at any instant in time.

Illustration 31.5: Power and Reactance

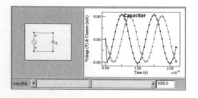

Assume an ideal power supply. The graph shows the voltage **(red)** across the source and the current **(black)** through the circuit as a function of time (voltage is given in volts, current is given in milliamperes, and time is given in seconds).

Resistive circuit: Look at the plot of voltage and current. Power is given by $P = VI$, but the current and voltage vary in time. It is more useful to think about the average power, which is $P = V_{rms}I_{rms} = I_{rms}^2 R = V_{rms}^2/R$. Notice that the current and voltage are always in phase, and so the product VI is always positive.

Capacitive circuit: Look at the plot of voltage and current. Notice that when the voltage is going from 0 to a more positive number, the current is going from a maximum value toward 0, but then when the voltage is going from its max back toward 0, the current has changed direction and is going from 0 down to a negative value. Thus, on the average over many cycles, the current and voltage are out of phase by $\pi/2 = 90°$. When the voltage is positive, the current is negative as much as it is positive, and the same applies when the voltage is negative. This means that the average power is 0. Compare this with the resistive load. When the voltage is positive, the current is positive, and when the voltage is negative, the current is negative. The resistor is always drawing current away from the source. Another way to think about this is that the capacitor simply stores charge. As the voltage changes direction, the current goes back and forth between the source and the capacitor, so the capacitor does not dissipate any energy over time (it simply stores the energy briefly).

Inductive circuit: Compare the plot of voltage and current for the inductor to that of the capacitor. Can you explain why the average power is 0 for this inductor just as it is for the capacitor?

If you have a circuit with a combination of resistive, capacitive, and inductive loads, calculating the average power dissipated now requires calculating $V_{rms}I_{rms}\cos\varphi$, where φ is the phase shift between the current and the voltage (see **Exploration 31.4**).

Illustration 31.6: Voltage and Current Phasors

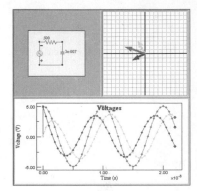

Assume ideal components. The bottom graph shows the voltage as a function of time across the source **(red)**, the resistor **(blue)**, and the capacitor **(green)**. Current is shown in **black** (voltage is given in volts, current is given in milliamps, and time is given in seconds).

You cannot simply use $V = IR$ when working with AC circuits, because you must account for the phase differences in the voltages and currents. Notice that when you look at **Voltages Only**, the voltages across the power supply, the resistor, and the capacitor are not in phase. One way to account for the phase differences is to describe the voltage with phasors as shown in the animation in the top right box. The voltage of each component is represented by a vector that rotates at the frequency of the source. The angle between the vectors represents the phase difference between the voltages, while the length of the vectors represents the peak voltage across each circuit element. **Illustration 31.7** and **Explorations 31.5** and **31.6** develop this idea further. We can also use this phasor representation to describe the current. Look at **Voltage and Current** and notice that the voltage from the source is out of phase with the current. So, instead of using $V = IR$, Ohm's Law becomes $V = IZ$, where Z is the impedance and includes the frequency response and phase shift associated with the various circuit components.

Illustration 31.7: RC Circuits and Phasors

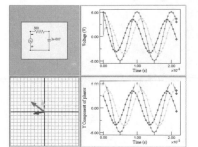

Assume an ideal power supply. The top graph shows the voltage as a function of time across the source **(red)**, the resistor **(blue)**, and the capacitor **(green)** (voltage is given in volts and time is given in seconds).

In order to analyze circuits with impedances that change as a function of frequency, we can use a phasor representation for the voltages across and the current through the various circuit elements. This allows us to take into account the phase difference between the voltages across the capacitor, the resistor, and the power supply.

To begin analyzing the circuit in the animation, we should first notice that Kirchhoff's law holds at any instant of time. Pause the animation and pick a time and find the voltage across the source, the resistor, and the capacitor. Verify that the voltage across the resistor plus the voltage across the capacitor equals the source voltage at that time. Notice, however, that if you add up the peak voltages, the peak resistor voltage plus the peak capacitor voltage is not equal to the peak source voltage.

You must account for the phase difference in the voltages. One way to account for the phase differences is to describe the voltage and current with phasors. Below the circuit are an animation and graph that show the phasor representation of the circuit elements (this allows us to show the phase difference), where the capacitor voltage is $\pi/2$ behind the resistor since its voltage lags the current. Notice that the phasors rotate at an angular speed of $\omega = 2\pi f$. The projection of the phasor vector on the y axis is plotted in the lower graph. Pause the animation. Note that the phasor animation matches the circuit graph. Try another frequency and verify that the two plots are the same (except at $t = 0$ because of the initial condition of the capacitor). Therefore, we can use phasor diagrams to show the phase angle between the source voltage, the resistor voltage, and the capacitor voltage. For more on phasors, see **Illustration 31.6** and **Explorations 31.5** and **31.6**.

Illustration 31.8: Impedance and Resonance, RLC Circuit

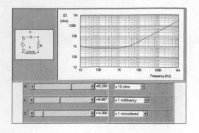

The impedance of a circuit is the relationship between the voltage and the current, $V = IZ$, where Z is the impedance. In a purely resistive circuit, $Z = R$, and the voltage and current are in phase. When capacitors and inductors are included, the relationship between voltage and current is more complicated. Calculating the impedance means taking the phase shift between voltage and current into account. In a series RLC circuit, the impedance is given by

$$Z = [R^2 + (\omega L - 1/\omega C)^2]^{1/2}.$$

Notice that the impedance will be the smallest when $\omega L = 1/\omega C$. If the impedance is smallest, what does that mean for the current at a given voltage? The frequency associated with this condition is called the resonant frequency. In the graph above you can see how the impedance changes as a function of frequency for different values of the resistance, the capacitance, and the inductance. If you change R, does the resonant frequency change? What if you increase C? What if you increase L? Using the equation, you should be able to predict what will happen as you change the values. At resonance, the power dissipated is the greatest and the impedance is said to be purely ohmic because $Z = R$ (since $\omega L = 1/\omega C$).

Exploration 31.1: Amplitude, Frequency, and Phase Shift

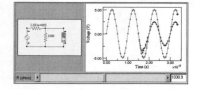

We characterize the voltage (or current) in AC circuits in terms of the amplitude, frequency (period), and phase. The sinusoidal voltage of this function generator is given by the equation

$$V(t) = V_0 \sin(\omega t - \varphi) = V_0 \sin(2\pi ft - \varphi),$$

where V_0 is the amplitude, f is the frequency ($\omega = 2\pi f$ is the angular frequency), and φ is the phase angle (voltage is given in volts and time is given in seconds).

To begin with, keep the resistance of the variable resistor equal to zero. Pick values for the voltage amplitude (between 0 and 20 V), frequency (between 100 and 2000 Hz), and phase angle (between -2π and 2π).

a. What does the amplitude on the graph correspond to?

b. If you increase the amplitude, what do you expect to happen? Try it.

c. Measure the time between two peaks (or valleys) on the graph. This is the period (T). What does $1/T$ equal?

d. What do you need to change to increase the time between two peaks? Try it.

e. Compare the plots when $\varphi = 0$ and when $\varphi = 0.5\pi$. (You can right-click inside the plot to make a copy.)

f. What happens when $\varphi = \pi$?

g. Pick a value of φ other than 0. Measure the time, t (measured from $t = 0$), it takes for the graph to cross the horizontal axis with a positive slope (going up). φ should be equal to $2\pi ft$. So, the phase (or phase shift) tells you how much the graph is shifted from a straight $\sin 2\pi ft$ curve.

h. Note that when $\varphi = 0.5\pi$, the plot is a cosine curve. Why?

Now, change the variable resistor. The plot shows both the voltage across the 1000 Ω resistor **(blue)** and the voltage supply **(red)**. Kirchhoff's laws hold for any instant of time in an AC circuit.

i. Use the techniques you learned for DC circuits to calculate the current in the circuit at several different points.

j. Verify that this circuit is simply a voltage divider.

k. What value does the variable resistor need to have for the maximum voltage across the 1000-Ω resistor to be one-third of the value of the source?

Exploration 31.2: Reactance

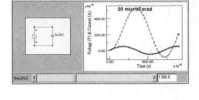

Assume an ideal power supply. The reactance, X, of a circuit element is the ratio between peak voltage and current so that $V = IX$. For a resistor, $X_R = R$. This exploration shows that for an active load like a capacitor or inductor, the reactance depends on the frequency as well (voltage is given in volts, current is given in amperes or milliamperes [note graph labels], capacitance is given in farads, inductance is given in henries, and time is given in seconds).

a. For a **capacitive load**, vary the frequency and observe what happens to the current. How is this result related to the formula for capacitive reactance? The graph shows the voltage across **(red)** and current from the power supply **(black)** as a function of time. Note that you may need to wait for transient effects to decay if you change the frequency.

b. Double the capacitance and try it again. What happens? Explain your observations in terms of the capacitive reactance.

c. Repeat (a) and (b) with an **inductive load**. What happens if you **double the inductance**? Explain your observations in terms of the inductive reactance.

If we take this to the limit of $f \to 0$ (DC circuits), then a capacitor is essentially an open circuit. At high frequencies, the capacitor is essentially a short circuit (acts like a wire with little or no resistance).

d. Explain these limits in terms of what a capacitor does (stores charge) and how it works.

e. At low frequencies, is an inductor essentially an open circuit or a short circuit? What about at high frequencies?

f. Explain in terms of what an inductor does (in terms of induced current).

Exploration 31.3: Filters

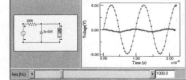

Since the reactance varies with frequency, we can use capacitors (or inductors) to filter out different frequencies. The voltage of the source is **red**, while the oscilloscope voltage is the **blue** plot on the graph (voltage is given in volts and time is given in seconds).

a. At very low frequencies does the capacitor have a high or a low reactance?

b. Therefore, at low frequencies, will the current through the capacitor be large or small?

Capacitor filter: **Try filter 1**.

c. Will this circuit allow high or low frequency signals to reach the oscilloscope? Explain.

d. Try it.

e. Is the amplitude of the voltage measured by the oscilloscope bigger at low or high frequencies?

If it is bigger at higher frequencies, then it "allows" high frequencies through more readily than lower frequencies and it is called a high-pass filter. If it "allows" low frequencies through, it is a low-pass filter. Look at the circuit for **filter 2**.

f. Do you think this is a high-pass or low-pass filter? Why?

g. Try it and determine which kind of filter it is.

Many signals are not simply made up of one single frequency. They are a combination of frequencies, and this is where filters are useful. Try a wave function composed of **two different frequency waves with the low-pass filter**. Try **this wave function with the high-pass filter**. (Note: You cannot change the frequency of this wave function with the slider bar in this animation.)

h. What is the difference between the oscilloscope signals in the two cases?

i. Explain.

Exploration 31.4: Phase Angle and Power

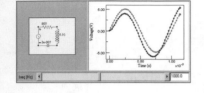

Assume an ideal power supply. The graph shows the voltage **(red)** as a function of time across the source and the current **(black)** through the circuit (voltage is given in volts, current is given in milliamperes, and time is given in seconds). (Note: The initial current vs. time graphs are not centered about 0 because of the initial state of the capacitor and inductor.)

To calculate the power dissipated by an **RLC series circuit** you cannot simply use $I_{rms}V_{rms}$ because the power supply current and voltage are not in phase (unlike with a purely resistive load). This is due to the different phase shifts between voltage and current associated with the capacitors and inductors. The current through all elements must be the same, so the voltages across each are phase shifted from each other (see **Illustrations 31.4** and **31.5**). Here the equations for calculating power are

$$P = I_{rms}^2 R = I_{rms}V_{rms}\cos\varphi = I_{rms}^2 Z\cos\varphi,$$

where Z is the impedance of the series circuit (V_{rms}/I_{rms}) and φ is the phase shift between current and voltage defined as ($\omega = 2\pi f$):

$$Z = [R^2 + (\omega L - 1/\omega C)^2]^{1/2} \quad \text{and} \quad \cos\varphi = R/Z.$$

a. Pick a frequency value. Find the impedance from V_{rms}/I_{rms}.

b. Compare this value with the calculated value found using the equation above.

c. Calculate the phase shift.

d. Compare this calculated value with the value of phase shift measured directly from the graph. To measure the phase angle, since one period ($1/f$) represents a phase shift of 2π, measure the time difference between the peaks of the voltage and current plots and divide by the period (the time between the peaks of the voltage or the current) to find the percentage of 2π by which the current is shifted.

e. What is the power dissipated?

Exploration 31.5: RL Circuits and Phasors

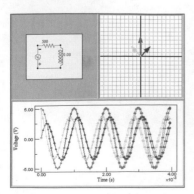

Assume an ideal power supply. The graph shows the voltage as a function of time across the source **(red)**, the resistor **(blue)**, and the inductor **(green)** (voltage is given in volts and time is given in seconds).

To analyze the currents and voltages in this circuit, notice that you cannot simply use $V = IX$ along with the peak current values. You must account for the phase differences between the voltages and the currents. One way to account for the phase differences is to describe the voltage, current, and reactance with phasors. Next to the circuit is an animation that shows the phasor representation of the circuit elements (this allows us to show the phase difference) in which the inductor voltage **(green)** is

$\pi/2$ ahead of the resistor voltage **(blue)**. The magnitude of each vector is the peak voltage across the element.

a. Since there is no phase shift between the resistor current and the voltage across the resistor, what would a phasor for the current look like in the phasor diagram?

b. Why is the phasor for the inductor $\pi/2$ ahead of the resistor voltage (i.e., the current through the circuit)? (Hint: Does the inductor current lead or lag the voltage [see **Illustration 31.4**]?)

c. How does the length of the inductor's phasor change as you change the frequency?

d. Why does the length change as a function of frequency?

Notice that the phasors rotate at an angular speed of $\omega = 2\pi f$. The projection of any given voltage phasor on the *y* axis is the voltage across the circuit element at that time.

e. Pause the animation and explain how you can tell that the phasor diagram matches the voltages across the different circuit elements shown in the graph. In other words, verify that the *y* component of the vector in the phasor animation matches the value of the voltage shown on the graph.

From the vectors on the phasor diagram, we can develop a connection between the peak (or rms) voltage and the peak (or rms) current, where $V_0 = I_0 Z$, and the phase difference between the voltage and current is given by φ. On the phasor diagram V_0 (**the source voltage–red**) is the vector sum of the two voltage vectors (**resistor-blue** and **inductor-green**), and φ is the angle between V_0 and the current (in the same direction as the resistor voltage phasor on the diagram). **Exploration 31.6** develops the use of phasors for RLC circuits.

Exploration 31.6: RLC Circuits and Phasors

Assume an ideal power supply. The graph shows the voltage as a function of time across the source **(red)**, the resistor **(blue)**, the capacitor **(green)**, and the inductor **(yellow)**, as well as the current through the circuit **(black)** (voltage is given in volts, current is given in milliamperes, angles are given in degrees, and time is given in seconds).

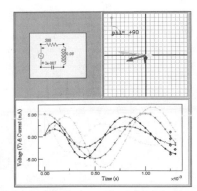

From the vectors on the phasor diagram, we can develop a connection between the peak (or rms) voltage and the peak (or rms) current, where $V_0 = I_0 Z$ and the phase difference between the voltage and current is given by φ. On the phasor diagram V_0 (**the source voltage–red**) is the vector sum of the three voltage vectors (**resistor-blue, inductor-yellow,** and **capacitor-green**) and φ is the angle between V_0 and the resistor phasor (since resistor current and voltage are in phase). The *y* components of the phasor vectors are the voltages across the various circuit elements. See **Illustrations 31.6** and **31.7** as well as **Exploration 31.5**.

a. Explain the phase difference between the blue, yellow, and green vectors in the phasor animation.

b. Pick a frequency and pause the animation. Verify that the red vector is the vector sum of the other three vectors.

c. Pick a frequency and measure φ on the phasor diagram using the pink protractor.

d. Explain how you can tell that the phasor animation matches the voltage and current vs. time graph for the circuit.

e. Also measure the phase angle on the voltage and current graphs. To measure the phase angle, since one period $(1/f)$ represents a phase shift of 2π, measure the time difference between the peaks of the voltage and current plots and divide by the period.

f. Measure Z for this same frequency $(Z = V_0/I_0)$. Check your answers by using the equations for impedance and the phase shift between the voltage and the current, $Z = (R^2 + (\omega L - 1/\omega C)^2)^{1/2}$ and $\cos\varphi = R/Z$.

Exploration 31.7: RLC Circuit

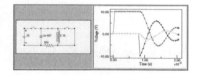

Assume ideal components. The graph shows the voltage across the source **(red)** and the current from the source **(black)** as functions of time (voltage is given in volts, current is given in milliamperes, and time is given in seconds).

An RLC circuit is similar to an oscillating spring or child on a swing. If you push the swing at exactly the same frequency as the natural frequency of oscillation (the most common way to push a swing), it quickly goes higher and higher. But if you push (or pull) part way through the swing at times that do not match the natural rhythm of the swing, the swing will not go as high as quickly and might even swing lower (in a fairly jerky fashion). When the current is the largest, this is called the resonance.

a. What is the resonant frequency of this circuit?

b. As you move the frequency of the driving source closer to the natural frequency of oscillation, what happens to the voltage and current?

c. Pick a new value for the variable resistor. What is the resonant frequency of this circuit?

d. What are the differences in the resonances with different values of R?

e. Compare the resonant frequency to $(1/2\pi)(1/LC)^{1/2}$. It should be the same.

Exploration 31.8: Damped RLC

Assume ideal components. The graph shows the voltage across the capacitor **(red)**, the voltage across the inductor **(blue)**, and the voltage across the resistor **(green)** as functions of time (voltage is given in volts).

Change the switches as you explore the behavior of this circuit.

a. Pick a specific time, measure the voltages on the graph, and verify that Kirchhoff's law holds when the switch is open and when the switch is closed.

b. What determines the time between peaks of the voltage when you close the switch?

c. Change the value of the variable resistor. What happens to the time for the oscillations if the resistor is large? When the resistor is small? Explain.

Problems

Problem 31.1

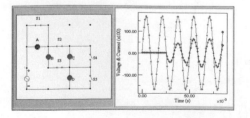

Assume an ideal power supply. The graph shows the voltage across the power supply **(red)** and current (* 100) from the power supply **(blue)** as a function of time (voltage is given in

volts, current is given in hundredths of amperes, and time is given in seconds). A measurement of a current of 50 on the graph would actually be a current of 0.5 A. Find the power rating of each of the bulbs A, B, C, and D.

Problem 31.2

A transformer is connected to an outlet. The graph shows the input voltage (voltage across the primary) and output voltage (voltage across the secondary and resistor) as a function of time. You can read the resistor value by moving the mouse over the resistor.

a. If the primary has 400 windings, how many windings does the secondary have?

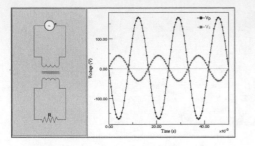

b. What is the rms current delivered to the primary?

c. How much power is delivered to the resistive load?

Problem 31.3

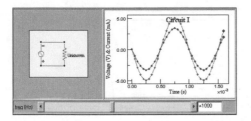

Assume an ideal power supply. The graph shows the voltage across the source **(red)** and the current through the circuit **(blue)** as functions of time (voltage is given in volts, current is given in milliamperes [10^{-3} Amps], and time is given in seconds). For each circuit,

What is the rms voltage? rms current? What is the value of the unknown resistor? What is the average power dissipated?

a. **Circuit I**

b. **Circuit II**

c. **Circuit III**

Problem 31.4

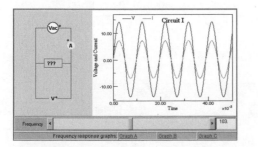

Identify the following unknown circuit elements (resistor, capacitor, or inductor) and match the frequency response graphs (Graphs A—C) to the current and voltage vs. time graphs shown for the circuits below. Explain your answers.

a. **Circuit I**

b. **Circuit II**

c. **Circuit III**

Problem 31.5

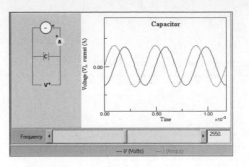

Determine the value of the circuit elements (voltage is given in volts, current is given in amperes, and time is given in seconds).

a. **Capacitor**

b. **Inductor**

Problem 31.6

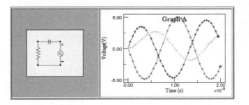

Assume an ideal power supply. The voltage of the source is **red**, the voltage across the resistor is **blue**, and the voltage across the capacitor is **green**. Identify the correct graph for the RC series circuits shown (voltage is given in volts and time is given in seconds).

Problem 31.7

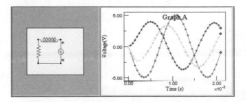

Assume an ideal power supply. The voltage of the source is **red**, the voltage across the resistor is **blue**, and the voltage across the inductor is **green**. Identify the correct graph for the RL series circuits shown (voltage is given in volts and time is given in seconds).

Problem 31.8

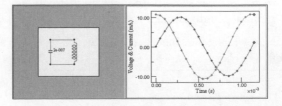

Assume ideal circuit components (voltage across the capacitor [blue] is given in volts, current [red] is given in milliamperes, and time is given in seconds). A capacitor is charged and connected to an inductor. Watch what happens to the voltage across the capacitor and to the current.

a. What is the value of the inductor?

b. What is the average power dissipated? Why?

Problem 31.9

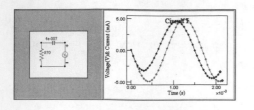

Assume an ideal power supply. The voltage of the source as a function of time is the **red** plot on the graph, and the current from the source is **black** (voltage is given in volts, current is given in milliamperes [mA = 10^{-3} A], and time is given in seconds). What is the power dissipated in each circuit?

a. **Circuit I:** RC circuit

b. **Circuit II:** RL circuit

c. **Circuit III:** RLC circuit

Problem 31.10

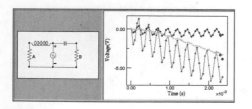

Assume an ideal power supply. The voltage of the source is the **red** plot on the graph.

a. Which plot shows the voltage across resistor A?

b. Which is the voltage across resistor B?

This is a basic design of a loudspeaker system in which the resistors represent speakers: a tweeter and a woofer. The woofer, a large speaker, wants low-frequency signals (and generates low-pitched sound), while the tweeter, the small-diameter speaker, should get high-frequency signals (to generate high-pitched sound).

c. Which resistor represents the tweeter and which represents the woofer? Explain.

Problem 31.11

Assume an ideal power supply. The graph shows the voltage across the source **(red)** and the current **(black)** through the

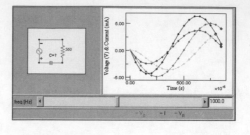

circuit as a function of time, as well as the voltage across the resistor **(blue)** and the voltage across the capacitor **(green)** (voltage is given in volts, current is given in milliamperes [mA = 10^{-3}A], and time is given in seconds).

a. What is the value of the unknown capacitor?

b. What is the average power dissipated when the reactance of the capacitor equals the resistance of the resistor?

Problem 31.12

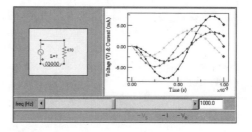

Assume an ideal power supply. The graph shows the voltage across the source **(red)** and the current **(black)** through the circuit as a function of time, as well as the voltage across the resistor **(blue)** and the voltage across the inductor **(green)** (voltage is given in volts, current is given in milliamperes [mA = 10^{-3} A], and time is given in seconds).

a. What is the value of the unknown inductor?

b. What is the average power dissipated when the reactance of the inductor equals the resistance of the resistor?

Problem 31.13

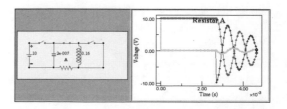

Assume an ideal power supply. Rank the resistors in the three switched RLC circuits (you will need to open the switches and start the graph by clicking on the "play" button). The graph shows the voltage across the capacitor as a function of time **(red)**, the voltage across the inductor as a function of time **(blue)**, and the voltage across the resistor as a function of time **(green)** (voltage is given in volts and time is given in seconds).

Problem 31.14

The voltage vs. time graph shows the voltages across circuit elements in this circuit. You can vary the frequency of the source voltage (voltage is given in volts, current is given in amperes, time is given in seconds, and frequency is given in hertz).

a. What is the resonant frequency of this circuit?

b. Which graph describes the frequency response of this circuit? Explain.

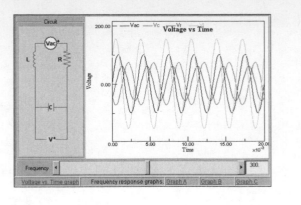

Electromagnetic (EM) Waves

Topics include electromagnetic waves, wavelength, and frequency.

Illustration 32.1: Creation of Electromagnetic Waves

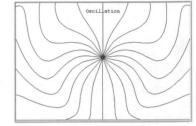

This Illustration shows the electric field lines due to a positive charge. Initially the charge is not moving. The slider can be used to vary the speed of the charge. When **translation** mode is selected, the slider controls the instantaneous velocity. When **oscillation** mode is selected, the slider sets the maximum speed.

How are the changing fields in an electromagnetic wave created? Electromagnetic waves such as heat, light, and radio waves are created by a charge that is *accelerating*. The magnitude of the electric field is related to the acceleration of the charge.

Play the animation in **translation** mode and move the velocity slider. Notice how the electric field lines form a disturbance that moves away from the charge when the velocity changes. Because a changing electric field will give rise to a changing magnetic field, and a changing magnetic field will cause a changing electric field, a traveling electromagnetic wave is created. Move the slider in translation mode and note that abrupt changes in the velocity produce very complex wave patterns.

Play the animation in **oscillation** mode and notice the sinusoidal appearance of the electromagnetic disturbance.

You may wonder where the energy that goes into the electromagnetic wave comes from. Is energy being created from nothing? The answer is no, of course. A charge will not oscillate on its own; it must be driven by some force. For example, the charge might be part of an AC current, oscillating as the voltage source oscillates. Although some energy is being radiated from the charge, energy is being put into the charge to keep it oscillating.

Illustration 32.2: Wave Crests

In the 19th century it was discovered that a moving charge produces an electromagnetic wave. Rapidly oscillating charged particles (such as electrons in an atom) produce visible light, while slowly oscillating charges (such as those in an antenna) produce radio waves. Although waves with different frequencies produce different effects when they interact with matter, their propagation through space is quite similar. These similarities are the subject of this Illustration.

Electromagnetic waves have regions of high and low field strengths that are analogous to the high- and low-pressure regions of a sound wave. Analogies between electromagnetic waves and sound waves can be useful, but they should not be pushed too far. This Illustration shows one such analogy. An oscillating charge within the back circle produces a wave, and this wave is seen to propagate away from the source. The wave crests and troughs moving away from the source represent regions of strong electric field. The troughs are regions where the electric field is also strong, but the field is pointing in the opposite direction from the field at the crests. As waves propagate away from the source, their amplitude decreases. The red wave traveling to the right illustrates this.

Electromagnetic waves are different from sound waves, and this Illustration does little to point out this difference. Sound requires a medium for propagation, whereas electromagnetic waves do not need a medium: They can propagate in a vacuum. Furthermore, electric fields cannot propagate energy without a complementary magnetic field. The magnetic field associated with an electromagnetic wave is perpendicular to the electric field and is not shown. But the wavelength, frequency, and amplitude of the electric field are correctly illustrated and provide clues to the following questions:

- Do the frequency and period of the electromagnetic wave depend on the distance from the source?
- How does electric field amplitude depend on the distance from the source?

Illustration 32.3: Electromagnetic Plane Waves

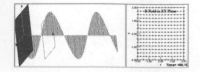

Electromagnetic waves (such as radio waves) can sometimes be approximated as plane waves if the observer is located far from the source. But what, exactly, does this plane wave look like? Before you begin, we should point out that plane waves are (like point masses) an idealization. Typical electromagnetic waves are not plane waves, not because they are curved (although they usually do have some curvature), but because they contain many frequencies and because they originate from more than one source. Although radio waves approximate a plane wave, visible light usually does not, unless it is produced by a laser. Because waves can be constructed by adding together multifrequency plane waves, understanding this Illustration is a good place to start.

The animation shows a plane electromagnetic wave's electric field. The magnetic field is not shown. Click-drag inside the large panel before you play the animation. What do you see? The lines pointing away from the z axis represent the electric field as measured along the axis. Move the slider. The slider controls the position (given in meters) of the transparent square. The transparent square represents the plane (hence the name plane wave) in which you are viewing the electric field in the right panel. Use the slider to estimate the wavelength. Play the animation. Note the animation time (given in nanoseconds) in the right-hand panel. What is the frequency of the wave? In what portion of the electromagnetic spectrum is the wave? Since the period is 6.68×10^{-8} s, the frequency is one over this, or about 1.5×10^7 Hz, or 15 MHz. Since $c = 3 \times 10^8$ m/s $= \lambda f$, this means that $\lambda = c/f = 20$ m, which is a radio wave.

The vectors along the z axis show the electric field along this path. What does the electric field in the xy plane look like for a particular value of z? Remember it is a plane wave. Move the square and notice that all points within the square have the same electric field, hence the name electromagnetic plane wave.

Notice that the wave equation for a pressure wave, $P(x, t) = A \sin(k x - \omega t)$, traveling in the x direction could be changed to describe this electromagnetic plane wave (traveling in the z direction) as $\mathbf{E}(z, t) = E_{max} \sin(k x - \omega t)\mathbf{i}$. Why does the electric field vector have a component in the x direction but not in the z direction? Maxwell's equations tell us that the electromagnetic wave is a transverse wave. Therefore, unlike the pressure wave, the electromagnetic wave cannot have a component in the direction of propagation.

Note that $k = 2\pi/\lambda$ and $\omega = 2\pi f$ so that $v = \omega/k = \lambda f$, where v is the wave speed, λ is the wavelength, and f is the frequency.

Illustration 32.4: Electromagnetic Waves: $E \times B$

This Illustration shows a *representation* of a simple sinusoidal electromagnetic wave that includes both the electric and magnetic fields versus z (the direction of travel).

When the wave is traveling in the positive z direction, these two fields can be written as

$$E_x = E_0 \sin (k z - \omega t), \quad E_y = 0, \quad E_z = 0,$$

$$B_x = 0, \quad B_y = B_0 \sin (k z - \omega t), \quad \text{and} \quad B_z = 0,$$

where $E_0 = c B_0$ and c is the speed of light in MKS units.

The electric field is measured along the z axis and drawn as a red line; the magnetic field is drawn as a green line. The length of these lines represents each field's magnitude. Although we have not drawn an arrowhead, each line represents a vector whose direction is away from the z axis along the line. Click-drag to the right or left to rotate the animation about the z axis. Click-drag up or down to rotate in the xy plane.

Many students misunderstand what is represented in this Illustration and think that the fields extend in the x and y directions in the same way a wave on a rope would. In other words, students often believe that the fields only extend a finite distance in the xy plane like the wave on a string. The representation actually tells you a field's strength only at different points along the z axis. However, with an electromagnetic plane wave traveling in the z direction, the field is uniform in the xy plane. You may want to review the field representations in **Illustration 32.3** to clarify this important concept.

Here we draw electric and magnetic vectors of equal length at points along the z axis. This is a bit of a misrepresentation. Electric and magnetic fields are measured in different units and their numeric values are, in fact, not equal in the MKS system. However, the energy carried by the electric field is equal to the energy carried by the magnetic field, and so most textbooks draw the vectors with equal lengths.

This Illustration also points out an important relationship between electric and magnetic fields. The **E** and **B** fields in an electromagnetic wave are in phase. Because one of these fields determines the other, we often only discuss one of them. We usually choose **E**.

Finally, there is an important relationship between **E** and **B** and the direction of propagation. The direction of travel is determined by **E** $\times$ **B**. In other words, **E** and **B** are perpendicular and the direction of propagation is given by the right-hand rule. You should confirm this cross-product relationship for yourself in the two animations.

Exploration 32.1: Representation of Plane Waves

Move the slider and observe the animation on the left-hand panel of your screen. The animation shows the electric field in a region of space. The arrows show the field-vector representation of the electric field. The amplitude of the field is represented by the brightness of the arrows. The slider allows you to move along the z axis. Notice that the electric field is always uniform in the xy plane but varies along the z axis (position is given in meters and time is given in nanoseconds).

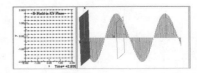

a. Construct a graph that represents the electric field along the z axis at $t = 0$ ns.

Now **view a representation** of the electric field. Click-drag inside the animation on the right to view the electric–field representation from different points of view. This representation should closely match the graph you drew for (a). Click on "play" to see a traveling wave. The representation on the right is often used to show a field like that on the left. Remember that the representation on the right is actually a graph of amplitude along the direction of propagation (z axis).

b. Keeping that in mind and looking at the graph on the right, rank the amplitude of the field at $t = 0$ ns for the following locations, from smallest to largest.

Location	x coordinate	y coordinate	z coordinate
I	1	0	−1.5
II	1	1	−1.5
III	0	0	−1.5
IV	0	1	−1.0
V	1	1	−0.5

c. Now, push "play" to see the traveling wave. At position $z = -0.5$ m, rank the amplitude of the field at the following times (approximately), from smallest to largest.

Time (ns)	x coordinate	y coordinate	z coordinate
$t = 0$	1	1	−0.5
$t = 1.7$	1	1	−0.5
$t = 3.3$	1	1	−0.5
$t = 5.0$	1	1	−0.5
$t = 6.7$	1	1	−0.5

d. What is the wavelength (distance between peaks) of the wave?

e. What is the frequency of the wave (the period $T = 1/f$ is the time it takes for the wave to repeat itself at a given location)?

f. What is the speed of the wave?

Exploration 32.2: Plane Waves and the Electric Field Equation

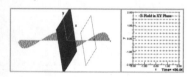

You can change the position of the square (that shows you the field–vector representation of the electric field), as well as the maximum value of the electric field and the wavelength (position is given in meters and time is given in nanoseconds).

The electromagnetic plane wave in the animation above is described by the equation

$$\mathbf{E}(z, t) = E_{max} \sin(k z - \omega t) \, \mathbf{I},$$

where $k = 2\pi/\lambda$ (λ is the wavelength) and $\omega = 2\pi f$ (f is the frequency).

a. Explain why the equation is a function of z and t for this wave.

b. Why is this equation a vector equation with a component in the x direction?

c. What is the associated equation for the magnetic field (check in your book if needed)?

d. What do you predict will happen in both representations (the vector field view to the right and the wave view to the left) if you increase the amplitude? Change the amplitude to check your prediction. Did the frequency change? Why or why not?

e. What do you predict will happen in both representations if you increase the wavelength? Try it. This time did the frequency change? Why or why not?

f. Pick a value of the wavelength (λ) and measure it.

g. Measure the frequency (f) at this wavelength.

h. What is the value of λf? (It should be 3×10^8 m/s.)

Note: When you change the wavelength, you need to let the animation play long enough for the old wavelength to disappear from the axis by letting the animation run for 100–200 ns before making any measurements.

Problems

Problem 32.1

The animation represents a traveling electromagnetic wave in the z direction. You can click-drag to the right or left to rotate about the z axis. Click-drag up or down to rotate in the xy plane. Which wave is the electric field and which one is the magnetic field?

Note that both the electric and magnetic fields in this animation are drawn with equal magnitude so you can see both fields. In reality, the magnetic field would be a factor of c smaller than the electric field in MKS units.

Problem 32.2

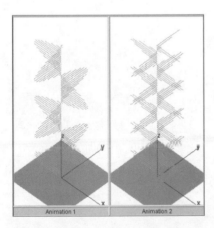

Two animations represent traveling electromagnetic waves in the z direction. You can click-drag to the right or left to rotate the animations about the z axis. Click-drag up or down to rotate in the xy plane. If Animation 1 represents green light, what is represented by Animation 2?

Problem 32.3

Three animations represent traveling electromagnetic waves in the z direction. You can click-drag to the right or left to rotate the animations about the z axis. Click-drag up or down to rotate in the xy plane. If green light, red light, and violet light are represented by the three animations, which of the animations represents which color?

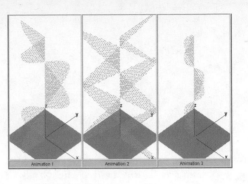

Problem 32.4

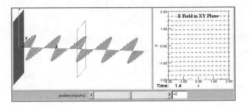

By moving the slider, you can change the position of the square in the electromagnetic wave animation. Then you can view the graph on the right, which shows you the field-vector representation of the electric field (position is given in microns [10^{-6} meters] and time is given in femtoseconds [10^{-15} seconds]). The animation only shows you the electric field.

a. In what direction is the magnetic field?

b. What are the wavelength, frequency, and speed of the electromagnetic wave represented here?

Problem 32.5

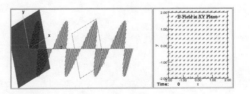

By moving the slider, you can change the position of the square in the electromagnetic wave animation. Then you can view the graph on the right, which shows you the field-vector representation of the electric field (position is given in millimeters and time in picoseconds [10^{-12} seconds]). The animation only shows you the electric field.

a. In what direction is the magnetic field?

b. What are the wavelength, frequency, and speed of the electromagnetic wave represented here?

Problem 32.6

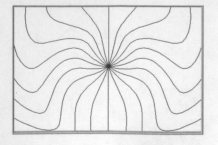

The animation shows how static electric field lines from a positive point charge produce a radiation field if the charge is caused to oscillate. Initially the charge is stationary.

Start the Charge Oscillating: Notice how the electric field lines form waves that move away from the charge when the charge is oscillating.

a. What is the direction of the magnetic field associated with the waves to the right of the charge?

b. In what direction is the radiation field a maximum?

Mirrors

Topics include mirrors, ray diagrams, focal point, and real and virtual images.

Illustration 33.1: Mirrors and the Small-Angle Approximation

Shown is an optics bench that allows you to add various optical elements (lens, mirror, and aperture) and light sources (beam, object, point source) and see their effect. Elements and sources can be added to the optics bench by clicking on the appropriate button and then clicking inside the applet at the desired location. Moving the mouse around in the applet gives you the position of the mouse, while a click-drag will allow you to measure angle (position is given in centimeters and angle is given in degrees).

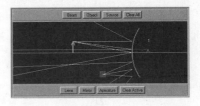

Add a mirror to the optics bench by clicking on the mirror button and then clicking inside the animation to place the mirror. Adjust the focal point of the mirror by dragging on the round hotspots. Notice that you can make the mirror either concave or convex. Make the mirror concave with a focal length of 0.5 cm and place it near the right-hand side of the applet. Now add a source of light by clicking on the object button and then clicking inside the animation. You can later add other sources of light.

Notice the rays emanating from the object, their reflection from the mirror, and the resulting image. Click the head of the arrow (the object) and move it around. First note that there are three rays that emanate from the head of the arrow. One ray comes off parallel to the principal axis (the yellow centerline) and is reflected through the focal point, one ray comes off at an angle to hit the mirror on the principal axis and is reflected, and one ray passes through the principal axis at the focal point of the mirror and is reflected parallel to the principal axis.

Do the rays always behave as you expect them to? Probably not. As you drag the head of the source and change its height and position, what do you notice about the rays when they reflect from the mirror? The rays reflect from the vertical line tangent to the mirror's surface. If you click on the mirror, you will see this line in green. This applet uses what is called the small-angle approximation. This approximation assumes that the object is of a much smaller dimension than the mirror. For larger focal length mirrors, you may barely notice the approximation, but for smaller focal length mirrors it becomes increasingly noticeable. In this Illustration a focal length of $f < 1$ cm will yield a noticeable difference between the rays you expect and the result of the small-angle approximation. Click on the mirror and drag the round hotspots to change the mirror's focal length to see the effect.

The optics bench allows you to try many different configurations to see how light will interact with a mirror. Take some time to play with the applet. You may also find it helpful to refer back to this Illustration as you develop your understanding of optics. A brief description of the three sources is given below.

- The "Beam" button adds a beam of parallel light rays. The angle of the light rays can be changed by dragging the hotspot after clicking on the beam.
- The "Object" button adds an arrow as an object. A ray diagram is drawn for the object if an optical element is present.

- The "Source" button adds a point source of light. The spread of the light rays can be adjusted by dragging the hotspot after clicking on the source.

Illustration 33.2: Flat Mirrors

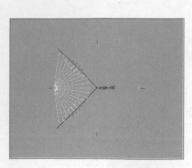

This animation shows images in two flat mirrors placed at an angle to each other. You can adjust the angle between the mirrors by click-dragging the green dot, and you can change the object size by click-dragging the red dot. The gray dots are the images. If you double-click in the animation window, you can see the path of some of the light rays from the source.

The yellow rays show the actual path of the light, while the gray "rays" show where it looks like the reflected yellow rays come from. When we look at objects, we assume light travels in straight lines (this is how our brain interprets the input it gets from our eyes). So when we look in a mirror, the images we see are behind the mirror because it looks like the light comes from the point behind the mirror. Since the light rays do not actually pass through the image points, the images are virtual ones. Try adjusting the angle between the mirrors to get more than two images. Why are there multiple images? Double-click to see the light rays and identify the images that are a result of light reflecting off the mirror more than once. Identify points where multiply reflected light rays cross. If you were located there, you would see multiple images. Follow each ray straight back to the virtual image to check. Note that since there are only a finite number of rays drawn, you may not get every ray you expect to see. As you decrease the angle between the mirrors, why are there more images?

Exploration 33.1: Image in a Flat Mirror

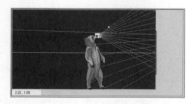

A bear stands in front of a plane mirror that is hanging on a wall. A point source of light is located near the mirror. You can drag this source to any location and can change the angle of its rays by click-dragging on the hotspot (position is given in meters and angle is given in degrees).

a. At what point on the mirror must the bear look in order to see her feet? For simplicity and ease, assume the bear's eye is located at the tip of her nose.

b. Move the bear to the position $x = 10$ m. If the bear looks at the same spot in the mirror found in (a), what will she see? Does this imply that she is able to see more, less, or the same amount of her body in the mirror when she moves away from the mirror?

c. In terms of the bear's height, how long must the mirror be for her to see her entire body?

Exploration 33.2: Looking at Curved Mirrors

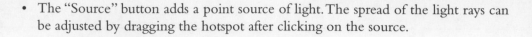

What is the difference between a real and a virtual image? What does your eye see when it looks into a mirror (position is given in meters and angle is given in degrees)?

a. Drag the object back and forth. In this animation when the image is on the left of the mirror it is a real image, but when it is on the right it is a virtual image. Why?

b. Place the object so that the image is to the right of the mirror (a virtual image). If your eye is where the eye is in the diagram, where does your eye/brain think the light is coming from? Because you think light travels in a straight line, when light diverges from a point, your brain assumes that the point it diverges from (the image point) is where the light originated. So, for a virtual image like this, your eye/brain sees an image and thinks it is behind the mirror.

c. What about a real image? Place the object at some point so that the image forms somewhere in front of the eye. Where does the eye think the light comes from? What does the eye see? (Is the image upright or inverted, bigger or smaller than the object?) What if the image point is beyond the eye? What would that look like? (Notice that for this case the light doesn't seem to have a convergence point so you'd see a blurry image.)

d. In which case is the light actually traveling through the image point? For real images, the light actually travels through the image point. If you put up a screen at the point that the rays cross, then a real image can be formed on the screen, whereas if you put up a screen at the point of a virtual image, you won't see anything on the screen (the screen is behind the mirror).

Exploration 33.3: Ray Diagrams

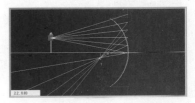

You will often use ray diagrams in order to determine where an image of an object will be, whether it will be real or virtual, and whether it will be inverted or upright. The animation shows an object arrow, a mirror, and a pink dot to show the focal point of the mirror. You can move the object using the slider (position is given in meters).

a. One point source is attached to the object in the animation. Move the object and notice where the light from the point source converges. Move the point source up and down and notice where on the image the light converges. In order to sketch a diagram of the object, in addition to the lens and the approximate position of the image, you need to know where the light from every point on the object converges. Instead of trying to draw a large number of the rays from many points on the object, we generally use three rays from the tip of the object.

b. As you move the object (with the slider) or move the point source, there is a ray that always passes through the focal point. Describe that ray. This is a ray generally included in a ray diagram.

c. Now switch to the "ray diagram" view. Describe the other two rays (compare them to the list in your textbook, if needed). As you move the object, describe what stays the same for each ray even when the object is in a different position and the image is changing position and size.

d. Move the object to a position between the focal point and the mirror. Compare the **object with point source** and **ray diagram** views.

Exploration 33.4: Focal Point and Image Point

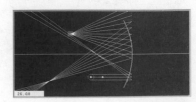

In the animation you have the option of adding both a parallel beam source and a point source of light, as well as mirrors (position is given in meters). Whenever you add a mirror, all the light sources will be cleared off the screen.

a. First, add a concave mirror and then add a parallel beam source. Where do the light rays converge? This is the focal point of the mirror.

b. Now, add a point source of light (move the parallel beam source to the right of the mirror). Where do the rays converge? Is it at the focal point?

c. Move the point source around. What happens to the point where the rays converge?

d. Now add an object. Put the object and point source at the same place. Where is the image in relation to the point where the rays converge? What is the difference between the focal point and an image point?

e. What happens when you move the object to the focal point of the mirror? Why?

f. What happens when the object is between the focal point of the mirror and the mirror itself?

g. What is the difference between images when the object is inside the focal point or outside the focal point? (If you find that the screen is too cluttered to see what is happening, you can clear the screen and add only a concave mirror and an object.)

h. Which images are real images and which ones are virtual? How can you tell?

i. Clear the screen and add a convex mirror and an object. Describe (and explain) the image formed.

Exploration 33.5: Convex Mirrors, Focal Point, and Radius of Curvature

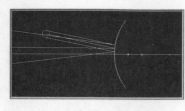

You can add a parallel beam source, a point source, and an object (position is given in meters and angle is given in degrees). How do you find the focal point of a convex mirror?

a. First, add a parallel beam source. Move it around so that one of the beams leaves the mirror parallel to the axis. This beam acts as if it came from the focal point. Why? So, to find the focal point, you need to extend the original path of this beam to the right side of the mirror. The easiest way is to use the "protractor" to click-drag an angle measure. You can move the protractor around as well as click-dragging to change the angle. If the mirror were not there, where would the original beam hit: the blue, green, red, or pink dot?

b. Now, move the parallel beam source until one of the beams bounces back on itself. This time where does an extension of the incoming beam originate: the blue, green, red, or pink dot? This is the radius of curvature of the mirror (radius of the circle that the lens would make). The radius of curvature should be twice the focal length.

c. Add a point source. How would you devise a method to determine the focal point of the mirror with a point source? Describe your method.

d. Finally, add an object and develop a method to determine the focal point with an object source.

Problems

Problem 33.1

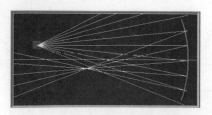

A point source is located to the left of a mirror. You can drag this point source to any position (position is given in meters and angle is given in degrees). Find the focal length of the mirror.

Problem 33.2

A beam of light is incident upon a mirror. You can click-drag both the position and angle of this beam (position is given in

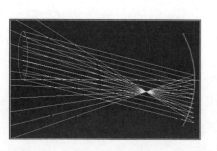

meters and angle is given in degrees). Find the focal length of the mirror.

Problem 33.3

The animation shows an object and its image (position is given in meters and angle is given in degrees). You can click-drag the top of the arrow-shaped object.

Which, if any, of the following statements are true for this mirror?

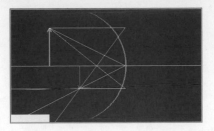

a. The magnification changes sign when the tip of the arrow is dragged below the principal axis.

b. Light rays leaving the tip of the arrow always pass through the image.

c. Real images are always inverted, while virtual images are always upright.

Problem 33.4

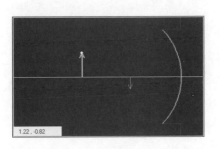

A dragable object is located to the left of a mirror. You can click-drag both the position and height of the object (position is given in meters and angle is given in degrees). Find the focal length of the mirror.

Problem 33.5

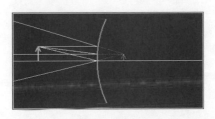

A dragable object is located to the left of a mirror (position is given in meters and angle is given in degrees). Find the focal length of the mirror.

Problem 33.6

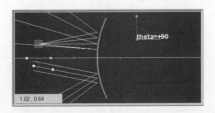

This animation has a point source, a beam source, and a mirror. You can also find an angle or create a line with an angle by moving the pink protractor and click-dragging to change the angle (position is given in meters and angle is given in degrees). Which dot is sitting at the focal point of the mirror?

Problem 33.7

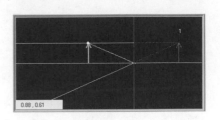

Many side mirrors on cars have a label that says "objects in mirror are closer than they appear." What type of mirror is it? The animation has an object and a mirror (position is given in meters and angle is given in degrees). You can choose the type of mirror you want to try.

a. To determine what type of mirror a side mirror on a car is, first consider (from your experience) whether the image in the mirror is upright or inverted? Does your answer change as the object changes distance from the mirror? Therefore, which type of mirror can't it be?

b. For objects to be "closer than they appear," could it mean (1) that the image distance is longer than the object distance (so you perceive the object to be farther away) and/or (2) that the image is smaller (since we judge distances by relative sizes of objects)? Therefore, which type of mirror can't it be (which mirror does not have a smaller image or a longer image distance)?

c. For the type of mirror that the car's side mirror is, then, why are "objects closer than they appear"?

Problem 33.8

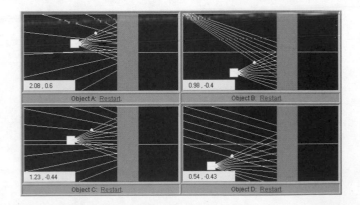

Four regions are hidden by a curtain. You can drag the source of light to any location and adjust the beam width (position is given in meters and angle is given in degrees).

a. What is behind each curtain?

b. Rank the objects in terms of their focal lengths, from smallest to greatest.

Refraction

Topics include refraction, total internal reflection, Snell's law, and dispersion.

Illustration 34.1: Huygens' Principle and Refraction

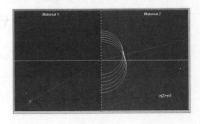

Huygens' principle states that all points on a wave front serve as point sources for secondary spherical waves that propagate outward. The position of the wave front at some later time is determined by the tangent to the surface of the secondary wave fronts. Huygens' principle can be used to predict observed optical phenomena such as refraction. Although the principle may seem strange and contrived, it is a direct consequence of the differential wave equation. This Illustration shows you Huygens' principle applied to light passing between two mediums.

The animation begins with the $n_1 = n_2$ animation. Click "play" to begin. You will see a wave front, represented by a white line, moving at an angle across the screen. Huygens' principle applies to all points along the path of the wave front. However, to make it simple, the visualization of the creation of the secondary wave fronts is only shown for the points down the center of the applet. In this case the medium on the right and left of the points is the same. As the applet plays, carefully watch as the secondary wave fronts are formed at the points in the center. Notice how the wave front, now defined as the tangent to the surface of the secondary wave fronts, is exactly the same as it was before. View the $n_1 = n_2$ animation several times until you feel comfortable with what it represents.

Now initialize and play the $n_2 > n_1$ animation. In this animation the wave front passes from one medium to another. Because $n_2 > n_1$, the waves slow down in the second medium. Carefully watch as the wave front passes from one medium to another. Since the wave fronts are traveling slower in the second medium, you see the primary wave front bend downward. This is particularly apparent if you pause the applet just as the wave front reaches the medium and then step forward as the wave front passes into the new medium.

Finally, initialize and play the $n_2 < n_1$ animation. In this case the waves speed up in the second medium and you see the wave front bend upward.

Illustration 34.2: Fiber Optics

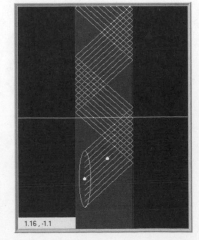

When you carry on a telephone conversation or watch cable television you are likely utilizing fiber optic technology to transmit and receive information. A fiber optic provides a less expensive, higher capacity alternative to copper wire with less signal degradation. You may think of fiber optics as being high tech, but the physics behind it is actually quite simple.

When light is incident on a medium of lower index of refraction at an angle greater than the critical angle, all of the light will be reflected. In the animation, a source of parallel rays of light is embedded in a medium of higher index of refraction than its surroundings. When the animation begins, the light rays strike the interface at an angle less than the critical angle for these two substances. Adjust the angle of the rays by clicking on the beam and then click-dragging the hotspot. At some point the angle is increased beyond the critical angle and the rays are entirely reflected back into the medium.

A fiber optic cable is a thin strand of glass surrounded by a material with an index of refraction less than glass. Light will travel through a fiber optic cable just as the light in the animation was transmitted through the blue region by reflecting off the boundary between the two materials.

Illustration 34.3: Prisms and Dispersion

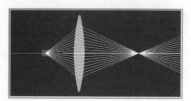

The index of refraction of a given material depends on the wavelength (or frequency) of the incoming light. Hence, the speed of light in that material also depends on the wavelength or frequency of light.

When the index of refraction of a material is given, therefore, it is really true for only one particular wavelength or color of light. This slight variation in the index of refraction leads to what are called chromatic aberrations in lenses (where the focal point is different for different colors). It is what allows for the separation of white light into colors using a prism (or drops of water). This phenomenon is called dispersion. When the speed of a wave in a particular medium is a function of frequency, the medium is dispersive. Note that for this Illustration we consider the dispersive qualities of glass ($1.6 < n < 1.68$), but air itself is also dispersive ($1.45 < n < 1.47$).

Change the wavelength of light (in air) and therefore change the color of the light entering the prism. Notice the angle at which the different colors exit the prism and the different index of refraction associated with each color.

When white light enters the prism, what happens? This is a very nice example of dispersion in glass. You see a rainbow of colors, both inside and outside of the prism, because each light ray refracts differently depending on its wavelength (or frequency). A raindrop can also refract sunlight. The result of the dispersion of light in water droplets during a passing rainstorm is often a rainbow.

Exploration 34.1: Lens and a Changing Index of Refraction

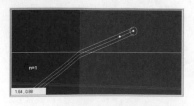

Light rays from a point source, initially in air, are shown incident on a lens.

a. How, if at all, would the path of the rays change if the source and lens were placed in another medium with an index of refraction of $n = 1.2$, which is less than the index of refraction of the lens? Make a prediction and then test your prediction by using the slider to increase the index of refraction of the surrounding medium.

b. Now, what happens if the index is increased to $n = 2.0$, so that it exceeds the index of refraction for the lens? Make a prediction and then check your prediction using the slider.

Exploration 34.2: Snell's Law and Total Internal Reflection

Light rays from a beam source, initially in air ($n = 1$), are shown incident on material with an index of refraction that you can vary by moving the slider (position is given in meters and angle is given in degrees). You can move the beam source and change the angle of the light from the source by clicking on the beam and click-dragging the hotspot.

a. Verify that Snell's law holds. Measure the incident angle and refracted angle. You can use the pink protractor to measure angles. You can drag the protractor around and click-drag to adjust the angle. Calculate the value of the index of refraction of the material. Theoretically, what is the maximum angle of incidence (the animation limits the angle of incidence to 45°, but that is not the maximum)? Given the maximum incidence angle, what is the maximum angle of refraction? This angle is sometimes called the critical angle. Develop a general expression for the critical angle as a function of the indices of refraction of the two materials.

b. Move the light source inside the material and change the beam so it leaves the blue material and goes into the air (black). Measure the angles of incidence and refraction and calculate the index of refraction of the material. What happens if the angle of incidence (from inside the material) is greater than the critical angle of refraction found in (a) above? Why? This is called total internal reflection.

c. Change the index of refraction. Calculate the new critical angle. Measure the critical angle and compare it with your calculated value.

d. Why is it only possible to have total internal reflection when light travels from a medium of higher index of refraction to one of lower index of refraction?

Exploration 34.3: Toward Building a Lens

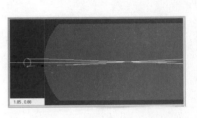

Light rays from a beam source, initially in air, are incident on a material of different index of refraction (position is given in meters). You can change the curvature of the surface of the material as well as the index of refraction.

a. Move the slider to decrease the curvature of the blue material. What happens when the edge is curved more (the radius gets smaller)? When the curvature is 1, where is the point at which all the rays converge (a focal point)?

b. Increase the index of refraction. When the curvature is 1, where is the point at which all the rays converge? If the index of refraction is 1, what happens? Why?

c. Mathematically, the relationship between this focal point inside the curved material, the curvature of the surface, and the radius of curvature of the surface is given by $f = nR/(n - 1)$. Verify this expression with the animation.

How a surface focuses light, then, depends both on the index of refraction as well as on the curvature of the material.

Exploration 34.4: Fermat's Principle and Snell's Law

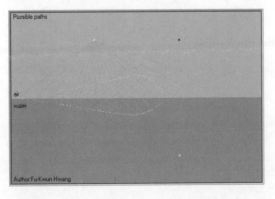

This animation demonstrates Fermat's principle: Light travels along the path that takes the shortest time. You can click–drag the source (white dot) and the end-points (reflected light, blue, and refracted light, green). The animation will show you the possible paths for the light to take. The white path is the path that takes the shortest time. You can also click on the words at the interface ("air/water") to switch between a light source in air or one in water. Notice that for the reflected light, the angle of incidence equals the angle of reflection. Once the path is completed, you can click in the animation to show the angles for the angles of incidence and refraction for the refracted light.

a. Verify that the angles obey Snell's law.

b. Click on the upper left-hand corner (on the words "Possible paths") to switch to "Real paths." (If you click on the words again, it will switch back to the "Possible paths.") What does the animation show in this mode, and how is it different from the "Possible paths" mode?

(Calculus required): Using the diagram below (and the hints that follow), prove that you can derive Snell's law using Fermat's principle.

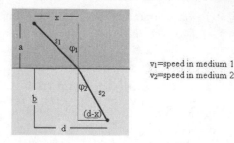

v_1=speed in medium 1
v_2=speed in medium 2

c. Since the time for the light to travel through the two media (along any arbitrary path) is $t = s_1/v_1 + s_2/v_2$, show that you can rewrite the time as

$$t = \frac{\sqrt{a^2 + x^2}}{v_1} + \frac{\sqrt{b^2 + (d-x)^2}}{v_2}$$

d. To find the path that minimizes the travel time between the two points, solve for $dt/dx = 0$. Why?

e. When you solve for $dt/dx = 0$, show that you get

$$\frac{x}{v_1\sqrt{a^2 + x^2}} = \frac{d-x}{v_2\sqrt{b^2 + (d-x)^2}}$$

f. Show (make the necessary substitutions) that this is the same as Snell's law, $n_1 \sin\theta_1 = n_2 \sin\theta_2$.

Exploration 34.5: Index of Refraction and Wavelength

Light rays from a beam source, initially in air, are shown incident on a sphere of water. You can change the wavelength of light by moving the slider.

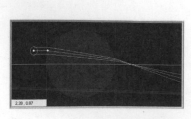

a. Move the slider to change the color of the light (by changing the frequency). As you move the slider to the right, does the frequency of light increase or decrease (look up the frequencies of different colors in your book if you need to)?

b. Where does the red light converge? Where does the blue light converge?

c. If the index of refraction of the circular blue region was 1, where would the point of convergence be? Therefore, explain why a higher index of refraction means a convergence point closer to the sphere.

d. For which color light, then, is the index of refraction higher? For which color is it lower?

Problems

Problem 34.1

A beam of parallel rays is shown passing through a medium surrounded by air. Both the position of the source and the angle of the source rays can be adjusted by click-dragging the circular hotspots. You can move the pink protractor around and use it to measure angles (position is given in meters and angle is given in degrees).

a. What is the index of refraction of the inner medium?

b. What is the angle of total internal reflection for the inner medium?

Problem 34.2

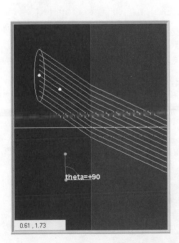

A light beam source, in air, is incident on a substance of unknown index of refraction. You can click-drag both the position and the ray angle of the beam. You can move the pink protractor around and use it to measure angles (position is given in meters and angle is given in degrees). What is the unknown substance?

Problem 34.3

A source of light is shown in a medium. Angles can be measured by click-dragging the movable green protractor (position

is given in meters and angle is given in degrees). Where is the image located as seen by someone looking into the substance from the right?

Problem 34.4

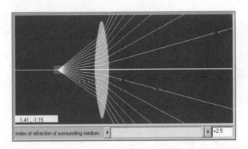

What is the index of refraction of the lens?

Problem 34.5

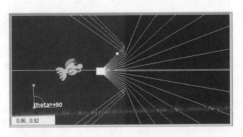

A fish in a tank is shown. Both the position of the source and the angle of the source rays can be adjusted by click-dragging the circular hotspots. You can use the movable green protractor and click-drag to measure angles (position is given in centimeters and angle is given in degrees).

a. When the fish's eye is 1 cm from the edge, how big is the circle through which the fish sees the world outside the fish tank when he looks straight ahead (to the right in the animation)?

b. Find an expression for the size of the circle through which the fish sees the outside world as a function of how far away the fish is from the right edge of the tank.

Problem 34.6

Rank the media from smallest index of refraction to largest. You can change the angle of the beam by clicking on the beam

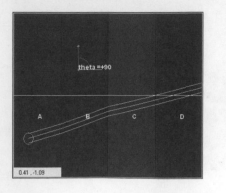

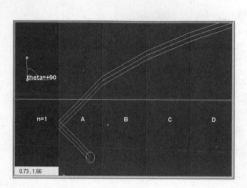

source and then click-dragging its hotspot. You can use the movable pink protractor and click-drag to measure angles (angle is given in degrees).

Problem 34.7

Four materials are next to each other, and the change in the index of refraction from one to the next is the same. (In other words, if the index of refraction of region A is 1.5 AND the index of refraction of region B is 2, then the index of refraction in region C is 2.5, and so forth.) You can move the beam source, but you cannot change the angle of the light. You can use the movable green protractor and click-drag to measure angles (angle is given in degrees).

What is the index of refraction of region D? Explain your observations as you drag the source through each of the different materials. (Why is there total internal reflection between some interfaces, but not others?)

Problem 34.8

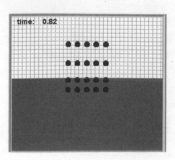

The animation models wave fronts in air entering a region of unknown index of refraction. Assume the wave fronts are traveling at the speed of light (in air) before entering the unknown region (position is given in units normalized to the speed of light in air).

a. Why are the wave fronts closer together in the unknown medium?

b. What is the index of refraction of the unknown medium?

c. In terms of the speed of light, how fast are the wave fronts traveling inside the medium?

Problem 34.9

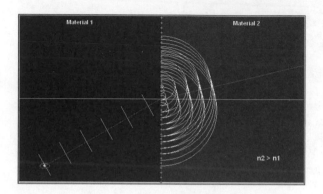

The animation shows a wave front passing from one medium to another. Huygens' principle is applied at the boundary between the mediums.

a. What happens to the wavelength for the refracted light wave when $n_2 > n_1$?

b. What happens to the wavelength for the refracted light wave when $n_2 = n_1$?

c. What happens to the wavelength for the refracted light wave when $n_2 < n_1$?

d. What happens to the frequency of the refracted light wave in each of the given situations?

Problem 34.10

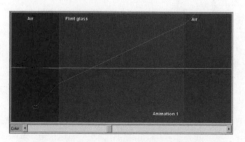

Light rays from a beam source, initially in air, are shown incident on a piece of glass. You can change the wavelength of light by moving the slider. Which animation is correct? Explain.

Lenses

Topics include lenses, ray diagrams, focal point, real and virtual images, and the lens maker's equation.

Illustration 35.1: Lenses and the Thin-Lens Approximation

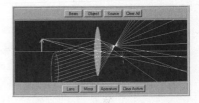

Shown is an optics bench that allows you to add various optical elements (lens, mirror, and aperture) and light sources (beam, object, point source) and see their effect. Elements and sources can be added to the optics bench by clicking on the appropriate button and then clicking inside the animation at the desired location. Moving the mouse around in the animation gives you the position of the mouse, while a click-drag will allow you to measure angle (position is given in centimeters and angle is given in degrees).

Add a lens to the optics bench by clicking on the "Lens" button and then clicking inside the animation to place the lens. Adjust the focal point of the lens by dragging on the round hotspots. Notice that you can make the lens either converging or diverging. Make the lens converging with a focal length of 1 cm and place it in the middle of the animation ($x = 2.5$ cm). Now add a source of light by clicking on the "Object" button and then clicking inside the animation. Place the object at $x = 0.1$ cm and give it a height of 0.5 cm. You can add other sources of light later.

Notice the rays emanating from the object, their refraction (bending), and the resulting image. Click the head of the arrow (the object) and move it around. First note that there are three rays that emanate from the head of the arrow. One ray comes off parallel to the principal axis (the yellow centerline) and is bent at the vertical centerline of the lens and then travels through the focal point on the far side of the lens, one ray comes off at an angle to hit the vertical centerline of the lens on the principal axis and is unrefracted, and one ray passes through the principal axis at the near focal point of the lens and is bent at the vertical centerline of the lens and comes off parallel to the principal axis.

Do the rays always behave as you expect them to? Probably not. As you drag the head of the source and change its height and position, what do you notice about the rays when they refract through the lens? The rays refract from the vertical centerline of the lens. If you click on the lens, you will see this line in blue. This animation uses what is called the thin-lens approximation. This approximation assumes that the lens is thin in comparison to its radius of curvature. In fact, in the thin-lens approximation, we take the thickness of the lens to be zero (this is why the refraction takes place at the vertical centerline of the lens). In a real lens, the rays from the object would form a real image to the right of the lens, but also a virtual image behind the object on the left of the lens.

The optics bench allows you to try many different configurations to see how light will interact with a lens. Take some time to play with the animation. You may also find it helpful to refer back to this Illustration as you develop your understanding of optics. A brief description of the three sources is given below.

- The "Beam" button adds a beam of parallel light rays. The angle of the light rays can be changed by dragging the hotspot after clicking on the beam.

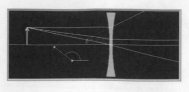

- The "Object" button adds an arrow as an object. A ray diagram is drawn for the object if an optical element is present.
- The "Source" button adds a point source of light. The spread of the light rays can be adjusted by dragging the hotspot after clicking on the source.

Illustration 35.2: Image from a Diverging Lens

An object (the arrow) is in front of a diverging lens. You can drag the lens around, but the object is fixed. Moving the mouse around in the animation gives you the position of the mouse, while a click-drag will allow you to measure angle (position is given in centimeters and angle is given in degrees). Notice the rays emanating from the object, their refraction (bending), and the resulting image. In this case the image is virtual.

To determine where the image would be located and that it is indeed virtual, place the lens at $x = 2$ cm. Note that there are three rays that emanate from the head of the arrow.

- One ray comes off parallel to the principal axis (the yellow centerline), is bent at the vertical centerline of the lens, and then travels to the right on a line that, if continued backward to the left, would pass through the focal point on the near side of the lens.
- One ray comes off at an angle to hit the vertical centerline of the lens on the principal axis and is unrefracted.
- One ray, which if continued to the right would pass through the focal point on the far side of the lens, is bent at the vertical centerline of the lens and comes off parallel to the principal axis.

Given these statements, can you determine the focal length of the lens? On one (or both) of the lines that if continued would pass through a focal point, click-drag the center circle of the compass tool into position over one of those lines. Now drag the green line until it is parallel to the ray and passes through the principal axis (the yellow centerline) so that the animation looks like the following image.

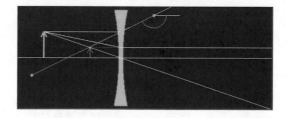

If the lens is at $x = 2$ cm, you should get the position at which the green line crosses the principal axis to be $x = 1$ cm or $x = 3$ cm, depending on which ray you choose. In either case, since the lens is at $x = 2$ cm, the focal length of the lens is 1 cm.

How do we determine where the image will be located? Notice that the three rays diverge. The dotted lines continue the three rays on the right of the lens backward on the left of the lens to a point where all three dotted lines converge. This is the location of the image. It is a virtual image: A screen placed at the image location would not be illuminated by the image of the object.

Exploration 35.1: Image Formation

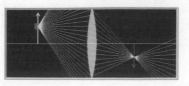

An object is placed in front of a converging lens outside of the focal point.

a. Draw a ray diagram to locate the image. You can check your answer below.

b. <u>View object and image</u>. Now consider a point at the top of the object. Light must leave this point and travel in all directions (otherwise everyone in a room would not be able to view an object at the same time). Draw the rays that leave the top of the object and travel through the lens. Once you have your drawing, check your answer by clicking on the link below. Were you correct? If not, why?

c. Initialize part (c) and then move the light source to different points on the object. As you drag the point source up and down, notice that the rays from one point on the object all converge at the same point on the image. Are all the rays leaving a point on the object blocked if half of the lens is cut off? **Add a screen**. How would the image appear if the top half of the lens were blocked?

Check (a) Check (b)

Exploration 35.2: Ray Diagrams

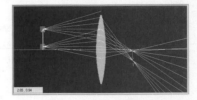

You will often use ray diagrams in order to determine where an image of an object will be, whether it will be real or virtual, and whether it will be inverted or upright (position is given in meters). The animation shows an object arrow, a lens, and pink dots to show the focal point of the lens.

a. Two point sources are attached to the object in the animation. Move the object and notice where the rays from the point sources converge. In order to sketch a diagram of the object, the lens, and the approximate position of the image, you need to know where the light from these sources (from the object) converges. As you move the object around, what do you notice about the rays that are parallel to the principal axis (either before entering the lens or after leaving the lens)? Why do they always cross the axis at the same place?

b. Instead of trying to draw a large number of the rays from many points on the object, we generally use three rays from the tip of the object (sometimes called principal rays) to sketch a ray diagram. Change to the "ray diagram" view. Describe the three rays (compare them to the list in your textbook, if needed). Which one goes from the object through the lens and then through the focal point? Which one seems to be undeflected as it goes through the lens? Which one goes through the focal point (on the object side) and then through the lens?

c. Look at a diverging lens with a point source. Try sketching a ray diagram for a diverging lens. Check it by looking at the ray diagram.

Exploration 35.3: Moving a Lens

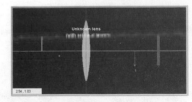

In this animation the lens is movable, but the object is not (position is given in meters). Initially, you have a lens of an unknown focal length (that you cannot adjust using the slider).

a. What are the object and image distances for the lens? Find the focal length of the lens.

b. There is another spot where you can put the lens that will give an image at the same position (on the blue screen). Move the lens until an image appears at the same spot (on the blue screen). What are the object and image distances this time?

c. For a given distance between an object and a screen, develop an equation for the two spots where you can place a lens to get a clear image on the screen. Verify your expression for a **lens with an adjustable focal length** (use the slider to change the focal length). Note that when you click or drag this lens, the focal length (f.l.) appears on the screen.

Exploration 35.4: What is Behind the Curtain?

A dragable source of light is shown along with an optical element that has been hidden behind a pink curtain (position is given in meters).

a. What is behind the curtain? Do not read the rest of the question until you have answered this part.

b. AFTER you have determined what is behind the curtain, **remove the curtain** to see if you were right. Surprised?

c. Most students predict there is a converging lens with a focal length around 1 m behind the curtain. Without removing the curtain, could you have known this prediction was incorrect?

d. Find the focal length of the individual lenses. Remember that when an object sits at the focal point of a lens, the light exits the lens parallel to the axis, and when light enters a lens parallel, it converges at the focal point. So, move the object to the focal point of the first lens, and the light should converge at the focal point of the second lens.

Exploration 35.5: Lens Maker's Equation

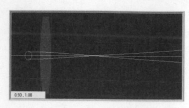

Light rays from a beam source, initially in air, are incident on a material of different index of refraction (position is given in centimeters). You can change the curvature of the surface of the material as well as the index of refraction.

a. Build a plano-convex lens. Decrease the radius of curvature of the left side while keeping the right at 30 cm. As you decrease the radius of curvature, what happens to the beam? When the curvature of the left side is 1 cm, where is the point at which all the rays converge? How far is the point where the rays converge from the center of the "lens" you are making? This is the focal point of the lens.

b. What happens if you keep the left side essentially flat (radius = 30 cm) and decrease the radius of curvature of the right side? What is the focal point when this radius is 1 cm? What happens to the focal point if you increase the index of refraction of the material? What happens if you decrease it?

c. Build a double-convex lens. Decrease the radius of curvature of both sides of the lens. What is the focal point when the radius of curvature is 1 cm for both sides? How does the focal point change with a different index of refraction?

d. Analytically, the focal length is described by the lens maker's equation: $1/f = (n - 1)(1/R_1 + 1/R_2)$, where R_1 and R_2 are the radii of curvature, f is the focal length, and n is the index of refraction. Verify that your earlier measurements are consistent with this equation.

e. For lenses made from glass $(n = 1.5)$, show that the radius of curvature of a double-convex lens (where the radii of both sides is the same) is equal to the focal length.

Problems

Problem 35.1

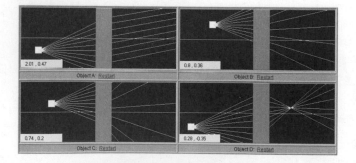

Four regions are hidden by curtains. You can drag the source of light to any location (position is given in meters).

a. What is behind each curtain?

b. Rank the objects in terms of their focal lengths from smallest to largest (most negative to most positive).

Problem 35.2

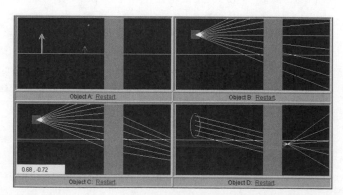

Four regions are hidden by curtains. The light sources are dragable. The arrow represents an object, and the fainter arrow represents its image (position is given in meters).

a. What is behind each curtain?

b. Rank the objects in terms of their focal lengths, from smallest to largest (most negative to most positive).

Problem 35.3

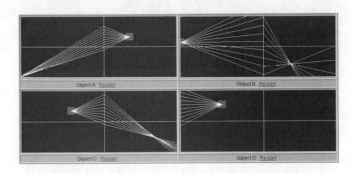

Four lenses are represented by vertical lines. You can drag the light sources to any location (position is given in meters). Rank the lenses in terms of their focal lengths, from smallest to largest (most negative to most positive).

Problem 35.4

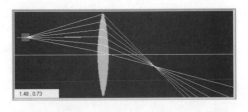

A lens is shown with a light source. The light source is constrained to move along the line $y = 0.4$ m (position is given in meters). What is the focal length of the lens?

Problem 35.5

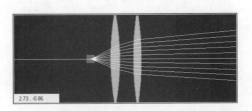

A dragable light source is shown with two lenses (position is given in meters).

a. What is the focal length of the lens system?

b. What is the focal length of each individual lens?

Problem 35.6

A lens is shown with a source of parallel light rays. You can drag the source and change the orientation of its rays by dragging its hotspots (position is given in meters).

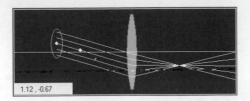

a. If an object were placed at $x = 0$ m, where would the image of that object be located?

b. Would the image be upright or inverted?

Problem 35.7

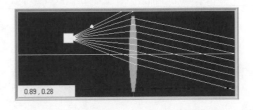

A lens is shown with a point source of light. You can drag the source and change the orientation of its rays by dragging its hotspots (position is given in meters).

a. If an object were placed at $x = 0$ m, where would the image of that object be located?

b. Would the image be upright or inverted?

Problem 35.8

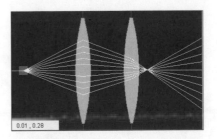

A light source is located to the left of two lenses (position is given in centimeters). You can click-drag either lens through a limited range of x positions. Find the focal length of each lens.

Problem 35.9

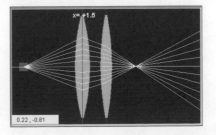

A light source is located to the left of two lenses (position is given in centimeters). You can click-drag the left lens through a limited range of x positions. The right lens will move in tandem. Find the focal length of each lens.

Problem 35.10

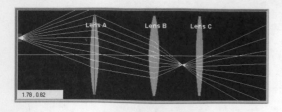

Three lenses are shown with a point source of light. You can drag the source and lenses with some restrictions. (Position is given in meters.) Find the focal length of each lens.

Problem 35.11

Light rays from a point source, initially in air, are incident on a lens made from a material with an index of refraction of 2.5 (position is given in centimeters). The radius of curvature of the left side of the lens is −10 cm (10-cm radius and concave, so rays diverge upon entering it). You can change the radius of

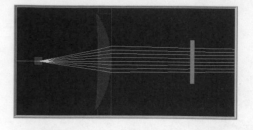

the right side. Build a lens so that the light rays from the point source will converge on the green screen.

a. What is the focal length of this lens?

b. Calculate the radius of curvature of the right side of the lens, then test your calculation.

Optical Applications

Topics include lenses, mirrors, the eye, cameras, microscopes, telescopes, and laser cavities.

Illustration 36.1: The Human Eye

The animation depicts a simplified model of the eye in which the front of the eye is a single converging lens (position is given in arbitrary units and angle is given in degrees).

Initialize the healthy eye and then add a **far source** of light. Notice how the parallel rays of light from the faraway source converge at the back of the eye on the retina. The retina is to the eye what film is to a camera. The retina is made up of nerves that convert the light energy into an electrical signal that is sent to the brain. So in order for an object to be "seen," its image must be FOCUSED on the back of the retina.

Now remove the **far source** and add a **near source**. Notice that the light from the nearby source is focused behind the retina. In this case the person would see a blurry image. As evolution would have it, our eyes have the ability to accommodate. You can change the focal length of your eye by using the muscles of your eye to change the curvature of the lens. Try looking at a faraway object and then at something close by, such as your finger. You will feel the muscles in your eye respond as you change your focus. In the animation, accommodation is accomplished by using the slider at the bottom to vary the focal length of the lens. Now vary the focal length of the lens, using the slider, until the image of the light source is focused on the retina.

People with normal vision focus on faraway objects with their eyes relaxed. Notice that the far source in the animation was focused when the focal length was at its maximum, one unit. As you use your muscles to accommodate, you shorten the focal length of your eye.

Put your finger in front of your eyes about an arm's length away. You should be able to see a clear image of your finger. Now slowly bring your finger toward you. At some point, you will no longer be able to focus on your finger and it will become blurry. This is your near point. It is the closest distance at which you can focus on an object. If you have not already done so, initialize a healthy eye with a near source of light focused on the retina. Now move the source of light toward the eye. At some point you will no longer be able to accommodate (using the slider) to focus the source. That is the near point for the eye in the animation. Notice that the eye in the animation is not to scale relative to a real eye. If we had made it to scale you would need a much larger computer screen.

The far point is just like the near point, except it is the farthest point an eye can focus on. For people with normal vision, the far point is at infinity.

Initialize the nearsighted eye and add a **far source**. Notice that the light does not focus on the retina when the eye is relaxed. Instead, it focuses in front of the retina. Use the slider to try to focus the light. Notice that accommodation does not help in this situation. Now remove the **far source** and add a **near source**. Notice that the nearsighted person has no trouble focusing on the nearby source. A person who is nearsighted can clearly see near objects but not faraway objects.

Now **initialize the farsighted eye** and investigate it as you did with the nearsighted eye. Notice that a farsighted person can see faraway objects but has difficulty focusing on nearby objects.

Initialize a nearsighted eye with a **far source**. Unaided, this eye cannot focus on the far source. Now add an eyeglass lens. Notice that you can change the focal length (power) of the eyeglass lens by clicking on it and then dragging on the hotspots. You can make the lens either converging or diverging.

Since light is focused in front of the retina in a nearsighted eye, nearsightedness is corrected using a diverging lens. Can you find the correct focal length to correct this eye? In the same way, farsightedness is corrected using a converging lens.

Illustration 36.2: Camera

This animation can be used to demonstrate the basic operation of a camera (position is given in arbitrary units and angle is given in degrees). Various lenses and light sources can be added by clicking on the appropriate links.

Initialize a **normal lens** and a **near source**. The camera is "focused" by dragging the lens to change the lens-to-film distance until the rays from the source all converge on the film. An object at the point of the source will be in focus on the film with this film-to-lens separation. Now add an **object source**. Notice that when it is focused, the image falls directly on the film.

This Illustration models a camera with one lens. A camera is actually comprised of several lenses that work together as a unit. Multiple lenses are necessary to correct for aberrations. For example, the bending of light by a lens is actually somewhat dependent on the color of the light. This property of nature leads to chromatic aberration (misalignment of the colors in an image), which is corrected by using several carefully chosen lenses.

Illustration 36.3: Laser Cavity

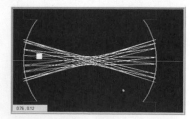

Under some circumstances, an atom in an excited state can be stimulated to drop to a lower energy state when hit by a photon (particle of light). When the atom drops to the lower energy state, a photon identical to the incident photon is released. If nearby atoms are also in an excited state, a chain reaction will be set off, with released photons going on to stimulate the release of even more photons. All of the photons will be identical, meaning they will have the same wavelength, phase, polarization, and direction of travel. If the chain reaction can be maintained, a beam of laser light is obtained.

It is crucial to the operation of a laser that emitted photons are retained to stimulate more emissions. The purpose of the laser cavity (or resonant cavity) is to confine the emitted photons. The laser cavity consists of two mirrors. One of the mirrors is highly reflective and the other is partly reflective. The one that is partly reflective will allow some of the produced laser light to pass, which is the source of the laser beam, and will reflect the rest to maintain the chain reaction.

A model of a laser cavity is demonstrated in the animation. Light is reflected off of two mirrors. If the circumstances are right, the cavity has stability. That is, light will reflect off the mirrors in such a way that it is confined within the cavity. When the animation is initially loaded, the cavity is stable.

- Click on the mirror on the right and drag it to increase the separation of the mirrors. At what point does the cavity become unstable?

- You can change the focal length of the mirrors by dragging the focal point when the mirror is selected (click on it). Drag the mirrors so that a stable condition exists. What happens to the stability when the focal length of one mirror is increased? What about when the focal length is decreased?

Exploration 36.1: Camera

This animation can be used to demonstrate the basic operation of a camera (position is given in arbitrary units and angle is given in degrees). Various lenses and light sources can be added by clicking on the appropriate links. The camera is "focused" by dragging the lens to change the lens-to-film distance.

a. Click on the link for "Normal Lens" and add a near source. What is the closest position an object could be and still be focused on the film?

b. Remove the near source and add an object source. When the object source is at its original location ($x = 2.3$, $h = 1.2$), where must a normal lens be placed to focus this object? Where should a telephoto lens be placed? Where should a wide-angle lens be placed?

c. Rank the height of the images from (b), from smallest to largest.

d. Based on your answer for (c), which lens would you use if you wanted to take a picture in which the object took up most of the photographed area (zoom in)? Explain.

e. Based on your answer for (c), which lens would you use if you wanted to take a picture of the object and much of its surroundings (zoom out)? Explain.

f. Rank, from smallest to largest, the focal lengths of the three lenses.

Exploration 36.2: Telescope

In order to understand the magnification of a telescope, it is helpful to understand the idea of the angle that an object or image subtends. So, before exploring a telescope, we need to understand the idea of "the angle that an object subtends."

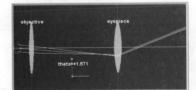

a. If an eye is located where the image of the eye is (position is given in centimeters and angle is given in radians), the object subtends an angle of 6°. You should check this by moving the protractor along the optical axis, putting its vertex at the front of the eye, and measuring the angle a light ray from the top of the arrow would make with the optical axis (remember that your protractor measures in radians).

b. Now, **add a simple magnifying glass**. The angle that the image subtends is the angle at which the light exits the lens crossing the optical axis. What angle does the image subtend? The magnification is the ratio of the height of the image to the height of the object, but with small enough angles, it is also the ratio of the angles the image and object subtend.

c. Two lenses, an eyepiece and an objective, are used to make a **telescope**. What good is the telescope if it takes essentially parallel light (from essentially infinity) and turns it back into essentially parallel light (for the relaxed eye)? The answer is in the difference in the angle the object and image subtend. Consider the angle that the object (very far away) subtends. Measure the angle between the beams from infinity and the optical axis. What is that angle?

d. Now, measure the angle the light exiting the eyepiece makes with the optical axis (the angle the image subtends). What is that angle?

e. The ratio of these angles is the magnification. What is the magnification of this telescope?

Problems

Problem 36.1

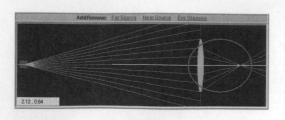

This animation models the eye (position is given in arbitrary units and angle is given in degrees). The slider can be used to simulate the eye's ability to accommodate. When the eye is in a relaxed state, the focal length of the lens system of the eye is taken to be one. Notice that as the eye accommodates, focusing on nearby objects, the focal length of the eye decreases. The location of the near source can be changed by dragging the source. The focal length of the eyeglass can be altered by clicking on the eyeglass and then dragging the hotspots.

a. Is the eye represented in the animation normal, nearsighted, or farsighted?

b. If the eye is nearsighted, what power eyeglasses should be used for normal vision? (The eye can see faraway objects when relaxed and also focus on an object at $x = 2.0$.)

c. If the eye is farsighted, what power eyeglasses should be used to allow the person to see an object at $x = 2.0$? Would a farsighted person with these glasses still be able to focus on a faraway object? Why are farsighted individuals often prescribed bifocals?

Problem 36.2

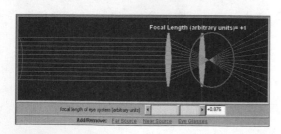

This animation models the eye (position is given in arbitrary units and angle is given in degrees). The slider can be used to simulate the eye's ability to accommodate. When the eye is in a relaxed state, the focal length of the lens system of the eye is taken to be one. Notice that as the eye accommodates, focusing on nearby objects, the focal length of the eye decreases. The location of the near source can be changed by dragging the source. The focal length of the eyeglass can be altered by clicking on the eyeglass and then dragging the hotspots.

a. Is the eye represented in the animation normal, nearsighted, or farsighted?

b. If the eye is nearsighted, what power eyeglasses should be used for normal vision? (The eye can see faraway objects when relaxed and also focus on an object at $x = 2.0$.)

c. If the eye is farsighted, what power eyeglasses should be used to allow the person to see an object at $x = 2.0$? Would a farsighted person with these glasses still be able to focus on a faraway object? Why are farsighted individuals often prescribed bifocals?

Problem 36.3

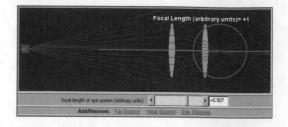

This animation models the eye (position is given in arbitrary units and angle is given in degrees). The slider can be used to simulate the eye's ability to accommodate. When the eye is in a relaxed state, the focal length of the lens system of the eye is taken to be one. Notice that as the eye accommodates, focusing on nearby objects, the focal length of the eye decreases. The location of the near source can be changed by dragging the source. The focal length of the eyeglass can be altered by clicking on the eyeglass and then dragging the hotspots.

a. Is the eye represented in the animation normal, nearsighted, or farsighted?

b. If the eye is nearsighted, what power eyeglasses should be used for normal vision? (The eye can see faraway objects when relaxed, and also focus on an object at $x = 2.0$.)

c. If the eye is farsighted, what power eyeglasses should be used to allow the person to see an object at $x = 2.0$? Would a farsighted person with these glasses still be able to focus on a faraway object? Why are farsighted individuals often prescribed bifocals?

Problem 36.4

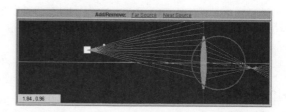

This animation models the eye (position is given in arbitrary units and angle is given in degrees). The slider can be used to simulate the eye's ability to accommodate. When the eye is in a relaxed state, the focal length of the lens system of the eye is taken to be one. Notice that as the eye accommodates, focusing on nearby objects, the focal length of the eye decreases. The location of the near source can be changed by dragging the source. What is the location of the near point of the eye?

Problem 36.5

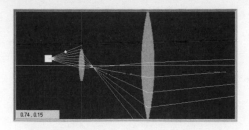

Two lenses, an eyepiece, and an objective are used to make a microscope (position is given in centimeters and angle is given in degrees). You may focus the microscope by click-dragging the object into position. Where should the object be placed for optimal viewing by a relaxed eye?

Interference

Topics include interference, double slits, and thin films.

Illustration 37.1: Ripple Tank

This applet calculates seven frames and then runs continuously. For a large number of sources, or for very small wavelengths, this calculation can take some time, so let the applet finish calculating all seven frames.

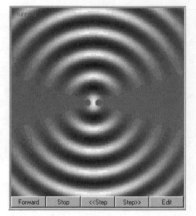

This animation depicts a ripple tank in which you can see interference effects with two or more sources. When two or more waves encounter each other, they interfere. Waves can interfere constructively, resulting in a larger amplitude at a particular point, or they can interfere destructively, resulting in a smaller amplitude.

Sources of waves are shown by red dots. The waves are represented in black and white. The location of the sources can be changed by click-dragging on the source in the animation. New sources can be added using the text boxes and the "add source" button, and the wavelength of the sources can be changed by using the wavelength text box and the "set wavelength and play" button.

In the amplitude view (select the "show amplitude view" button) the greatest amplitude is represented by white, negative amplitudes are represented by black, and areas with zero amplitude are represented by gray. Note that in a real ripple tank, this is the view you would see.

In the intensity view, where the intensity is proportional to the square of the amplitude, the greatest magnitude of the amplitude (positive or negative) is represented by white, while black shows regions of zero amplitude. This is precisely because the intensity is related to the square of the amplitude. Note that when we are looking at light waves on a screen, this is the view that you would see. Since the energy of the wave is proportional to the square of its amplitude, we could also interpret the intensity mode as the energy mode.

There are several important features we need to understand about this animation and the two representations. First let's consider one source. Clear sources and add a source at the origin with an amplitude of 1 and 0 phase. Also set the wavelength to 1. Look at the amplitude view by clicking the button and waiting for all seven frames to load. Measure the wavelength. Obviously you should get 1. This is the distance between adjacent white or black regions. Now change the view to the intensity view and again wait for all seven frames to load. Again measure the wavelength. Again you should get 1. Did you? You may not have. In the intensity view the wavelength is not the distance between adjacent white or black regions. You need to include one more white region or black region to get the wavelength. This is because in the amplitude mode the series of white and black regions represent amplitudes of $+ - + - + - + -$, etc. However, in the intensity mode the series of white and black regions represent intensities of $+ 0 + 0 + 0 + 0$, etc. but correspond to amplitudes of $+ 0 - 0 + 0 - 0$, etc., since the intensity is related to the square of the amplitude. If this is confusing (or even if it is not), consider the pictures below and notice that they give the same wavelength as long as you realize the difference in interpretation.

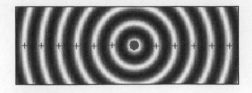 

amplitude mode: + represents an amplitude of +1

intensity mode: + represents an amplitude of +1

Now clear your source and add two sources with the same phase and amplitude 1 position unit apart at $y = 0$ ($x = -0.5$ and $x = 0.5$). Set the wavelength to 2 position units and play the animation. Notice how you get dead spots due to interference to the right and the left of the sources on the y axis. Why does this happen? Since the separation between the sources is one half of a wavelength for any position on the x axis, the two waves will always be 180° out of phase and will destructively interfere. Also note that on the y axis the two waves are equidistant from any position on the y axis and, therefore, the two waves constructively interfere (the two waves are always in phase). Hence, when we have interference, we are seeing that path difference creates a phase difference between the two waves.

You can also use the animation to further explore the properties of waves.

- What happens if the distance between the sources is increased or decreased?
- What happens if the phase between the sources is changed?
- What happens if the wavelength of the emitted waves is increased or decreased?
- What happens if more sources are added?

Illustration 37.2: Dielectric Mirrors

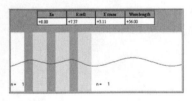

For applications that require mirrors with very high reflectance (such as a laser mirror), several dielectric layers are used to build the mirror. A typical mirror uses alternating indices of refraction to increase the reflectance to more than 98%. In the following example the mirror is made of alternating layers of zinc sulfide ($n = 2.3$) and magnesium fluoride ($n = 1.35$) film. The table shows the electric field. The intensity of the wave is $n * E * E$, which is proportional to the square of the electric field (and the energy of the wave).

When the animation is in its initial state, the incident light encounters only empty space. The data table shows you that, of the incident light (represented by E), all is transmitted ($E_{trans} = E$) and none is reflected ($E_{ref} = 0$). This is an obvious result because there is nothing to reflect from. Now add a layer of film by clicking on "add film." A layer of zinc sulfide and a layer of magnesium fluoride are added. What happens to the intensity of the transmitted and reflected light?

Now add several more layers. Notice that as each layer is added, more light is reflected and less escapes. By building a substance of alternating layers of carefully chosen materials you have constructed a mirror!

Exploration 37.1: Varying Numbers and Orientations of Sources

Two sources of light waves of equal frequency and amplitude are shown. In the amplitude view the greatest amplitude is represented by white, negative amplitudes are represented by black, and areas with zero amplitude are represented by gray. In the intensity view the greatest magnitude of the amplitude (positive or negative) is represented by white, while black shows regions of zero amplitude (position is given in nanometers).

a. What would the pattern look like (in both the amplitude and intensity views) if one source were removed? Answer: **view with one source**.

b. What is the wavelength? (Check both the amplitude and intensity views.)

c. In which view do you measure the wavelength by measuring the distance from the middle of the white band to the middle of the adjacent white band, and in which view do you have to measure the distance across two white bands (or black bands)? Why?

d. What would the pattern look like if the two sources were in phase with each other but rotated 90° to lie on the *x* axis? Answer: **rotate sources**.

e. Explain why the pattern looks the way it does.

Exploration 37.2: Changing the Separation Between Sources

Two sources of light waves of equal frequency and amplitude are shown. The magnitude of the electric field is represented by the light and dark areas. The lighter the spot, the greater the magnitude of the electric field at that spot (position is given in nanometers).

Begin with the **0.5 wavelength separation** animation. The sources are separated by one half the wavelength of the light.

a. Predict what pattern would be seen if the source separation was increased to one wavelength. AFTER you have made your prediction and written down your reasoning, check to see if you were correct. If you were incorrect, reexamine your reasoning by looking at the **one wavelength separation** animation.

b. When you feel confident in your understanding, test it by predicting the pattern if the source separation is 1.5 wavelengths. Check your prediction with the **1.5 wavelength separation** animation.

c. As a final test, predict the pattern for separations of 2 and 2.5 wavelengths. Check your prediction with the **two wavelength separation** and **2.5 wavelength separation** animations.

d. If a screen is placed on the right-hand side of the viewing window, how would the interference pattern change as the distance between the sources is increased?

Problems

Problem 37.1

Two sources of light waves of equal frequency and amplitude are shown. The magnitude of the electric field is represented by

the light and dark areas. The lighter the spot, the greater the magnitude of the electric field at that spot. A blue screen is shown on the right-hand side of the animation (position is given in arbitrary units).

a. What illumination pattern appears on the screen?

b. Does the appearance of the screen change with time?

Problem 37.2

Two sources of light waves of equal frequency and amplitude are shown. The magnitude of the electric field is represented by the light and dark areas. The lighter the spot, the greater the magnitude of the electric field at that spot (position is given in arbitrary units). At the points indicated (A, B, C, D), is the interference of the waves from the two

sources maximally constructive, completely destructive, or somewhere in between?

Problem 37.3

Two sources of light waves of equal frequency and amplitude are shown. The magnitude of the electric field is represented by the light and dark areas. The lighter the spot, the greater the magnitude of the electric field (could be positive or negative amplitude) at that spot (position is given in nanometers). Are the two sources in phase or 180° out of phase? Explain.

Problem 37.4

Two sources of light waves that are in phase, of equal frequency, and have the same amplitude are underneath the red rectangle. The magnitude of the electric field is represented by the light and dark areas. The lighter the spot, the greater the magnitude of the electric field at that spot (position is given in

arbitrary units). How far apart are the two sources in terms of their wavelength?

Problem 37.5

Two sources of light waves of equal frequency and amplitude are shown. The magnitude of the electric field is represented by the light and dark areas. The lighter the spot, the greater the magnitude of the electric field at that spot (position is given in nanometers).

a. What wave pattern would result if the phase of the bottom source was increased by 180°? Explain.

b. How would the pattern be different if, instead, the top source was changed?

Problem 37.6

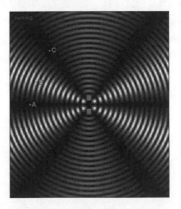

Two sources of light waves of equal frequency and amplitude are shown. The magnitude of the electric field is represented by the light and dark areas. The lighter the spot, the greater the magnitude of the electric field (could be positive or negative amplitude) at that spot (position is given in nanometers).

a. Are the sources in phase or out of phase?

b. In terms of the wavelength of the light produced by the sources, is the distance between the sources 0.5 wavelength, 1 wavelength, 1.25 wavelengths, 1.5 wavelengths, 1.75 wavelengths, or 2 wavelengths? Explain.

c. If the wavelengths of both sources were halved, would point A remain a point of complete destructive interference, change to a point of maximally constructive interference, or be somewhere in between? Would point B remain a point of maximally constructive interference? If not, how would point B change? Finally, would point C remain a point of complete destructive interference? If not, how would point B change?

Problem 37.7

The animation shows the amplitude of a wave pattern. The greatest amplitude is represented by white, negative amplitudes are represented by black, and areas with zero amplitude are represented by gray. The mouse can be used to make coordinate measurements (position is given in nanometers). A double slit is hidden underneath the red bar in the animation. What is the slit separation?

Problem 37.8

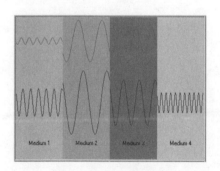

The animation shows a right-traveling light wave (shown in blue) incident on a region composed of different mediums. The left-traveling wave (shown in red above it) represents the sum of all reflections of the incident wave at the boundaries between mediums (position is given in arbitrary units). Notice that the wave amplitude varies with the medium in order to conserve energy.

a. Rank the four mediums in terms of their indices of refraction, from smallest to greatest.

b. Compare the amplitude of incident light and reflected light in medium 1 when there are **four layers** and **two layers**. Why is the amplitude of reflected light (shown in red) in medium 1 so much smaller when there are multiple layers than when there are only two layers?

Problem 37.9

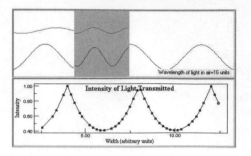

The animation shows a right-traveling light wave (shown in blue) incident on a region composed of an unknown index of refraction embedded in air (position is given in arbitrary units). The left-traveling wave (shown in red above it) represents the sum of all reflections of the incident wave at the boundaries. Notice that the wave amplitude varies with the medium in order to conserve energy.

a. Why are there numerous transmission peaks?

b. What is the index of refraction of the material?

Problem 37.10

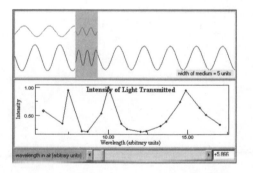

The animation shows a right-traveling light wave (shown in blue) incident on a region composed of an unknown index of refraction embedded in air (position is given in arbitrary units). The left-traveling wave (shown in red above it) represents the sum of all reflections of the incident wave at the boundaries. Notice that the wave amplitude varies with the medium in order to conserve energy.

Observe how the intensity of light transmitted through the medium changes by using the slider to change the wavelength of incident light. Note: Because the data points are connected, you must move the slider slowly to obtain a smooth curve.

a. Why are there numerous transmission peaks?

b. What is the index of refraction of the material?

c. Mathematically prove that the transmission peaks must be spaced farther apart as the wavelength increases.

Problem 37.11

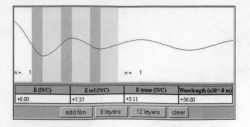

The animation demonstrates how layers of alternating materials can be used to form a mirror (position is given in arbitrary units). For more discussion of this process, see the dielectric mirror Illustration (**Illustration 37.2**) in this chapter. By itself, each material will allow light to be transmitted through it. But when layered in a particular way, the materials act as a mirror, reflecting most of the incident light.

How does such a mirror work? In other words, conceptually explain the physics behind this phenomenon.

Diffraction

Topics include diffraction, single slits, and gratings.

Illustration 38.1: Single Slit Diffraction

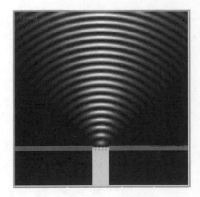

This applet calculates seven frames and then runs continuously. For a large number of sources, or for very small wavelengths, this calculation can take some time, so let the applet finish calculating all seven frames.

To model diffraction from a single slit, we can think of the light entering the slit as point sources for the light exiting (this is effectively Huygen's principle; see **Illustration 34.3**). The light from these point sources interfere with each other, and diffraction is due to this interference.

So, in the animation **small slit** you see five point sources generating the light passing through the slit and the interference pattern from the point sources (diffraction is due to the interference of the waves). Notice that the waves spread out from the slit and that the width of the light (waves) leaving the slit is wider than the slit. It looks as if light "bends" around the corner. Without diffraction, the light exiting the slit would be of the same width as the slit itself.

- Now look at light passing through a slightly **wider slit**. What is the difference in the effect of the small slit and the wider slit on the light passing through?

- If you change the wavelength of the source, the diffraction pattern also changes. Look at a source with a **longer wavelength** (represented by red color). Then observe a **shorter wavelength** (represented by blue color). What is the effect of the longer wavelength on the width of the light leaving the slit? In diffraction, as waves pass through a slit, the size of the slit and the wavelength determine how much the waves appear to "bend" around the slit.

For an example of the effect of wavelength and slit size on diffraction, think of the door to your room as a slit. If the wavelength is much smaller than the slit (visible light passing through a door, for example), there is no noticeable diffraction (you see a straight shadow of the door frame). But if the wavelength is much larger than the slit (like a sound wave), there is noticeable diffraction (sound from down the hall bends into your room. It also reflects into your room, so it is hard to separate the effects of diffraction from reflection.). If the wavelength of light were the size of the door, you would see a fuzzy shadow of the door frame.

Illustration 38.2: Application of Diffraction Gratings

This animation models a diffraction grating that is a series of parallel slits. As you change the wavelength, notice where the bright spots are. These are the spots where light traveling through different paths interferes constructively. This is due to the diffraction of the light from the different slits, which creates path differences.

The central line is from the light rays that constructively interfere at the center. The lines above and below the central lines are spots of constructive interference where light from one slit has traveled one complete wavelength farther than light from an adjacent slit. These points are called the first-order maxima. Similarly, the rays at the top and bottom of the screen, second-order maxima, are rays where the light from one slit has traveled two complete wavelengths of light farther than light from its neighbor. This model is a bit misleading because the light is dimmer for higher-order diffraction peaks, and here the brightness is the same for the two orders.

Diffraction gratings are used to study the spectrum of light from different elements. When an atom gets extra energy (is excited from its ground state), it releases energy in the form of an electromagnetic wave. The wavelength of the light released depends on the energy levels inside the atom. Each atom has its own unique, discrete light spectrum (different wavelengths of light emitted when the atom releases its extra energy). So, if the light from excited atoms goes through a diffraction grating, you can see the spectrum for that element. Look at what light from excited hydrogen atoms would look like with this diffraction grating. This is one way to determine what elements are in an unknown substance. White light sources have a continuous spectrum (like the white light spectrum). However, as the light from the interior of the Sun or other stars passes through the gas in the outer atmosphere of the star, that gas absorbs light at its own unique spectral wavelength. Hydrogen, would, for example absorb light at the wavelengths in its discrete spectrum. Looking at the light from the sun and other stars through a diffraction grating, astronomers can determine what elements are in the sun and stars by the spectral lines that are missing from the white light spectra.

Exploration 38.1: Modeling Diffraction from a Slit

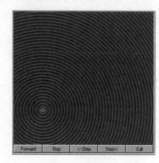

The animation simulates waves from a point source of light. Sources can be added by entering a position and clicking on "add source" and then "set wavelength and play." The position of sources can be changed by dragging in the applet (position is given in arbitrary units).

a. Use the animation to model diffraction from a slit. Turn in a screen shot showing your model along with an explanation of your model. Discuss any limitations of your model.

b. As the width of the slit opening is increased, the diffraction pattern should narrow. Confirm that your model is correct by testing this property. Turn in a screen shot of your test as evidence.

c. As the wavelength of light through the slit is decreased, the diffraction pattern should narrow. Confirm that your model is correct by testing this property. Turn in a screen shot of your test as evidence.

Exploration 38.2: Diffraction Grating

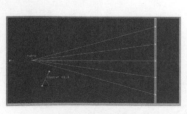

This animation models a diffraction grating that is a series of parallel slits in a material. You can change the wavelength of the light as well as the slit spacing and see the first- and second-order maxima (position is given in centimeters and angle is given in degrees).

First, consider different colors of light passing through the grating.

a. What happens when you increase the wavelength?

b. What happens when you decrease the wavelength?

c. Why do you see the results in (a) and (b)? Explain in terms of the interference between waves passing through the grating.

Now, consider the effect of the spacing between the slits in the grating.

d. What happens when you increase the number of slits per millimeter (decrease the spacing between slits)?

e. What happens when you decrease the spacing between slits?

f. Why do you see the results in (d) and (e)? Explain in terms of the interference between light waves passing through the grating.

g. Using the movable protractor, verify the relationship found in your textbook between the location of the maxima, the wavelength of the light, and the spacing between the slits.

Problems

Problem 38.1

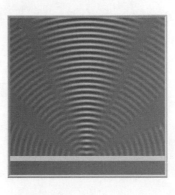

with zero amplitude are represented by gray. Determine the size of the slit.

Problem 38.4

This animation models light hitting a single slit. You can change the width of the slit and use the protractor to measure angles (position is given in centimeters and angle is given in degrees). What is the wavelength of the light?

We have made the higher-order diffraction patterns easier to see by brightening them (since the higher-order terms are very dim).

The animation shows waves from a single slit (which is underneath the red, blue, or green stripe). Rank the animations from the smallest slit to the largest slit.

Problem 38.2

The animation shows waves from a single slit (which is underneath the red, blue, or green stripe). Rank the animations from the smallest slit to the largest slit.

Problem 38.3

The animation shows the interference (diffraction) pattern from light exiting a single slit located underneath the green strip (position is given in microns, [10^{-6} meters]). In this animation the greatest amplitude of the wave is represented by white, negative amplitudes are represented by black, and areas

---

Problem 38.5

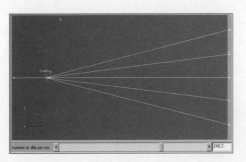

This animation models a diffraction grating that is a series of parallel slits in a material. You can change the slit spacing and

see the first- and second-order maxima (position is given in centimeters and angle is given in degrees). What is the wavelength of the light passing through the diffraction grating?

Problem 38.6

This applet models a diffraction grating that is a series of parallel slits in a material. As you change the wavelength, notice where the bright spots are. You can move the protractor around to measure angles (position is given in centimeters and angle is

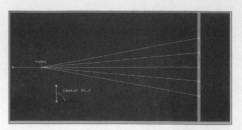

given in degrees). How many slits per millimeter does the diffraction grating have?

Polarization

Topics include polarization, linearly and circularly polarized light, and polarizers.

Illustration 39.1: Polarization

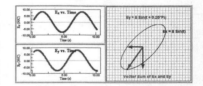

The animation shows the result of adding two perpendicular electric fields together. Each field is part of an electromagnetic wave traveling along the z axis. Each electric field is shown separately on the two graphs on the left. The graphs show the electric field at one point on the z axis for various times. On the right the animation shows both electric fields and their sum at the same point on the z axis and at the same times as the graphs on the left. It is as if you are looking down the z axis at the electric field. You can change the electric fields and the phase difference between the two fields and see the resulting waves.

The direction of polarization for an electromagnetic wave is described by the direction in which the electric field points. In Chapter 32 (Electromagnetic Waves) the electric field was always along either the x or the y axis (usually the x axis). An electromagnetic wave with this kind of electric field is called linearly polarized light. Light is linearly polarized when its electric field lies on a plane (linearly polarized light is often called plane-polarized light for this reason) defined by a line perpendicular to the propagation direction. To see this wave for numerous points along the z axis, revisit **Illustration 32.3**.

However, the electric field need not be on an axis. For a wave traveling in the z direction, the electric field pointing in the x or the y directions is not the only possibility. For example, the electric field could lie on a plane defined by a line off the x axis by 45° (or $\pi/4$ radians). If you are looking at just one point on the z axis, as we are for this animation, you see the electric field pointing along the 45° line. Such an electric field is shown when $E_x = 8$ N/C, $E_y = 8$ N/C, and there is a phase difference of 0 radians. Notice that the angle off of the x axis depends on the amount of the x and the y electric fields you have. So, for example, an electric field of $E_x = 8$ N/C, $E_y = 4$ N/C, and with a phase difference of 0 radians yields an electric field that is linearly polarized off of the x axis by 26.56° (or 0.464 radians).

Circular and elliptical polarization occurs when two or more linearly polarized waves add together such that the electric field rotates in a plane perpendicular to the direction of propagation. For circularly polarized light, the direction in which the electric field points rotates in a plane, but its magnitude stays the same. For elliptically polarized light both the magnitude and the direction of the electric field varies. If you enter the following values, $E_x = 8$ N/C, $E_y = 8$ N/C, and a phase difference of $0.5 * \pi$ radians, a wave that is right–circularly polarized will result. If you change E_y to 4 N/C, a wave that is right–elliptically polarized will result.

Illustration 39.2: Polarized Electromagnetic Waves

Light is composed of a traveling wave of changing electric and magnetic fields. Click on the link for a **linear wave** to see an example of the electric field component of an electromagnetic wave. Click-drag to the right or left to rotate about the

z axis. Click–drag up or down to rotate in the xy plane. The wave in the animation is x polarized, which means that its electric field oscillates in the x direction.

Some materials, called polarizers, will only transmit light with its electric field in a particular direction. To see an example of a linear wave with a polarizer, click on the link **linear wave with polarizer**. In this example light that is xy polarized is passed through a polarizer that only transmits the component of the electric field in the x direction.

The **circular wave** link shows an example of a circularly polarized wave, and the **circular wave with polarizer** link shows the effect of an x-direction polarizer on this circular wave. **Exploration 39.1** deals more extensively with circularly polarized light, while **Exploration 39.2** discusses polarizers.

Exploration 39.1: Polarization Tutorial

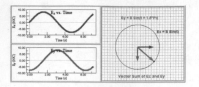

The animation shows the result of adding two perpendicular electric fields together. Each field is part of an electromagnetic wave traveling along the z axis. Each electric field is shown separately on the two graphs on the left. The graphs show the electric field at one point on the z axis for various times. On the right the animation shows both electric fields and their sum at the same point on the z axis and at the same times as the graphs on the left. It is as if you are looking down the z axis at the electric fields.

You can change the electric fields and the phase difference between the two fields and see the resulting waves.

Enter the following values: $E_x = 8$ N/C, $E_y = 0$ N/C, and phase difference = $0 * \pi$ radians. You have created a light wave, traveling along the z axis, with its electric field in the x direction.

a. What kind of polarized light did you create?

b. What is the vector equation of the wave you just created?

c. The wave you just specified is polarized in the x direction. What equations for E_x and E_y would result in light that is linearly polarized along a plane 45° above the +x axis?

d. What equations for E_x and E_y would result in light that is linearly polarized along a plane less than 45° above the +x axis?

Light is linearly polarized when its electric field is in a plane. Circular and elliptical polarization occurs when two or more linearly polarized waves add together such that the electric field of the wave rotates in the plane perpendicular to the direction of propagation. For circularly polarized light, the direction of the electric field rotates but its magnitude stays the same. For elliptically polarized light, both the magnitude and the direction of the electric field vary. For example, if you enter the following values, $E_x = 8$ N/C, $E_y = 8$ N/C, and phase difference = $0.5 * \pi$ radians, a wave that is right–circularly polarized will result. Now consider the values needed to answer the following questions.

e. What equations for E_x and E_y would result in light that is left–circularly polarized?

f. What equations for E_x and E_y would result in light that is right–elliptically polarized?

Exploration 39.2: Polarizers

This animation shows a traveling wave incident on a polarizer. The direction of the polarizer is indicated by the black line. Click–drag to the right or left to rotate about the z axis. Click–drag up or down to rotate in the xy plane.

a. Since the light is linearly polarized along the x axis, explain why, when the light is incident on a polarizing film that allows light polarized along the y axis, this light does not pass through.

b. Suppose this light is incident on a polarizing film polarized along a plane 45° from the x axis. What do you predict will happen? After making your prediction, **try it**.

c. Notice that for (b), the amplitude of the wave passing through the polarizing film is reduced since only the waves with components in the direction of the polarizing film pass through. Now, suppose the light is incident first on the polarizing film of (b) and then on the film of (a). What do you predict will happen? After making your prediction, **try it**.

d. Explain what you see with two polarizing films set up in this way. This explains why, sometimes, two polarizing films let more light through than just one.

Problems

Problem 39.1

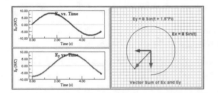

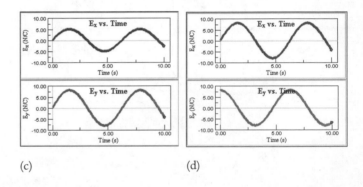

(c) (d)

The animation shows the result of adding two perpendicular electric fields together. Each field is part of an electromagnetic wave traveling along the z axis. Each electric field is shown separately on the two graphs on the left. The graphs show the electric field at one point on the z axis for various times. On the right the animation shows both electric fields and their sum at the same point on the z axis and at the same times as the graphs on the left. It is as if you are looking down the z axis at the electric field. You can change the electric fields and the phase difference between the two fields and see the resulting waves.

You are given several graphs of the x and y components of the electric field. Produce a representation using the animation, like that shown in the right-hand animation, for the sum

of the two fields. Use your representation to identify the type of polarization (include direction) that results.

Problem 39.2

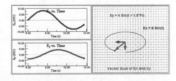

The animation shows the result of adding two perpendicular electric fields together. Each field is part of an electromagnetic wave traveling along the z axis. Each electric field is shown separately on the two graphs on the left. The graphs show the electric field at one point on the z axis for various times. On the right, the animation shows both electric fields and their sum at the same point on the z axis and at the same times as the graphs on the left. It is as if you are looking down the z axis at the electric field. You can change the electric fields and the phase difference between the two fields and see the resulting waves.

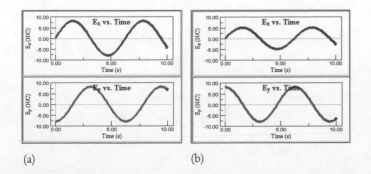

(a) (b)

For each of the following, write an equation for the net electric field, which is polarized as shown.

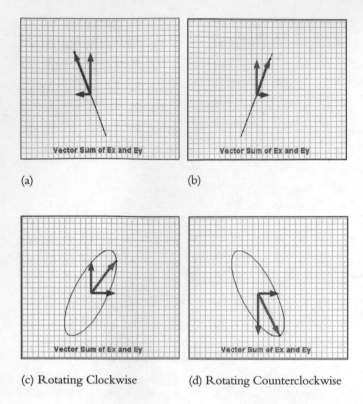

(a)

(b)

(c) Rotating Clockwise

(d) Rotating Counterclockwise

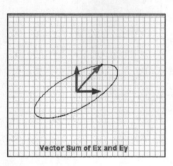

(e) Rotating Clockwise

Problem 39.3

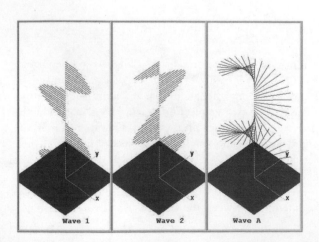

These animations show the electric field component of a traveling wave. Click and drag in each animation to see an alternative view of the wave.

Which wave represents the sum of Wave A and Wave B?

Problem 39.4

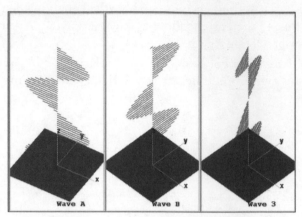

These animations show the electric field component of a traveling wave. Click and drag in each animation to see an alternative view of the wave.

Which wave represents the sum of Wave A and Wave B?

Problem 39.5

The animation shows the electric field of a light wave. Click-drag to the right or left to rotate about the z axis. Click-drag up or down to rotate in the xy plane.

a. In which direction is the light traveling?

b. Describe the polarization state of the wave. Include in your description the type of polarization and the direction of polarization.

Problem 39.6

The animation shows the electric field of a light wave. Click-drag to the right or left to rotate about the z axis. Click-drag up or down to rotate in the xy plane.

a. In which direction is the light traveling?

b. Describe the polarization state of the wave. Include in your description the type of polarization and the direction of polarization.

Problem 39.7

The animation shows the electric field of a light wave. Click-drag to the right or left to rotate about the z axis. Click-drag up or down to rotate in the xy plane.

a. In which direction is the light traveling?

b. Describe the polarization state of the wave. Include in your description the type of polarization and the direction of polarization.

Problem 39.8

The animation shows the electric field of a light wave. Click-drag to the right or left to rotate about the z axis. Click-drag up or down to rotate in the xy plane.

a. In which direction is the light traveling?

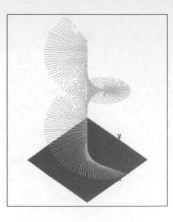

b. Describe the polarization state of the wave. Include in your description the type of polarization and the direction of polarization.

Problem 39.9

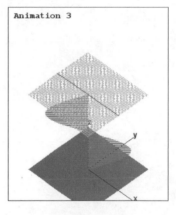

Initially, this animation shows a traveling wave. Click-drag to the right or left to rotate about the z axis. Click-drag up or down to rotate in the xy plane. The subsequent animations show this wave passing through a polarizer. The direction of polarization is indicated by the black line. Which, if any, of the animations are correct? Explain.

Problem 39.10

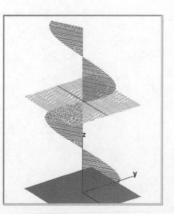

The animation shows a wave incident on a polarizer (with the direction indicated by the black line). Click-drag to the right or left to rotate about the z axis. Click–drag up or down to rotate in the xy plane. You can use the slider to change the polarization of the incoming linear wave. This also changes the amplitude of the exiting wave.

a. Develop an expression for the amplitude of the exiting wave as a function of the initial amplitude and the polarization angle.

b. Develop an expression for the energy transmitted through the polarizer as a function of polarization and the incident energy.

Optics Appendix:
What's Behind the Curtain?

Topics include lenses, mirrors, focal point, and curtains.

Problems

Problems 1–5

An object and its image are shown (position is given in meters). You can drag the object.

a. What optical element is hidden behind the curtain?

b. What is the focal length of the hidden element? Assume the curtain is centered on the optical element.

Problem 1

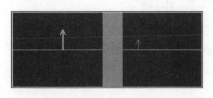

Problem 2

Problem 3

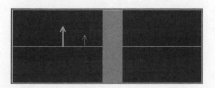

Problem 4

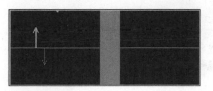

Problem 5

Problems 6–10

A source of parallel rays is shown (position is given in meters and angle is given in degrees). You can change both the position and the direction of the source and both the position and angle of the protractor by click-dragging.

a. What optical element is hidden behind the curtain?

b. What is the focal length of the hidden element? Assume the curtain is centered on the optical element.

Problem 6

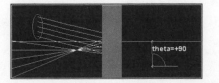

Problem 7

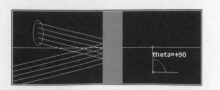

Problem 8

Problem 9

Problem 10

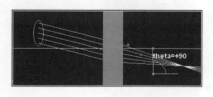

Problems 11–15

A point source of light is shown (position is given in meters and angle is given in degrees). You can change both the position and the direction of the source and both the position and angle of the protractor by click-dragging.

a. What optical element is hidden behind the curtain?

b. What is the focal length of the hidden element? Assume the curtain is centered on the optical element.

Problem 11

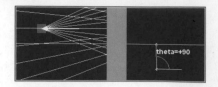

Problem 12

Problem 13

Problem 14

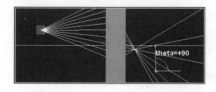

Problem 15

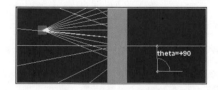

PHYSLET® PHYSICS CD-ROM

WOLFGANG CHRISTIAN AND MARIO BELLONI
PRENTICE-HALL, INC. A PEARSON COMPANY © 2004

YOU SHOULD CAREFULLY READ THE TERMS AND CONDITIONS BEFORE USING THE CD-ROM PACKAGE. USING THIS CD-ROM PACKAGE INDICATES YOUR ACCEPTANCE OF THESE TERMS AND CONDITIONS.

Prentice-Hall, Inc. provides this program and licenses its use. You assume responsibility for the selection of the program to achieve your intended results, and for the installation, use, and results obtained from the program. This license extends only to use of the program in the United States or countries in which the program is marketed by authorized distributors.

LICENSE GRANT

You hereby accept a nonexclusive, nontransferable, permanent license to install and use the program ON A SINGLE COMPUTER at any given time. You may copy the program solely for backup or archival purposes in support of your use of the program on the single computer. You may not modify, translate, disassemble, decompile, or reverse engineer the program, in whole or in part.

TERM

The License is effective until terminated. Prentice-Hall, Inc. reserves the right to terminate this License automatically if any provision of the License is violated. You may terminate the License at any time. To terminate this License, you must return the program, including documentation, along with a written warranty stating that all copies in your possession have been returned or destroyed.

LIMITED WARRANTY

THE PROGRAM IS PROVIDED "AS IS" WITHOUT WARRANTY OF ANY KIND, EITHER EXPRESSED OR IMPLIED, INCLUDING, BUT NOT LIMITED TO, THE IMPLIED WARRANTIES OR MERCHANTABILITY AND FITNESS FOR A PARTICULAR PURPOSE. THE ENTIRE RISK AS TO THE QUALITY AND PERFORMANCE OF THE PROGRAM IS WITH YOU. SHOULD THE PROGRAM PROVE DEFECTIVE, YOU (AND NOT PRENTICE-HALL, INC. OR ANY AUTHORIZED DEALER) ASSUME THE ENTIRE COST OF ALL NECESSARY SERVICING, REPAIR, OR CORRECTION. NO ORAL OR WRITTEN INFORMATION OR ADVICE GIVEN BY PRENTICE-HALL, INC., ITS DEALERS, DISTRIBUTORS, OR AGENTS SHALL CREATE A WARRANTY OR INCREASE THE SCOPE OF THIS WARRANTY.

SOME STATES DO NOT ALLOW THE EXCLUSION OF IMPLIED WARRANTIES, SO THE ABOVE EXCLUSION MAY NOT APPLY TO YOU. THIS WARRANTY GIVES YOU SPECIFIC LEGAL RIGHTS AND YOU MAY ALSO HAVE OTHER LEGAL RIGHTS THAT VARY FROM STATE TO STATE.

Prentice-Hall, Inc. does not warrant that the functions contained in the program will meet your requirements or that the operation of the program will be uninterrupted or error-free.

However, Prentice-Hall, Inc. warrants the CD-ROM on which the program is furnished to be free from defects in material and workmanship under normal use for a period of ninety (90) days from the date of delivery to you as evidenced by a copy of your receipt.

The program should not be relied on as the sole basis to solve a problem whose incorrect solution could result in injury to person or property. If the program is employed in such a manner, it is at the user's own risk and Prentice-Hall, Inc. explicitly disclaims all liability for such misuse.

LIMITATION OF REMEDIES

Prentice-Hall, Inc.'s entire liability and your exclusive remedy shall be:
1. the replacement of any CD-ROM not meeting Prentice-Hall, Inc.'s "LIMITED WARRANTY" and that is returned to Prentice-Hall, or
2. if Prentice-Hall is unable to deliver a replacement CD-ROM that is free of defects in materials or workmanship, you may terminate this agreement by returning the program.

IN NO EVENT WILL PRENTICE-HALL, INC. BE LIABLE TO YOU FOR ANY DAMAGES, INCLUDING ANY LOST PROFITS, LOST SAVINGS, OR OTHER INCIDENTAL OR CONSEQUENTIAL DAMAGES ARISING OUT OF THE USE OR INABILITY TO USE SUCH PROGRAM EVEN IF PRENTICE-HALL, INC. OR AN AUTHORIZED DISTRIBUTOR HAS BEEN ADVISED OF THE POSSIBILITY OF SUCH DAMAGES, OR FOR ANY CLAIM BY ANY OTHER PARTY.

SOME STATES DO NOT ALLOW FOR THE LIMITATION OR EXCLUSION OF LIABILITY FOR INCIDENTAL OR CONSEQUENTIAL DAMAGES, SO THE ABOVE LIMITATION OR EXCLUSION MAY NOT APPLY TO YOU.

GENERAL

You may not sublicense, assign, or transfer the license of the program. Any attempt to sublicense, assign or transfer any of the rights, duties, or obligations hereunder is void.

This Agreement will be governed by the laws of the State of New York.

Should you have any questions concerning this Agreement, you may contact Prentice-Hall, Inc. by writing to:

ESM Media Development
Higher Education Division
Prentice-Hall, Inc.
1 Lake Street
Upper Saddle River, NJ 07458

Should you have any questions concerning technical support, you may write to:

New Media Production
Higher Education Division
Prentice-Hall, Inc.
1 Lake Street
Upper Saddle River, NJ 07458

YOU ACKNOWLEDGE THAT YOU HAVE READ THIS AGREEMENT, UNDERSTAND IT, AND AGREE TO BE BOUND BY ITS TERMS AND CONDITIONS. YOU FURTHER AGREE THAT IT IS THE COMPLETE AND EXCLUSIVE STATEMENT OF THE AGREEMENT BETWEEN US THAT SUPERSEDES ANY PROPOSAL OR PRIOR AGREEMENT, ORAL OR WRITTEN, AND ANY OTHER COMMUNICATIONS BETWEEN US RELATING TO THE SUBJECT MATTER OF THIS AGREEMENT.

Program Instructions
- Windows:
To access the Physlet Physics curricular material content, follow the following steps:
- Insert the "Physlet Physics" CD into your CD-ROM drive.
- Browse to the "Physlet_Physics" CD-ROM using your Windows explorer window.
- Double-click the "start.html" file.
To access the Physlet Physics Exploration Worksheets, double-click the "exploration_worksheets" folder, choose a chapter folder to view the selection of associated Exploration PDF worksheets, and double-click a PDF file to view.

Minimum System Requirements
- Windows:
400MHz Intel Pentium processor
Windows 2000/XP
32 MB or more of available RAM
800x600 monitor resolution set to 16 bit color
Mouse or other pointing device (?)
4x CD-ROM Drive
Active internet connection (optional)

Requires a browser supporting the Java 1.4 Virtual Machine and JavaScript to Java communication. Internet Explorer 5.5 or higher with the Sun Java plugin 1.4 or higher, or Mozilla 1.3 or higher with the Sun Java plugin 1.4 or higher are recommended.

Adobe Acrobat Reader 5.0 © 2002 or above is required to view and print the pdfs of the Exploration Worksheets.

- Macintosh:
This CD-ROM is not supported on Mac OS.